BREWING TECHNIQUES IN PRACTICE

BREWING TECHNIQUES IN PRACTICE

An In-depth Review of Beer Production with Problem Solving Strategies

Werner Back, Martina Gastl, Martin Krottenthaler,
Ludwig Narziß, Martin Zarnkow

FACHVERLAG HANS CARL

IMPRINT

Disclaimer

All information in this book was compiled by the authors to the best oftheir knowledge and reviewed together with the publisher with the utmostcare. However, in terms of product liability law, content errors cannotbe completely ruled out. Therefore, the authors and the publisherare neither under any obligation nor provide any guarantee and also do not assume liability for any errors in content, for personal injury, propertydamage or financial loss.

Bibliographic Information in the Deutsche Nationalbibliothek (German National Library)

The Deutsche Nationalbibliothek lists this publication in the Deutsche Nationalbibliografie (German National Bibliographic Records); detailed bibliographic information is available on the Internet at http://dnb.ddb.de.

Cover: Christina Schönberger
Translation: McGreger Translations, Freising, Germany
Layout and typesetting: Komhus Agentur für Kommunikation, Essen, Germany; Communicate Design and Language, Mülheim an der Ruhr, Germany
Printed by paper&board solutions ug, Langenzenn
ISBN 978-3-418-00852-3

FOREWORD

There has been a desire within the international brewing community for some time to make this technical text available in English, and with the present translation of 'Brewing Techniques in Practice' this has finally been realized. In doing so, information concerning beer styles and brewing methods commonplace outside of Germany has been incorporated into this volume. Craft beer has also been taken into consideration. The existing chapters were revised and updated to reflect the current state of scientific findings and technical knowledge. Like the original text, the English edition has been organized along thematic lines, which address routine day-to-day tasks faced by brewing technologists.

The complex explanations found in this book provide a well-reasoned, articulate and comprehensive overview of each topic for brewers working in the industry as well as for students of brewing. The respective concepts are clearly elucidated in their entirety in each chapter, both at a scientific and a technical level. Particular emphasis is placed on the fundamental aspects of biochemical processes as well as production techniques, so that the information can be readily applied in practice. The organization of the book, which includes a thorough summary at the end of each chapter, has proven to be an indispensable tool, especially as it allows problems, causes and solutions to be explored with ease. Those in the industry will find the tables with standard values for substances typically found in wort and beer to be quite useful. A table has been added providing conversions of international units along with key indicators for technical processes. An overview of the requirements necessary for a beer to be acknowledged as having been "brewed according to the German *Reinheitsgebot*" has likewise been appended to this English edition.

Additionally, a section of the book has been set aside for the purpose of distinguishing the *Reinheitsgebot* (referred only to lager beer), the medieval purity law of 1516, from the current provisional beer law (*Vorläufiges Biergesetz*), which comprises the current regulations governing beer production in Germany. The *Reinheitsgebot* still serves as the core of the *Vorläufiges Biergesetz*, in that the four ingredients stipulated for brewing beer in the law from 1516 still apply under the existing regulations. The *Vorläufiges Biergesetz* regulates all further aspects of modern beer production.

I am grateful to my co-authors for their contributions to the current English edition, as well as to my former colleagues at the *Technologie der Brauerei I* at the TUM-Weihenstephan, who laid the groundwork for this technical brewing text. I owe all of them a debt of gratitude for their diligent and competent teamwork.

My thanks also go to the Fachverlag Hans Carl for the willingness to publish this revised and updated English edition of the handbook.

Werner Back
Freising, August 2019

List of authors of the German edition

Werner Back (editor)
Ingrid Bohak (†)
Felix Burberg
Torsten Dickel
Oliver Franz
Martina Gastl
Stefan Hanke
Klaus Hartmann
Markus Herrmann
Dietmar Kaltner
Matthias Keßler
Stefan Kreisz
Martin Krottenthaler
Florian Kühbeck
Ralf Mezger
Ludwig Narziß
Mark Schneeberger
Christina Schönberger
Cem Schwarz
Elmar Spieleder
Frithjof Thiele
Kornel Vetterlein
Sascha Wunderlich
Michael Wurzbacher
Martin Zarnkow
Joachim Zürcher

CONTENTS

MALT

1 INTRODUCTION

Malt quality is of great consequence in beer production and thus has a substantial impact on the quality of the finished beer. Individual production steps, e.g. lautering, fermentation and filtration, as well as attributes central to the character of beer, e.g. flavor, color, foam and stability, are heavily influenced by malt quality. The malt utilized in beer production is mainly produced from malting barley; however, for some specialty beers, e.g. Southern German-style wheat beer, malt is also produced from wheat or even other cereals, e.g. rye or oats (cf. Cereals and Pseudocereals). Barley is a natural product, making it subject to regional and seasonal fluctuations. The task of compensating for this variation, at least to the extent physically possible, falls to the maltster whose vocation it is to make homogeneous malt of a consistent quality available to breweries. However, biological and economic constraints limit the degree to which quality can be rectified in the malthouse. Maintaining high standards of quality for German malting barley and, in turn, for the malt created from the barley for the purpose of brewing beer, is the responsibility of the entire production chain, from the farmer to the brewer. Advancements in barley breeding and cultivation have resulted in malting barley of an extremely high quality, particularly spring barley varieties. Specifications define the quality of malt required for effortless processing and thus have become the standards used by malt producers and other processing companies.

Through selection of the barley variety and the level of malt quality and hence the standard values and thresholds for the quantifiable attributes described in the malt analysis, a brewer ultimately determines the quality of the raw materials required for a particular beer style. When deciding which attributes should receive the highest priority, the accuracy of the analyses as well as how these various attributes interact with one another should be taken into account. Meticulous attention must be exercised in obtaining analysis results. The procedures for conducting the brewing analyses established throughout Europe have been published in collections of brewing analysis methods by the *Mitteleuropäische Brautechnische Analysenkommission* (Central European Brewing Technology Commission or MEBAK) and by the European Brewery Convention (EBC) [1, 5].

The laboratory mashing method for the evaluation of malting barley varieties was changed prior to the 2012 harvest. The Congress mash method was replaced with an isothermal 65 °C mash (similar to hot water extract), allowing a more practically oriented assessment of new barley varieties to be carried out while also providing insights into processability [2]. However, when evaluating malt quality, the results obtained with the Congress mash method are not equivalent to those found with the isothermal 65 °C mash method. In this situation, comparative analysis is needed to find a common basis. Direct conversion factors will certainly never be generated for adapting the pool of data collected for the Congress mash method to the data for the isothermal 65 °C mash method [3]. Due to the considerably advanced proteolytic and cytolytic modification of grain accomplished in the malthouse, brewers can concentrate their efforts on degrading the starch in the mash vessel (cf. Mashing). Before discussing the individual attributes measured in malt analysis, one should understand that the quality of the assessment itself, as well as the quality of a particular lot of malt, largely depends on representative sampling. The importance of collections, along with the rules governing collection and the preparation of samples, have been described in numerous publications [1, 6, 4].

2 THE QUALITY ATTRIBUTES OF BARLEY MALT AND WHEAT MALT

2.1 QUALITY ATTRIBUTES OF BARLEY MALT

First and foremost, a barley malt analysis describes the three primary modification processes that have occurred in the kernel: cytolysis, proteolysis and amylolysis. The single most important task of the maltster, given the fact that modern brewhouse procedures often entail mashing in at temperatures above 60 °C, is to effect a homogenous and complete degradation of the cell walls and to attain a suitable level of protein modification. Thus, malt quality plays a key role in ensuring that the production process runs smoothly. In large breweries, where up to twelve batches of wort are brewed per day, modifying the temperatures and rests in the mash program to accommodate individual fluctuations in malt quality is not practicable if brewing operations are to remain on schedule. Thus, mashing is largely limited to amylolysis, that is, the degradation of amylose and amylopectin required for brewing (cf. Mashing).

2.1.1 CYTOLYSIS

Cytolysis describes the degradation of the substances providing structure and support to the cells that surround the starch in the endosperm. Structural proteins and polysaccharides in the cell wall, especially β-glucans, are subject to these degradation processes. If the processes are allowed to continue during malting until the support structures of the cells are largely broken down, the enzymatic degradation of the endosperm during mashing is much less arduous, resulting in higher brewhouse yields. Likewise, insufficient modification of these support structures not only brings about shortfalls in brewhouse yield but also causes large quantities of high molecular weight β-glucans to become soluble and enter the process of wort production.

Older sources attest to the favorable influence of high molecular weight β-glucans on foam and mouthfeel. As long as the β-glucans are not in gel form, quantities of up to approximately 350 mg/l do not pose a problem from a brewing standpoint (isothermal 65 °C mashing procedure)[7, 8, 9]. β-Glucan gel can lead to filtration issues even at concentrations as minute as 10–15 mg/l, a level only slightly above the reliable detection threshold (cf. Filterability – Issues with Turbidity). Wort produced using mash programs with mash-in temperatures above 60 °C are particularly susceptible to gel formation. At mash temperatures in the range from 60 to 65 °C, a substantial amount of β-glucans still bound to the cell walls is liberated by the enzyme β-glucan solubilase. However, degradation of this high molecular weight fraction can no longer take place, since the endo-β-glucanases – the enzymes responsible for breaking down these large β-glucans – are inactivated at temperatures as low as 52 °C. Thus, given a consistent malt quality, brewhouse procedures incorporating high mash-in temperatures will always lead to higher total β-glucan concentrations in wort and beer. For this reason, high mash-in temperatures require more extensive modification of the cell wall (cf. Mashing).

Various key metrics are used to describe the level of cytolytic modification. The value for friability has proven useful in this regard. The procedure is simple, and the value can also be ascertained rapidly. Cell wall modification is evaluated with a friabilimeter to determine the overall friability of a specific lot of malted grain and the percentage of kernels that are classified as completely glassy. This information is then used to draw a conclusion regarding how uniform the process of malting the barley has been. High values for friability are not necessarily an indication of over-modification, as long as they only apply to cell wall modification and not to protein modification. Therefore, breeders involved in developing malting barley varieties face the challenge of striking the right balance among the individual traits used to define modification, especially with respect to proteolytic and cytolytic processes.

Other attributes providing information about the degree of cytolytic modification include the viscosity and the β-glucan content of both the Congress wort and of the 65 °C mash as well as the homogeneity and modification of the malt. The values obtained for the 65 °C mash are a better gauge of cytolytic modification than the values obtained with the Congress mash.

Owing to the 45 °C rest, the Congress mash method promotes more β-glucan degradation. However, with this method, once the temperature of the mash reaches 45 °C, it is immediately heated to 70 °C, which does not allow enough time for an adequate β-glucan solubilase rest. Thus, variation in cytolytic modification among different lots of malt cannot be sufficiently characterized by means of the Congress mash method. The isothermal 65 °C mash, on the other hand, more clearly distinguishes this variation with its high mash-in temperature and intensive β-glucan solubilase rest. From time to time, the difference in the results between the Congress mash and the isothermal 65 °C mash is also employed as an assessment criterion. One should be mindful of the fact that both MEBAK and the EBC no longer include the method for determining the difference in extract between fine and coarse grist in their analysis collections. Based on statistical evaluation in combination with practical tests, the following analysis results and limit values have been shown to be useful for mash programs utilizing high mash-in temperatures.

Table 1: Cytolytic malt analysis attributes for mash programs utilizing high mash-in temperatures (isothermal 65 °C mashing procedure)

friability	> 85 %
whole glassy kernels	< 2 %
viscosity, isothermal 65 °C mash (adj. to 8.6 %)	< 1.6 mPa·s
β-glucans, isothermal 65 °C mash (adj. to 8.6 %)	< 350 mg/l
homogeneity	> 75 %

Practical experience has shown that it is prudent to determine the viscosity of the isothermal 65 °C mash in addition to the friability (including the glassy kernels) and homogeneity as part of a routine analysis program. Only in cases of considerable uncertainty (results exceed or fall below the limit values) would it be worthwhile to perform the remaining analyses in table 1. Interpretation of the results is recommended as follows: If the results for at least two of the analyses fall outside of the range for the limit values given, then mashing in at a high temperature with the malt in question can lead to difficulties during lautering and filtration. If, on the other hand, the results remain within the limit values and filtration issues arise, these can most likely be traced back to errors in brewing techniques (cf. Filterability – Turbidity Problems). If only one of the values exceeds the limit, then the analysis should be repeated. In some instances, the analyses to determine the percentage of whole glassy kernels and the concentration of β-glucans exhibit very poor reproducibility. The corresponding statistical data analysis can be found in the collection of analysis methods published by MEBAK or EBC. Low values for homogeneity and a disproportionate rise in the viscosity and the concentration of β-glucans in the Congress wort compared to the values from the 65 °C mash indicate that highly modified malt has been mixed with slightly modified malt [10].

2.1.2 PROTEOLYSIS

Proteolysis describes the degradation of the protein in the kernel and its resultant transformation into more soluble molecules, which can be of a low, medium or high molecular weight. While extensive cell wall modification does not impact beer quality in either a positive or negative manner, it does improve processability. Both excessive and negligible levels of protein modification are regarded as detrimental. Low protein modification carries the risk of depriving the yeast of assimilable nitrogen compounds. Negative outcomes include inadequate yeast reproduction and the formation of undesirable fermentation by-products (e.g. diacetyl). On the other hand, a high level of protein modification results in excessive degradation of high molecular weight proteins. A dearth of sufficient concentrations of high molecular weight proteins – but also a surplus of medium molecular weight proteins and certain amino acids (lysine, arginine and histidine) – has a negative effect on foam stability. Malt subjected to elevated levels of protein modification produces wort and beer that tends to be darker in color and contains an overabundance of certain amino acids. The presence of these amino acids leads to the production of uncharacteristic aromas in beer, along with reduced flavor stability (cf. Flavor Stability). Furthermore, beer containing higher concentrations of amino acids is more susceptible to beer spoilage microbes.
The Kolbach index (degree of protein modification), the soluble nitrogen content and free amino nitrogen (FAN) are key metrics for proteolysis. The Kolbach index is the most commonly used metric for assessing proteolysis in commercial breweries. It represents the percentage of the total protein converted into soluble form during malting and subsequently during the Congress mash (calculated value). The preferred range for pale, all-malt beers is between 38 and 42 % (Congress mash method). The degree of protein solubility limits the possibilities for theoretical combinations derived from the absolute values for the protein content of the malt (standard values: 9.5–11 %) and the soluble protein (standard values: 3.9–4.7 %). This is intended to ensure that the overall composition of soluble protein is balanced in the ranges of both the high molecular weight (foam stability) and the low molecular weight (yeast nutrition) proteins, independent of the total protein content of the malt. Soluble protein is calculated by determining the soluble nitrogen (conversion factor: 6.25) in most cases. Consequently, calculations based on the data above yield values of 650 to 750 mg of soluble nitrogen per 100 g of malt (dry matter) and 130 to 160 mg of FAN per 100 g malt (dry matter) which should account for approximately 21 % of the soluble nitrogen (Congress mash method). A curtailed mash program or the use of adjuncts, such as rice or corn, would mean that higher amounts are required (cf. Cereals and Pseudocereals).

2.1.3 AMYLOLYSIS

Of the metrics available for amylolytic activity, the following measurements are routinely performed: extract, limit of attenuation, β-amylase activity (expressed as diastatic power in WK units or β-amylase activity in BU) and α-amylase activity (ASBC or DU, using a Megazyme kit). The extract content indicates the percentage of finely milled malt (dry matter) that can be solubilized using the laboratory mash method and gives an indication of the expected yield during the brewing process. In the case of barley malt, the values are between 78.0 and 83.5 % (Congress mash method).

The limit of attenuation is a metric for evaluating the quality of wort in a laboratory and provides information on how well the extract can be metabolized by the yeast. The degree of final attenuation reached in the finished beer at a brewery should be compared to the laboratory results for the limit of attenuation. The final attenuation should be equal to or as close as possible to the limit of attenuation. The quantity of fermentable sugars and their relative proportion are the defining factors for the limit of attenuation; however, it is also affected by the gelatinization temperature of the starch (cf. Mashing). In addition, the role of trace elements and the nitrogen composition should also not be underestimated. As a measure of quality of the Congress wort, the following applies: the higher the limit of attenuation, the better (> 81 %). Final attenuation in the brewery is sometimes disproportionately high and can be difficult to control. Within this context, there is some discussion as to whether or not attempts should be made to regulate the limit of attenuation in the malt (cf. Acidification with Natural Lactic Acid). The β-amylase activity of malt is primarily of interest outside of Germany, e.g. in countries where large quantities of adjuncts are used. These adjuncts usually contribute very little enzymatic activity of their own. For an all-malt beer, values for β-amylase activity greater than 200 WK or 750 BU, respectively, are considered adequate. If β-amylase activity is too low, this can cause a shift in the sugar spectrum and may result in abnormal fermentations in extreme cases (diauxie).

The pace-setting enzyme during starch degradation is α-amylase. It degrades starch into fragments consisting of amylopectin and amylose, providing substrate for β-amylase. An α-amylase activity of more than 60 ASBC units or DU is desirable. If the mash is acidified, α-amylase activity may be inhibited to some extent (pH optimum: 5.4–5.6). If the gelatinization temperature of a given lot of malt is high, it would be beneficial to avoid acidification during mashing, since this could result in increased iodine values. Furthermore, undesirable consequences may arise as well, such as higher values for turbidity in the filtered beer. Furthermore, a lack of available substrate for β-amylase can adversely impact the limit of attenuation (cf. Mashing).

2.1.4 ADDITIONAL MALT SPECIFICATIONS

2.1.4.1 DMS precursor

The DMS precursor (DMS-P), otherwise known as S-methylmethionine (SMM), is an amino acid not found in barley in its unbound form, but it is present in malt.

Dimethyl sulfide (DMS) splits off of the larger DMS-P molecule at above approximately 70° C, essentially at every stage of malt and wort production in which thermal processes exceed that temperature.

The vast majority of DMS should be cleaved from the DMS-P molecules and eliminated during malt kilning and wort boiling. One should be cognizant of the fact that DMS will continue to be formed in the whirlpool (cf. Wort Boiling, Wort Boiling Systems). The temperature and duration of curing during kilning are the most effective means for influencing the level of DMS-P in the malt. In principle, the following applies: the higher the temperature or the longer the duration of curing, the lower the level of DMS-P in the malt. However, economic considerations and excessive thermal stress (see TBI) can be detrimental to pale malt and thus serve as arguments against curing longer at higher temperatures.

DMS can create flavor and aroma impressions in the finished beer that are reminiscent of boiled cabbage or cooked vegetables. The sensory threshold for free DMS in beer is approximately 50–100 µg/l. Therefore, depending on the intensity of the wort boiling process, the DMS-P content of malt should not exceed 5–7 ppm (cf. Wort Boiling, Wort Boiling Systems).

2.1.4.2 TBI (thiobarbituric acid index)

The TBI is a metric representing the sum of the thermal stress to which the malt has been subjected during the kilning process. Maillard products deepen the color of the malt and the finished beer and may, in part, have a negative impact on the beer's flavor stability. These compounds are formed in reactions brought about by thermal stress. For pale malt, the TBI measured in the Congress wort should not exceed 18, and the values for DMS-P must be reduced to below the recognized upper limit. As already stated above and is often the case in malting and brewing technology, a compromise must be reached between the lowest possible thermal stress on the malt and the cleavage of DMS-P along with the elimination of DMS when determining the intensity of the kilning process (fig. 1).

Figure 1: Relationship between the temperature/duration of kilning and the TBI/DMS-P content in malt according to FORSTER [8].
For example: a concentration of 5 ppm DMS-P is obtained under the following conditions 3.2 h/84 °C/TBI=14.5 or 5.5 h/82 °C/TBI=18.

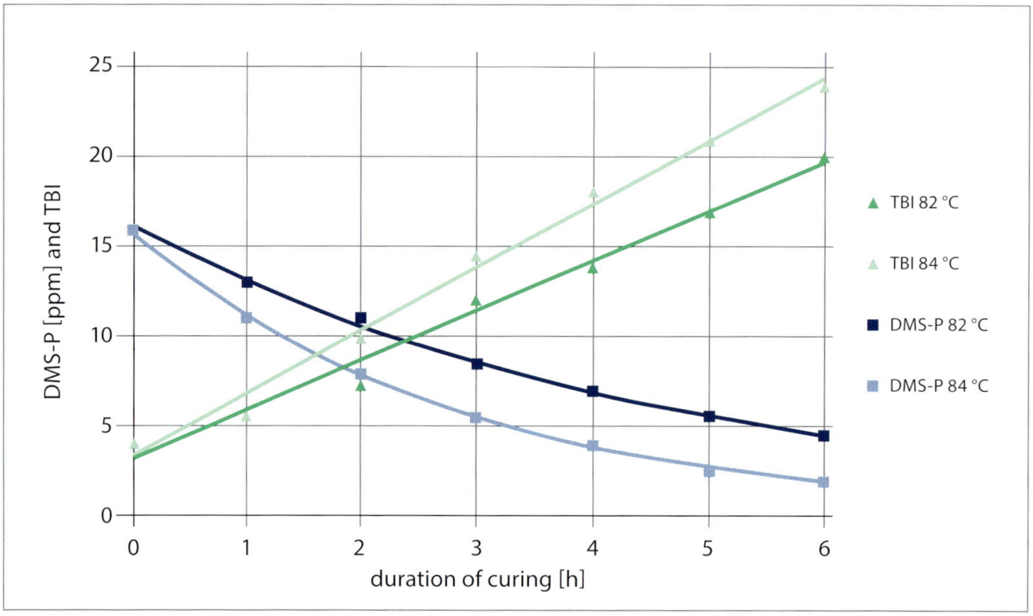

2.1.4.3 Malt color and boiled wort color

Obviously, the color of the malt significantly influences the color of the finished beer. Both are determined photometrically or visually from Congress wort samples (unboiled or boiled). In comparing analytical results, one should note which methods were used to determine the color. Pale malts used to produce Central European lager beers should possess a color between 2.5 and 3.5 EBC. The color itself is indicative of the malt and how it was produced. In principle, the color of the boiled wort can be used to predict the color of the finished beer if information is known about the brewing techniques employed. The thermal stress brought about by the brewing process, any potential oxidation processes (especially during mashing) and the extent of the pH drop during fermentation all have a considerable influence on the beer color, independent of the malt color. The application of less intense wort boiling methods and low thermal stress on both the malt and the wort can result in finished beer that is too light in color. In such cases, the desired color can be adjusted by adding of any number of specialty malt products [10, 12].

2.1.4.4 pH

The pH of the malt is determined by measuring the pH of the Congress wort. For pale barley malt, the pH should range from 5.80 to 5.95. Darker malts possess a larger quantity of Maillard products and thus a lower pH, which varies between 5.50 and 5.80. If the pH of pale malt is too low, this can be an indication that it has been overly modified or too intensely sulfured. One can expect that the mash produced using malt of a low pH will also exhibit a low pH.

2.2 QUALITY CRITERIA FOR WHEAT MALT

The criteria for assessing the quality of wheat malt were appropriated directly from those for judging barley malt. Since the technological requirements are different for the production of Southern German wheat beers, a critical view of these quality criteria is therefore necessary. The differences in barley and wheat as well as in clear, bottom-fermented beer and cloudy, top-fermented beer require a fundamentally different approach to the raw materials. The quality of wheat, the capacity to convert wheat to malt and the influence of the malt quality on that of the wheat beer, are not nearly as well researched as the same characteristics in barley and barley malt. Nevertheless, in each respective section of the discussion below, the properties of wheat malt are compared to the properties of barley malt and, where relevant, their similarities and differences are also examined [4].

2.2.1 CYTOLYSIS

The polysaccharide β-glucan, which is present in the cell walls of barley malt, is the primary focus of the analyses for characterizing the cytolytic processes involved in malt modification, and it also largely determines the viscosity of wort produced from barley malt. Staining β-glucans with calcofluor provides the analytical foundation for evaluating the homogeneity of malt or for directly determining the concentration of β-glucans in wort and beer. Wort brewed with wheat malt generally exhibits a higher viscosity (1.6–1.8 mPa·s in Congress wort) than wort brewed with barley malt. However, the viscosity is not attributable to β-glucans, but rather to other polysaccharides called pentosans. Little is known concerning their behavior during the processes of malting and brewing. It has been reported that if they are exposed to hot side aeration during mashing, the pentosans in rye malt can increase the viscosity of the wort quite dramatically, leading to difficulties in the lautering process [13]. Nevertheless, although gel formation has been documented with β-glucans, its formation in beer containing pentosans has yet to be demonstrated. With regard to the aforementioned analysis of barley malt and the evaluation of its cytolytic modification, one should keep in mind that not all of the tests based on β-glucans are applicable to wheat malt. In essence, the analyses used to assess the homogeneity of barley malt as well as the direct determination of β-glucans in wort and beer are not relevant for judging the quality of wheat malt. Moreover, the friabilimeter test is of little value in determining how friable the structure of the endosperm is in wheat malt. For this reason, only the viscosity remains as a means for assessing the level of cytolytic modification. Unfortunately, the results for the viscosity of Congress wort produced with wheat malt cannot be adequately correlated with lautering behavior or filterability in the production of *Kristallweizen* (filtered Southern German wheat beer) (cf. Filterability – Issues with Turbidity). However, given the fact that no other parameters exist for evaluating the processability of wheat malt, viscosity is currently the only reliable indicator for the degree of cytolytic modification in wheat malt.

2.2.2 PROTEOLYSIS

The analyses used to assess the proteolytic modification of barley malt are also used for wheat malt: total protein (conversion factor: 6.25), soluble protein, the Kolbach index (the quotient derived from these two values) and FAN. Protein modification in wheat malt has a significant impact on the aroma of weissbier [14]. Research results indicate that more extensive proteolytic modification results in the formation of fewer esters, thus leading to more neutral beers.

Furthermore, proteins play a central role in the consistent and permanent haze present in weissbier. Researchers are currently investigating the properties of these haze-forming compounds as haze is an important attribute of weissbier and other cloudy beer styles. Weikl has already shown that yeast cells are generally too large and sink too quickly to generate lasting turbidity [15]. Particles capable of doing so are between 0.1 and 1.0 μm in size. These particles have been identified as proteins and also, in part, as α-glucans. However, exactly how the variety of wheat and the malting technology influence the occurrence of these particles in beer remains largely unknown and is the subject of current research on weissbier.

Compared with bottom-fermented beers, the presence of proteinaceous particles appears to exercise a much greater influence on the filterability of *Kristallweizen* [16]. The desire for stable turbidity levels on the one hand and good filterability on the other suggests that two types of wheat malt with different specifications with respect to protein modification are required for weissbier and for *Kristallweizen*. Generally speaking, the key metrics for proteolytic modification of barley malt can be applied directly to wheat malt, although the overall protein content of wheat will be slightly higher, which in fact is desirable. Sacher has shown that a protein content of around 12 % for wheat malt and a moderate level of protein solubility are beneficial for the aroma of weissbier [14].

2.2.3 AMYLOLYSIS

Even with a significantly higher protein content, due to its lack of husks, wheat malt yields a higher amount of extract than barley malt, but in the end, weissbier exhibits a lower final degree of attenuation. This may also be due to a considerably lower level of α-amylase activity in wheat malt. As a rule, iodine tests show that saccharification turns out to be less extensive in a weissbier mash. For the most part, the α-glucans responsible for higher iodine values are considered favorable for achieving a stable turbidity, although a reliable confirmation is still lacking. In *Kristallweizen*, α-glucans are the primary cause of elevated values for turbidity, according to the 90° scattered light measurement. In any case, they are deemed undesirable in *Kristallweizen*. Measuring the α-amylase activity of wheat malt does not provide any valid insight into its processability. However, it has been shown that in the absence of any suitable alternatives, enzymatic activity can be helpful in estimating the quality of wheat malt. In this regard, the highest amount of enzymatic activity possible is preferable.

2.3 FOOD SAFETY

In light of a number of recent scandals and due to the globalization of the food industry, many consumers, merchants and legislators have focused their attention on the safety of food. Beer is largely protected against pathogenic microbes, due to many positive attributes, which include a low pH and compounds from hops (cf. Microbiology). Even the raw materials barley and wheat, and the malt produced from them, are foods that exhibit a low and manageable risk for consumers. Nevertheless, in recent years, a large number of safeguards have been created, strengthened by new legal regulations and the establishment of upper limits across the EU. The upper limits and analysis methods are the subject of a never-ending, highly politicized discussion. It is imperative to avoid any negative publicity arising from product safety issues which were initially reported as factual, only to be revealed later as untrue. For this reason, performing analyses for the detection of residues and toxins require the greatest care and expertise, and as noted previously, competent sample preparation is indispensable.

The risks inherent to the production chain – from farmers to merchants onto maltsters and finally to the delivery of malt to the brewer – can be impartially assessed. Risk analysis and risk assessment must be carried out independently by each company. Sources for additional information on current developments relevant to this topic are provided in the overview in table 2.

3 SUMMARY

Because malting barley is an agricultural product, the quality of barley is subject to varietal and seasonal fluctuations as well as disparities in cultivation practices. Maltsters endeavor to minimize this variation to an extent that it is both technologically feasible and economically reasonable, in order to best fulfill the quality criteria specified by brewers. A quality assessment of the malt attributes is generally carried out according to the methods developed by MEBAK or the EBC. These analyses describe, above all, the scope of the three main processes involved in modification, that is, proteolysis, cytolysis and amylolysis. They each profoundly influence the quality of finished beer. It is the task of brewers to determine upper and lower limits for the quality of the malt based on the type of brewhouse equipment and the brewing techniques at their disposal. In doing so, the various parameters should not be considered in isolation but rather as a whole. One must also be aware that individual attributes or sets of attributes can influence numerous production steps and thus the quality characteristics of the beer, sometimes in opposition to one another. Malt specifications, therefore, always represent a compromise and can be adjusted to adapt to the quality of a particular year's harvest and to the characteristics of a given variety of malting barley.

Table 2: Sources for information on food safety

German and European authorities, legal principles, limit values and regulations	Sources
Lebensmittel- und Bedarfsgegenständegesetz (LMBG) Regulation (EC) no. 178/2002 of the European Parliament and of the Council of 28 January 2002 Lebensmittelhygieneverordnung (LMHV) Neuartige Lebensmittel- und Lebensmittelzutatenverordnung (NLV) Trinkwasserverordnung (TrinkwV) Gesetz zur Regelung der Gentechnik (GenTG)	The *Bayerische Staatsministerium für Umwelt, Gesundheit und Verbraucherschutz (StMUGV)* is the government agency responsible for the consumer protection system VIS (Verbraucherschutzinformationssystem-http://www.visernaehrung.bayern.de/de/left/recht/recht-ix.htm). The website above is in German and provides detailed information concerning the applicable laws and regulations, including a link to the federal bureau for food safety.
European Food Safety Agency (EFSA). The EFSA offers independent, scientific guidance to all questions within the framework of food safety, including the well-being and protection of animals in addition to the health of crops as community legal regulations apply to them. They also address questions concerning nutrition. Brewers of Europe	http://www.efsa.europa.eu/ https://brewersofeurope.org
Information about upper and lower limit values specific to malt and beer	Deutscher Brauerbund e.V. http://www.brauer-bund.de/ Bayerischer Brauerbund e.V. http://www.bayrisch-bier.de/ Deutscher Mälzerbund https://www.deutscher-maelzerbund.de/

4 OVERVIEW

		Barley malt		Wheat malt*
Laboratory analysis	Unit	Various mashing methods	*"Hoch-kurz"* mashing procedure	
Amylolysis				
extract	%, d.m.	> 81	> 81	> 83
extract, isothermal 65 °C mash			> 81	
α-amylase	ASBC, moisture-free/ DU, moisture-free	> 60	> 60	> 28
diastatic power	WK	> 200	> 200	
β-amylase	BU, moisture-free	> 750	> 750	
limit of attenuation	%, apparent	> 80	> 80	
limit of attenuation, isothermal 65 °C mash	%, apparent		> 84	
Proteolysis				
raw protein	%, moisture-free	9.5–11	9.5–11	11–12.5
soluble nitrogen	mg/100 g malt, d.m	550–700	650–750	650–780
soluble nitrogen, isothermal 65 °C mash	mg/100 g malt, d.m		570–670	
Kolbach index	%	38–40	39–42	37–40
Kolbach index, isothermal 65 °C mash	%		34–38	
free amino nitrogen	mg/100 g malt, d.m	120–150	130–160	
free amino nitrogen, isothermal 65 °C mash	mg/100 g malt, d.m		> 140	

*Congress mash method

		Barley malt		Wheat malt*
Laboratory analysis	Unit	Various mashing methods	*"Hoch-kurz"* mashing procedure	
Cytolysis				
friability	%	> 82	> 85	
whole glassy kernels	%	< 2	< 2	
viscosity	mPa·s (adj. to 8.6 % w/w)	< 1.58	< 1.56	< 1.8
viscosity, isothermal 65 °C mash	mPa·s (adj. to 8.6 % w/w)	< 1.65	< 1.60	
β-glucans	mg/l	< 300	< 200	
β-glucans, isothermal 65 °C mash	mg/l	< 350	< 350	
modification	%	> 85	> 90	
homogeneity	%	> 75	> 75	
Further Analyses:				
DMS-P	ppm, a.d.	< 7	< 7	
DON	µg/kg	500	500	500
NDMA	µg/kg	< 2.5	< 2.5	< 2.5

*Congress mash method

5 REFERENCES

[1] Methner, F. (ed.): Collection of Brewing Analysis Methods. Raw Materials. Self-published by MEBAK, Freising 2018.

[2] Gastl, M., Rittenauer, M., Becker, T.: Kongressmaische – isotherme 65 °C-Maische: Ein Brückenschlag. Brauwelt 156 (2016), no. 45/46, pp. 1360–1363.

[3] Gastl, M., Knöfel, T., Becker, T.: Umstellung auf die isotherme 65 °C-Maische – Neubewertung der Qualitätsmerkmale. Brauwelt 154 (2014), no. 10, pp. 274–276.

[4] Narziß, L., Back, W., Die Bierbrauerei Bd. I, Technologie der Malzbereitung. Wiley-VCH, Weinheim 2012, pp. 673–700.

[5] European Brewery Convention: Analytica-EBC. https://analytica-ebc.com. Fachverlag Hans Carl, Nürnberg 1998 and 2017.

[6] Münzing, K.: Probenahme und Getreidevorbereitung für Rückstellmuster und Laboruntersuchungen, Veröffentlichungsnummer 7418 der Bundesanstalt für Getreide-, Kartoffel- und Fettforschung in Detmold und Münster. pp. 649–655.

[7] Lusk, L. T., Duncombe, G. R., Lay, S., Navarro, A., Ryder, D.: Barley β-Glucan and Foam Stability. Journal ASBC. 59 (2001), pp. 183–186.

[8] Kreisz, S.: Der Einfluss von Polysacchariden aus Malz, Hefe und Bakterien auf die Filtrierbarkeit von Würze und Bier. Dissertation, Technical University of Munich (Weihenstephan), 2003.

[9] Krebs, G.*, Müller, M.*, (both authors contributed equally), Becker, T., Gastl, M.: Characterization of the macromolecular and sensory profile of non-alcoholic beers produced with various methods. Food Research International. 116 (2019), pp. 508–517, doi: 10.1016/j.foodres.2018.08.067.

[10] Sacher, B.: Hohe Einmaischtemperaturen – Anforderungen an die Malzqualität. 40. Mälzereitechnische Arbeitstagung. Doemens, Gräfelfing 2001.

[11] Forster, C.: Der Einfluss der Darrtechnologie auf die Malz- und Bierqualität. Dissertation, Technical University of Munich (Weihenstephan), 1996.

[12] www.weyermann.de/eng/; www.bestmalz.de/en; www.ireks-malz.de/english.htm; www.castlemalting.com/Default.asp?Language=English.

[13] Braun, F.: Mälzungs- und Maischversuche zu Möglichkeiten der technologischen Einflußnahme auf ungünstige Fließeigenschaften bei Roggenmalzwürzen. Thesis, Technical University of Munich (Weihenstephan), 1997.

[14] Sacher, B.: Über den Einfluß von Sorte, Umwelt, agronomischen Maßnahmen und Mälzungstechnologie auf die wertbestimmenden Eigenschaften von Winterweizenmalzen. Dissertation, Technical University of Munich (Weihenstephan), 1997.

[15] Weikl, D.: Untersuchung der physikalischen Aspekte der Trübungsentstehung in Bier und ihrer messtechnischen Erfassung unter besonderer Berücksichtigung von dauertrüben Weizenbieren. Thesis, Technical University of Munich (Weihenstephan), 2003.

[16] Keßler, M.: Untersuchungen zur Filtrierbarkeit von Weizenbieren. Thesis Technical University of Munich (Weihenstephan), 2003.

HOPS

1 INTRODUCTION

The common hop *(Humulus lupulus)* belongs the botanical Order Rosales (nettles) and is classified as a member of the hemp family *(Cannabaceae)* [1].

Figure 1: Timeline of global change in beer production and the area under hop cultivation

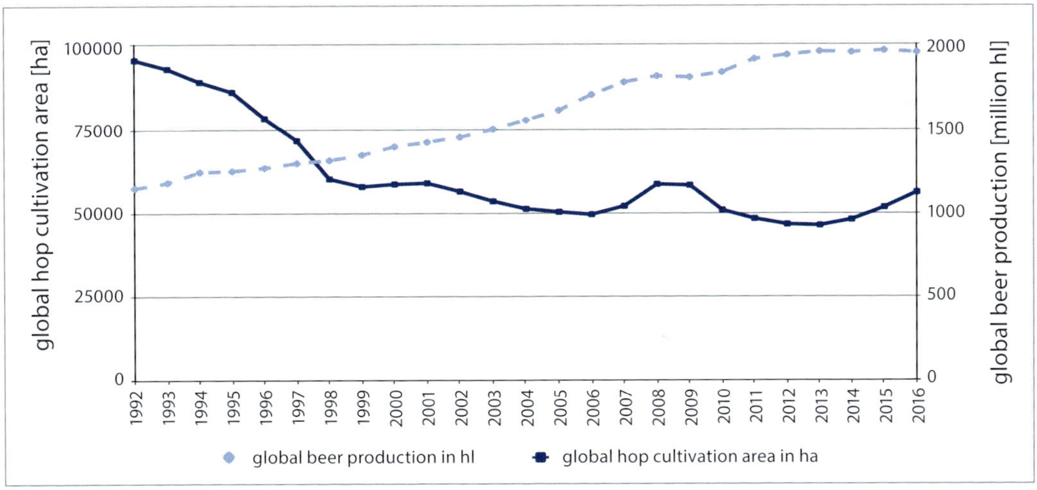

Figure 2: Change in the global average for α-acid concentrations in bitter hops and aroma hops [2, 3]

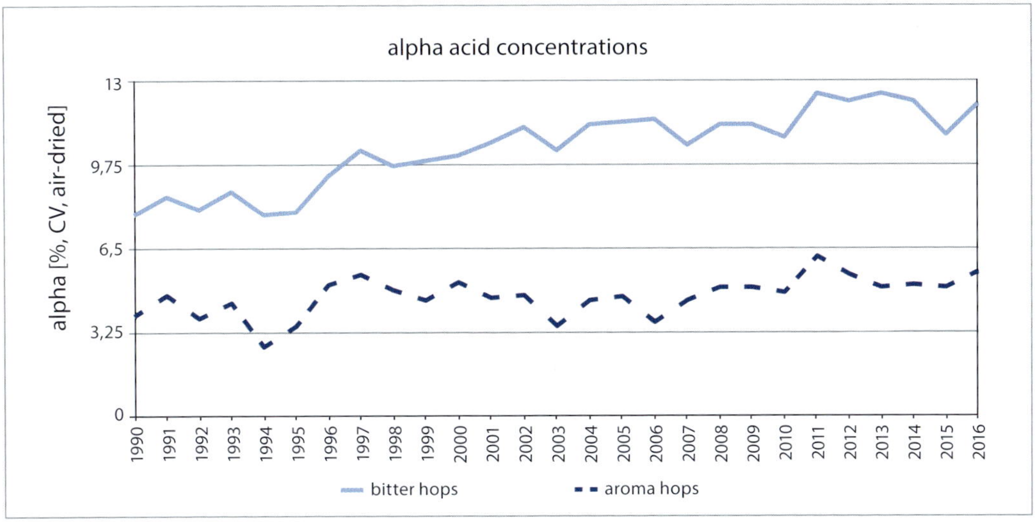

Hops are cultivated in more than 50 countries around the world, spanning an area of approximately 56,000 hectares. Of this, around 17,700 hectares are devoted to hop cultivation in Germany. Progress in hop breeding is evident in the rising α-acid concentrations in bitter hop varieties. Craft beer has generated global interest in beer with more hop character, which is now also reflected in more conventional beer styles. This has resulted in an increase in the cultivation of aroma hops, bringing the current global production of aroma hops to 60 %. Over the past six years, the amount of α-acids (g/hl) added in the production of beer has risen by 15 %. Presently, an average of 4.6 g of α-acids is added per hectoliter of beer [3].

More detailed information regarding German hop varieties such as the concentration of bitter substances, composition of essential oils, sensory characteristics, polyphenol content, pest resistance including agronomy and brewing quality [4, 5] are published in the CMA Hop Variety Portfolio [6]. Descriptions and specifications for all hop varieties (both German and international varieties) can be readily accessed on the internet [7, 8, 9, 10].

1.1 VALUABLE COMPOUNDS IN HOPS

Lupulin, a yellowish green substance that is sticky when fresh, is of great importance to brewers. Localized in cup-shaped glands located on the inside of the inner and outer bracteoles on a hop cone, lupulin contains bitter substances and aroma compounds that are considered essential to the brewing process. The composition of hops depends on the individual variety, the year of the harvest, their provenance and the timing of the harvest as well as how they were dried and stored [11, 12].

1.1.1 BITTER ACIDS

The total resins represent the sum of all bitter compounds in hops and can be divided into two groups based on their solubility in hexane: hard resins (insoluble) and soft resins (soluble). Bitter acids constitute the primary component of soft resins, which can be subdivided into α-acids (3–17 %) and β-acids (2–10 %). The α-acids occur as five homologs: humulone, cohumulone, adhumulone, prehumulone and posthumulone. The lupulone group also contains five homologs, which only differ from their humulone counterparts by one functional group on the side chain of the C3 atom (dimethyl allyl instead of OH). The bitter compounds impart a fine, pleasant hop bitterness to beer, improve drinkability, enhance foam stability and increase its shelf life through their antibacterial properties [13]. Studies have shown that bitter compounds exert a positive influence on individuals suffering from type 2 diabetes [14].

The bitterness potential and the value of different substances or compounds in brewing are variable [15]. In addition to solubility [16] in wort and beer, the quality of the bitterness is also important. Due to their low solubility in wort and beer, unisomerized α-acids do not make any notable sensory contribution to the bitterness [17]. The true soft resin fraction is expressed as "total soft resins – (α-acids + β-acids)". They are derived from their respective bitter acids and are soluble in wort and beer. They are formed through oxidative processes (e.g. from the α-acid humulinone and its derivatives, from the β-acid hulupone and its derivatives). They lend a mild bitterness [18, 19] to beer (refer to section 1.3). In accordance with the latest recommendations, soft resins are now classified as "auxiliary bitter compounds" [20].

During thermal isomerization, optically active six-ring compounds are converted to five-ring compounds. Degradation processes create the isohumulone derivatives (abeo-iso-α-acids, allo-iso-α-acids and anti-iso-α-acids) shown in figure 3. These compounds are complex for the most part, exhibit low volatility, good solubility in wort and are later found in beer [21]. The derivatives of β-acids, e.g. hulupones, are recognized as compounds that contribute to the bitterness of beer and are formed through oxidation reactions [22, 23, 24, 25].

While the α-acids show only a limited solubility at the pH values typically found in wort, they are transformed during wort boiling into more soluble forms, namely *cis*- and *trans*-iso-α-acids. The designations *cis* and *trans* refer to the spatial position of the hydroxyl group attached to the C4 atom with respect to the isoprene side chain on the C5 atom [26]. The *trans* form is less stable than the *cis* form due to their stereochemical properties [27, 28]. The intensity of the bitterness is described as the same for both isomers [28, 29]; however, some reports indicate that the *cis* form is somewhat more bitter [30, 31, 32].

Figure 3: Degradation products of *cis*- and *trans*-iso-alpha-acids formed during wort boiling. Humulinic acid <u>47</u>, allo-isohumulone <u>48</u>, allo-isohumulone hydroperoxide <u>49</u>, allo-isohumulone hydroxide <u>50</u> [33].

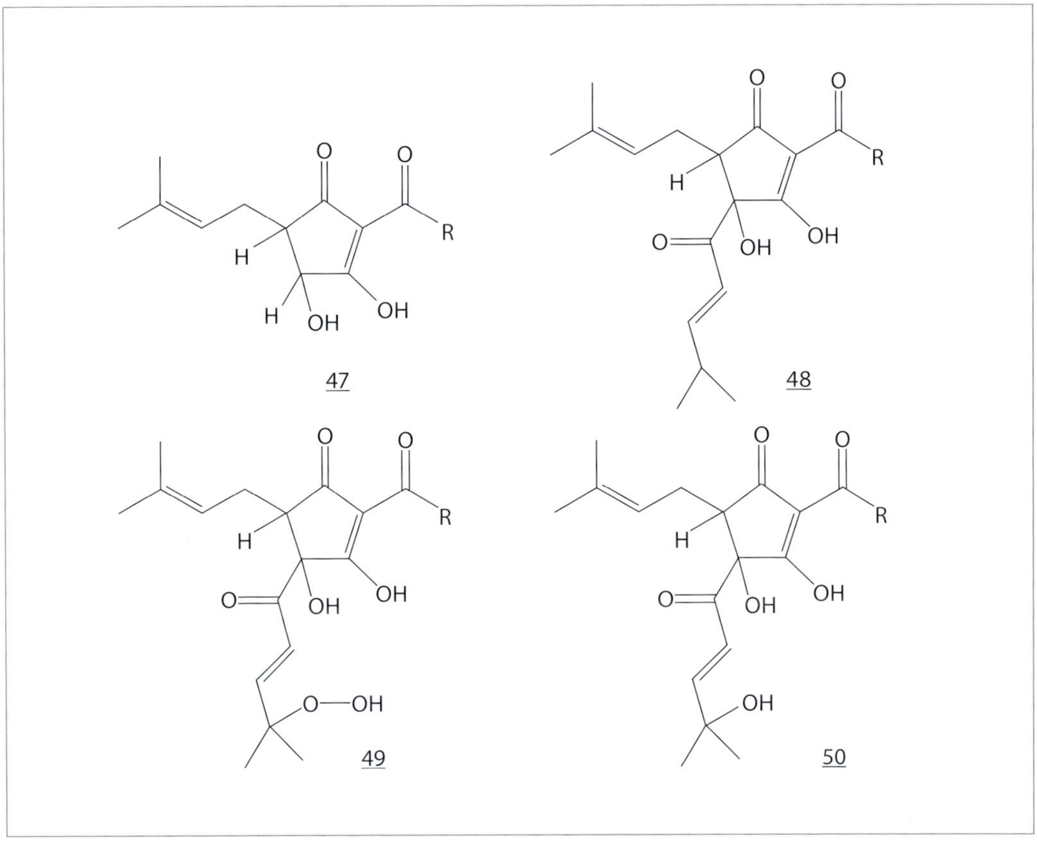

1.1.2 AROMA COMPOUNDS

To date, several hundred volatile substances have been identified in hops. These compounds are divided into a hydrocarbon fraction, which consists of the terpene hydrocarbons (70–75 %), oxygenated compounds (25–30 %, oxygenated fraction) as well as other aroma substances. The hydrocarbon fraction can be subdivided into monoterpenes and sesquiterpenes. Among others, monoterpenes include α-pinene, β-pinene, limonene and myrcene. Myrcene comprises anywhere from 15 to 60 % (by weight) of the total hop oils. Thus, it is the most important substance from a quantitative standpoint. Common sesquiterpenes include β-caryophyllene, α-humulene and farnesene [34]. Farnesene is typical for hop varieties from the Saazer cluster (Spalter, Tettnanger, Select) representing up to 30 % of the total hop oil content.

The oxygenated fraction includes alcohols, aldehydes, ketones, acids, esters and oxygenated heterocyclic compounds. The oxidized terpenes linalool (fig. 4) and geraniol exhibit floral and rose-like aromas, respectively. Linalool is a chiral compound occurring predominantly in the (R) configuration (92–94 %), which is considerably more aroma-active than the (S) configuration [35, 36, 37, 38].

Figure 4: Chemical structure of the chiral compound linalool

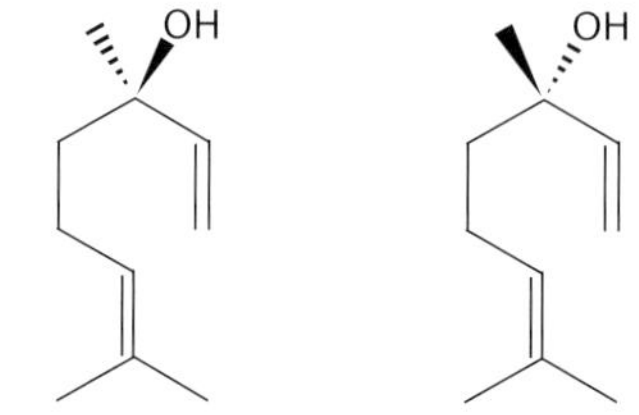

(R)- linalool (S)- linalool

Research has shown that the chiral distribution of linalool in aroma hops is approximately 94 % (R)-linalool [39]. This distribution remains consistent in unprocessed hops through to all conventional hop products. In a study spanning one year, the (R)-linalool content decreased from 92 % to 77 % in beer stored at 20 °C. Initially, (S)-linalool makes up only 10 % of the total amount of linalool present, but its concentration increases as the beer ages. However, this still does not explain the overall loss. Shifts in the linalool concentration result in a loss of aroma intensity [40].

During wort boiling, the nonpolar, hydrophobic terpenes and sesquiterpenes vaporize and exit the brewhouse in the steam from the kettle. Therefore, only trace amounts are found in beer. By contrast, compounds belonging to the oxygenated fraction occur in measurable quantities in beer and can be considered carriers for hop aroma. Although a portion of these compounds are also lost with the steam, they are formed at the same time during wort boiling [41].

Esters can contribute fruity aroma notes to beer. Fatty acids are generally the compounds responsible for a "cheesy aroma", especially in hops stored under unsuitable (warm) conditions. Heterocyclic compounds containing oxygen (epoxides) are formed through the autoxidation of sesquiterpene hydrocarbons. In addition, sulfur-containing aroma compounds, such as thioesters, sulfides and sulfur heterocycles, are found in hops [42, 43, 44].

Hop oil composition is dependent on the hop variety and can be strongly influenced by conditions during drying and the method of processing. The effects of oxidation can be minimized to a great extent by ensuring that the hops are continuously stored under cold conditions from start to finish, upon delivery of the fresh hops to the facility until processing is complete. Selected aroma hop varieties and the perceived intensity of their "fruity" and "floral" notes are listed in figure 5 [45].

Figure 5: Intensity of fruity and floral aromas in selected aroma hop varieties

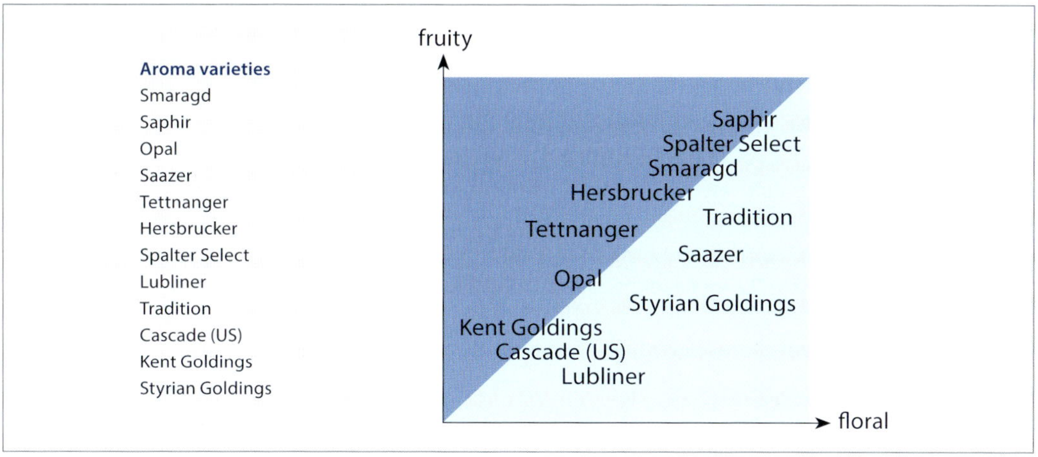

1.1.3 POLYPHENOLS

Depending on the hop variety and the type of hop product used, up to 30 % of the polyphenols in wort originate from hops, while the remainder comes from the malt. Around 100 individual phenolic compounds can be detected and measured analytically. Polyphenols consist of approximately 80 % condensable and 20 % hydrolyzable compounds. The former consist of polyphenol monomers and their glycosides. Lower molecular weight polyphenols can polymerize to form higher molecular weight products.

Polyphenols also include phenol carboxylic acids such as hydroxybenzoic acids and hydroxycinnamic acids. The primary prenylflavonoid in hops is xanthohumol [1].

Aroma hops possess higher quantities of low molecular weight polyphenols than bitter hops [46, 47, 48, 49]. Quantities vary depending on the provenance, the individual hop variety and the process of cultivation [50].

Proanthocyanidins composed of catechin or leucoanthocyandin units appear to be particularly effective at not only precipitating proteins but also creating turbidity [34, 51]. The term proanthocyanidin is derived from the capacity of these compounds to transform into their corresponding anthocyanidins when heated with acid in the presence of oxygen. In the brewing industry, they are also referred to as anthocyanogens [51].

Low molecular weight polyphenols are natural antioxidants. They increase the reducing capacity of the beer, thereby influencing its flavor stability [50]. Through their astringent characteristics, polyphenols can affect the bitterness, the mouthfeel and the body of a beer [24].

Furthermore, it is believed that polyphenols act as free radical scavengers in the human body and therefore could have an inhibitory effect on cancer cells (cf. Beer and Health). Xanthohumol has demonstrated an anti-carcinogenic effect both in animal tests and *in vivo* as part of clinical trials [48, 53, 54, 55, 56, 57]. The prenylflavonoid xanthohumol is localized in the lupulin glands of hops.

High molecular weight polyphenols can increase the color of the beer and create an astringent bitterness, especially if the wort is boiled for a long time. They reduce colloidal stability and lead to the formation of haze or turbidity in beer (cf. Filterability – Issues with Turbidity). In order to achieve a high degree of physical stability in beer, introducing polyphenols into the brewing process is generally avoided and efforts are even made to partially remove them. For instance, polyphenol-free hop extracts and barley varieties lacking procyanidins are available for this purpose. Alternatively, the polyphenol fractions responsible for turbidity can be removed during beer filtration by means of adsorption onto polyvinylpolypyrrolidone (PVPP) [48]. Partial elimination of polyphenols occurs during the separation of hot and cold break material, through the removal of whirlpools, tank bottoms and during filtration.

1.1.4 ANALYTICAL CHARACTERIZATION OF COMMONLY ENCOUNTERED HOP VARIETIES

Table 1 lists hop varieties cultivated in Germany. Of these, eleven are aroma hops with α-acid concentrations of 3–8 %, four are "flavor hops" with α-acid concentrations of 5.5–8.5 % as well as seven bitter hop and high α varieties with α-acid concentrations of 9–18.5 %. The ratio of ß-acids to α-acids is also provided: 0.7–2.4 for aroma hops, 0.5–0.7 for flavor hops and 0.2–0.5 for bitter hops and high alpha hops. Earlier, the

Table 1: Analytical characterization of commonly encountered hop varieties

Variety	Spalter	Tettnanger	Hallertauer Mfr.	Hersbrucker Spät	Saazer
Abbreviation	SP	TE	HA	HE	SA
Type	aroma	aroma	aroma	aroma	aroma
Bitter Substances					
α-acids (EBC 7.4)	4.1	4.0	4.1	3.1	3.2
β:α (EBC 7.7)	1.3	1.4	1.3	2.4	1.1
cohumulone (EBC 7.7)	24	25	21	20	25
Polyphenols					
polyphenols (EBC 7.14)	5.3	5.2	4.6	4.4	5.3
polyphenols:α	1.3	1.3	1.1	1.4	1.7
xanthohumol (EBC 7.7)	0.34	0.29	0.27	0.21	0.30
xanthohumol:α	0.083	0.073	0.066	0.069	0.094
Aroma Substances					
total oil (EBC 7.10)	0.60	0.60	0.85	0.75	0.55
β-caryophyl-lene:humulene	0.28	0.29	0.29	0.48	0.28
farnesene (EBC 7.12)	> 10	> 10	< 3	< 0.5	> 10
linalool (EBC 7.12; mg/100g)	4	4	6	5	3
linalool:α	1.0	1.0	1.5	1.6	0.9
yield (kg/ha)	1200	1300	1250	1750	1200

proportion of cohumulone was deemed important, but this belief is now under contention. In any case, this is difficult to assess in a qualitative manner. Previously, cohumulone concentrations were lower in conventional aroma varieties and higher in bitter varieties; however, over the past few decades, breeders have strived to lower the amount of cohumulone in newer varieties. Presently, higher values for cohumulone are more common. The polyphenol content is another point of interest, as no recognizable group characteristic exists with regard to their levels. As far as alpha acids are concerned, aroma hops introduce significantly more to wort or beer. This is true for the prenylflavonoid xanthohumol as well as for aroma compounds. As explained in 2.1.3, the ratio of ß-acids to α-acids serves as an indicator for the proportion of "auxiliary bitter substances" (referred to as "soft resins" in the past). They also reflect the ratio of EBC bittering units [IBU] to iso-α-acids in wort and beer and thus influence the harmonious character of hop bitterness in beer.

Perle	**Hallertauer Tradition**	**Spalter Select**	**Saphir**	**Opal**	**Smaragd**
PE	HAT	SE	SR	OL	SD
aroma	aroma	aroma	aroma	aroma	aroma
7.4	6.2	5.1	4.1	7.9	5.9
0.7	0.8	1.0	1.9	0.8	0.9
30	26	23	15	15	15
4.1	4.3	4.9	4.5	3.7	4.5
0.6	0.7	1.0	1.1	0.5	0.8
0.55	0.41	0.42	0.37	0.41	0.32
0.074	0.066	0.082	0.090	0.051	0.054
1.30	0.70	0.70	1.10	0.95	0.90
0.31	0.28	0.4	0.43	0.34	0.30
< 0.5	< 0.5	> 10	< 0.5	< 3	< 3
4	7	8	10	11	10
0.5	1.1	1.6	2.4	1.4	1.7
1850	1950	2000	2000	1900	1900

Table 1: Analytical characterization of commonly encountered hop varieties

Variety	Cascade	Mandarina Bavaria	Hallertauer Blanc	Huell Melon	Northern Brewer
Abbreviation	CA	MB	HC	HN	NB
Type	aroma	aroma	aroma	aroma	bitter
Bitter Substances					
α-acids (EBC 7.4)	6.0	7.9	8.5	5.8	9.2
β:α (EBC 7.7)	0.7	0.5	0.6	0.7	0.4
cohumulone (EBC 7.7)	31	30	24	29	27
Polyphenols					
polyphenols (EBC 7.14)	4.3	4.0	5.4	3.9	3.9
polyphenols:α	0.7	0.5	0.6	0.7	0.4
xanthohumol (EBC 7.7)	0.39	0.59	0.38	0.56	0.61
xanthohumol:α	0.065	0.075	0.045	0.097	0.066
Aroma Substances					
total oil (EBC 7.10)	1.00	1.20	1.10	1.10	1.50
β-caryophyl-lene:humulene	0.50	0.30	1.00	1.20	0.34
farnesene (EBC 7.12)	> 10	> 3	< 0.5	> 10	> 0.5
linalool (EBC 7.12; mg/100g)	4	5	5	3	4
linalool:α	0.7	0.6	0.6	0.5	0.4
yield (kg/ha)	2100	2100	2300	1900	1600

Nugget	Hallertauer Magnum	Hallertauer Taurus	Hallertauer Merkur	Herkules	Polaris
NU	HM	TU	MR	HS	PA
bitter	bitter	bitter	bitter	bitter	dual
11.3	13.9	15.9	13.3	16.7	18.6
0.3	0.2	0.3	0.5	0.3	0.3
29	27	23	20	36	26
3.4	2.6	3.1	4.2	3.8	4.0
0.3	0.2	0.2	0.3	0.2	0.2
0.68	0.47	0.89	0.37	0.80	0.79
0.060	0.034	0.056	0.028	0.048	0.042
1.70	2.40	2.00	2.20	1.70	3.20
0.47	0.28	0.29	0.29	0.28	0.40
< 0.5	< 0.5	> 0.5	< 0.5	< 0.5	< 0.5
10	8	19	13	8	8
0.9	0.6	1.2	1.0	0.5	0.4
2200	2000	2000	2000	2700	2200

1.2 HOP PRODUCTS

Figure 6: Classification of hop products

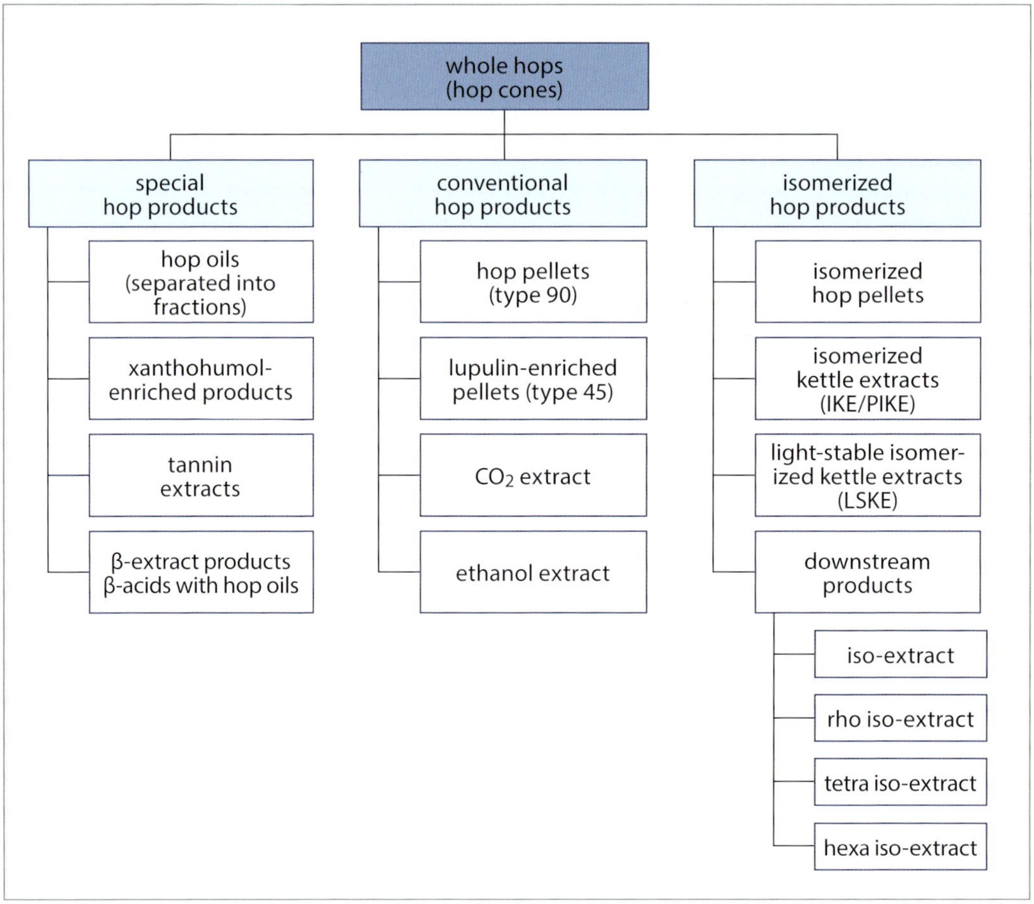

The different categories of hop products are shown in figure 6. The distinguishing feature of hop products is their reduced volume compared to whole hops. In particular, enriched hop products (type 45 pellets or hop extracts) offer a number of advantages. Hop products can easily be stored in refrigerated spaces. Transport and handling of hop products are greatly simplified; they also require less packaging material. The uniform distribution of α-acids in each production batch facilitates the precise dosing of hops in the brewery. Furthermore, hop products are better protected from oxidation through special packaging, which in turn preserves their quality over a longer period of time. The concentrations of pesticide residues found in whole hops are below the maximum permissible values stipulated by law. Even lower values are present in hop products, although these may vary greatly depending on the type [58, 59, 60]. Hop products can be differentiated on the basis of their intended application and at what point in the process they are introduced (bitterness, aroma, foam enhancement, light stability, etc.). If a hop product is added after wort boiling, that is, directly into the beer, it is referred to as a "downstream product" [61].

Brewers can choose from a broad range of hop products. Factors influencing this decision include not only cost and conditions specific to the brewery but also legal aspects such as whether a hop product is permitted in beer production in a particular country. For example, this is governed by legislation in the

European Union in addition to other federal laws. In Germany, a law known as the *Vorläufiges Biergesetz* (Provisional Beer Law) dictates that only conventional hop products may be added on the hot side of production. Whole hops and hop pellets are also permitted on the cold side (dry hopping).

1.3 ANALYSIS METHODS

Conventional hop additions in the brewery are usually determined on the basis of the conductometer value or according to the α-acid content of hop products (pellets, CO_2 extract) measured with HPLC. This approach has been shown to impart a consistent bitterness to beer. The conductometer value for bitterness is used for ethanol extract, since it correlates strongly with the sensory impression of perceived bitterness in beer [63].

Conventional standard analysis methods are often designed to measure the sum of specific groups of substances. However, the trend is moving toward a quantitative determination of individual compounds. The methods for determining α-acids can either be performed in an unspecific manner with the conductometer value (CV) or in a specific manner by means of HPLC [64, 65]. Figure 7 is a sample chromatogram showing the complete separation of the three primary homologs.

Figure 7: HPLC chromatogram of bitter acids [66]

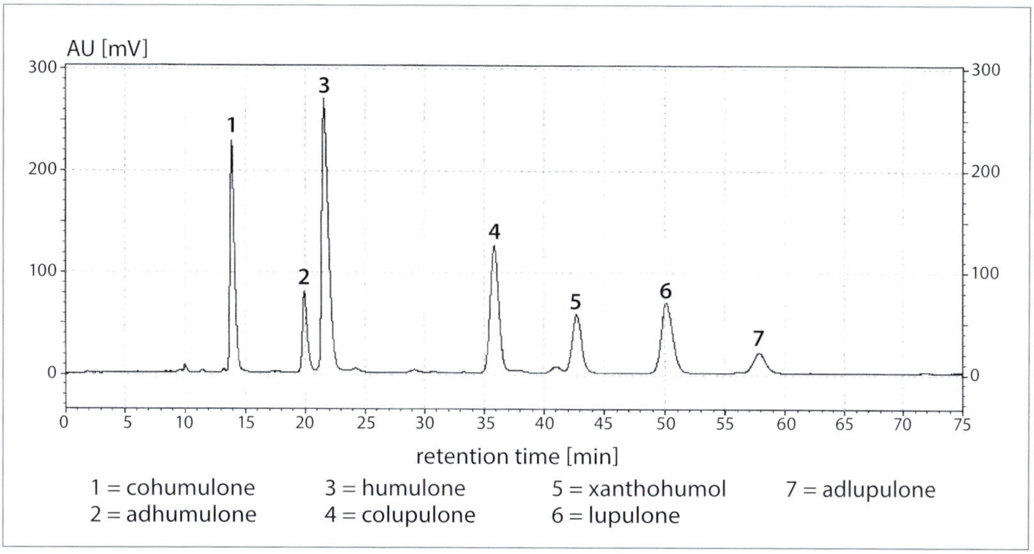

Considerable differences can be observed if the conductometer value is compared with the sum for α-acids measured by HPLC methods. Figure 8 shows the relative differences between results obtained using EBC methods 7.5 (CV) and 7.7 (HPLC). The conductometer analysis not only measures the α-acids but also other substances that can be precipitated with lead acetate. For this reason, the conductometer value is higher than the sum of the α-acids measured by HPLC. The difference between these values is dependent on hop variety and the year of the harvest, among others.

The sum of the bitter compounds in beer is expressed as the EBC bitterness value (IBU). Here again, the determination of iso-α-acids by HPLC generates lower values. This difference is referred to as "NIBC" or "non-isomerized bitter compounds". These substances are classified as "auxiliary bitter compounds". The proportion of auxiliary bitter compounds increases with the ratio of β-acids to α-acids and the point at which the hops are added in the brewing process (e.g. in partial additions) [21].

The total oil content is measured through distillation. Individual substances in the hop oil can be determined by means of capillary gas chromatography (GC). In some instances, gas chromatography is performed in conjunction with mass spectroscopy (GC-MS) or multidimensional GC-MS [43].

The sum of the polyphenols present can be measured using a wet chemistry method, while the analysis of individual compounds is performed with HPLC [43].

2 SPECIAL ASPECTS OF HOPS IN BREWING APPLICATIONS

2.1 HOP ADDITION

In brewing, hops are added to favorably influence beer aroma and flavor. Beer aroma is strongly impacted by the essential oils found in hops. Hop oils are volatile substances that contribute to the characteristic aroma of beer. Conversely, beer flavor is attributable to non-volatile substances. The greatest contribution to the bitter flavor is made by the iso-α-acids, which are responsible for more than 75 % of the bitterness perceived in beer [67, 68]. The quality of bitterness associated with cohumulone is a point of contention. Some researchers believe that cohumulone lends a harsh, unpleasant bitterness to beer [69, 70] although this finding has not been confirmed by others [31, 68, 70, 71, 72, 73]. The intensity of the bitterness of all three homologs is approximately the same [31, 68], although there is a tendency to rate the bitterness of iso-cohumulone as somewhat lower [30, 32, 70].

Figure 8: Differences in analysis results obtained with EBC methods 7.5 and 7.7

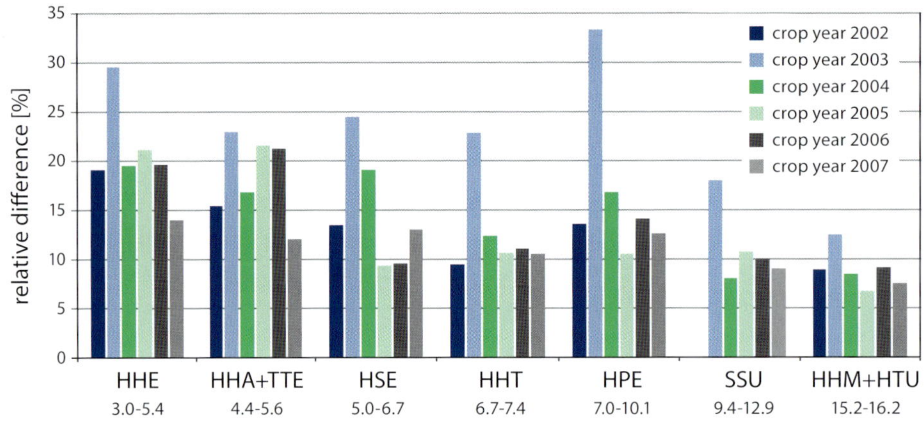

Alphabasis: (EBC 7.5)
2003 harvest year with a low alpha basis

Legend:

HHE: Hallertau Hersbrucker Spät
HHA: Hallertau Hallertauer Mittelfrüher
TTE: Tettnang Tettnanger
HSE: Hallertau Spalter Select
HHT: Hallertau Hallertauer Tradition
HPE: Hallertau Perle
SSU: Slovenian Super Styrian
HHM: Hallertau Hallertauer Magnum
HTU: Hallertau Taurus

In the list of hop varieties, the cultivation area is given first, followed by the hop variety.

The quantity of hops should be adjusted to achieve the desired flavor profile of the respective beer style. The flavor and aroma of the hops should pair well with the beer matrix.
The following parameters should be taken into consideration:

- composition of the raw materials (water, malt, adjuncts, additives)
- hop variety
- hop product
- quantity of hops added
- timing of the hop addition (the length of time hops are boiled after each partial addition)
- when dry hopping: the point at which the hops are added (fermentation, maturation, storage), the length of contact time and the amount of mixing
- wort boiling system
- amount of time spent in the whirlpool and the wort temperature in the whirlpool
- quality of break separation (hot break, cold break material)
- reuse of hot break in the brewing process
- fermentation (spectrum of aroma compounds in the beer)
- composition of the wort (e.g. original gravity, pH)
- final degree of attenuation
- filtration (adsorption of hop oils on filter aids, yeast and particulates)
- stabilization (polyphenols)
- container choice (clear glass, e.g. lightstruck flavor)
- consumer expectations (regional and cultural differences)
- declaration and labeling (legal regulations)

A pleasant, balanced flavor profile has a direct impact on the enjoyment of beer. Independent of this, targeted hop additions allow different beer styles to be developed, which occupy a distinct niche in the beer market. Hops are an excellent tool to help breweries develop beers that stand out from the competition.

In accordance with the German Provisional Beer Law, only conventional hop products (fig. 6) may be added to the wort, i.e. whole hops, pellets and extract [74]. With the exception of extract, this is also applicable to beer. If whole hops are used, a hop-straining device is necessary to remove the solid material from the wort. The solid material from the hop pellets is separated from the wort along with the hot break (e.g. in a whirlpool). An excessive amount of hops has a negative impact on whirlpool function. If larger quantities of hot break material are carried over into fermentation, a harsh bitterness can result. After the wort has left the whirlpool, the trub cone should not touch the walls of the vessel. Extraction devices can be used for hop addition to help preserve hop aroma and reduce losses.

Hops introduce nitrate to wort and beer. The amount of nitrate is lowered significantly through the use of hop extracts and enriched hop pellets. Given a nitrate concentration of 1000 mg/100 g of whole hops (9 % α-acids, air-dried) and a hopping rate of 8 g α-acids per hl of wort, this comes to a nitrate concentration of approximately 9 mg/l. If a hop extract made from the same hops is used instead, the resultant nitrate concentration is 0.05 to 1 mg/l [75]. Through extraction with CO_2 or ethanol, the nitrate concentration in the resultant hop extract is reduced by at least 98 % [76]. Since larger amounts of hops are used in dry hopping, their nitrate content takes on a special significance.

For beers produced in areas not subject to the German Provisional Beer Law, referred to in this chapter as "international beers", there are many different hop products available that have been processed to a greater degree (fig. 6). Isomerized hop products can be added to hot wort in the kettle. These include iso-pellets and isomerized "kettle extracts". Another alternative is to dose "downstream products", as they are known, into the beer immediately prior to beer filtration or in some cases, in the brewhouse. Iso-extracts are used to correct beer bitterness. Some reduced iso-extracts are employed to enhance foam stability or protect beer against developing a lightstruck flavor. An impact on the aroma and flavor profile of the beer cannot be ruled out since the isomerization of α-acids is performed at higher temperatures in non-conventional hop products [77]. Gentle processing is especially important when the isomerization of iso-pellets is carried out in a heating chamber.

Furthermore, different hop fractions have been enriched to create hop products designed to fulfill specific requirements.

In particular, reduced iso-extracts have different impacts on the sensory perception of bitterness and its intensity. When determining the sensory bitterness of identical quantities of tetrahydro-iso-α-acids and iso-α-acids in beer, tetrahydro-iso-α-acids exhibit a bitterness that is 1.0–1.4 times higher than that of iso-α-acids. Compared to iso-α-acids, rho-iso-α-acids possess 0.6–0.8 times the bitterness [78, 79, 80]. In some cases, if the dosage is too high, an unbalanced bitterness may result.

There are a multitude of reasons for adding hop products at different stages in the brewing process. Other than special legal requirements (e.g. the German Provisional Beer Law), the composition of the hop products dictate at which point they may be added. Hop products dosed into beer end up in the final product. The sensory characteristics of the hop product, its solubility in beer and many other aspects must comply with the requirements stipulated in the laws governing food and its production. If the hop product is added in the brewhouse, then it will certainly be subjected to further stages of processing. For instance, this ensures the elimination of hop compounds with low solubility or undesirable aroma compounds. Isomerized hop products are primarily chosen for economic reasons.

Bitter hops should be added at the beginning of the boil, not only to achieve the highest level of isomerization possible but also to drive out undesirable aroma compounds through evaporation. The later the hops are added, the less aroma is eliminated with the vapors, leaving more hop oils in the finished beer. Aroma hop additions intended to produce beers with a pronounced hop aroma should first be added towards the end of the boil or simply placed in the whirlpool prior to casting out the wort in the kettle. This helps achieve an optimal transfer of hop aroma and flavor to the beer [35].

Figure 9: How the point of hop addition influences beer character

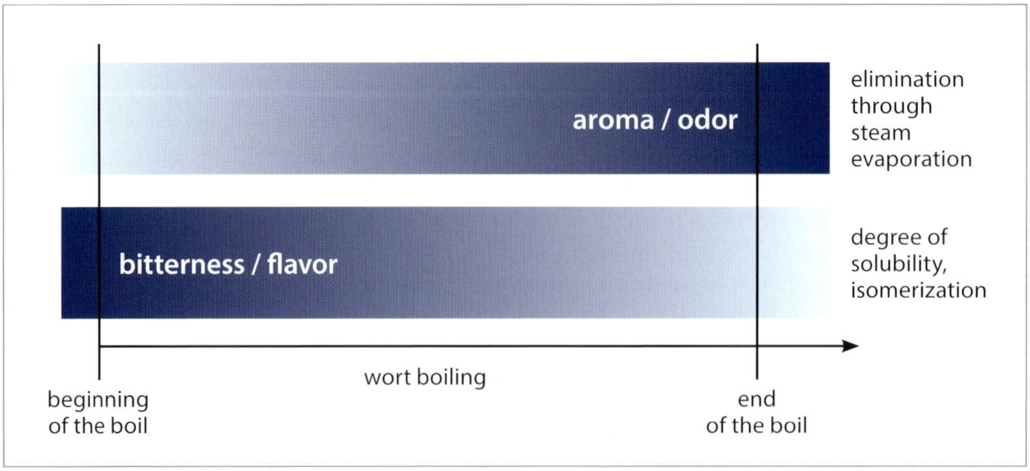

It is evident from figure 9 that early hop additions can positively influence both the bitterness and the flavor of beer. To impart hop aroma to beer, the hops should be added close to the end of the boil or directly to the beer, on the cold side.

2.1.1 BEER PRODUCED WITH A LOW HOPPING RATE

A "weakly hopped" beer or one produced with a low hopping rate typically contains around 15–20 bittering units and exhibits no perceptible hop aroma in the nose. The original gravity of beer is close to 11 % (bottom or top-fermented). The beer should possess an underlying bitterness that does not dominate the overall sensory impression. Descriptions, recommendations and brewing techniques for producing this type of beer are outlined below.

Provisional Beer Law:

- approximately 20 bittering units (IBU), due to the intense flavor profile of the beer from the use of 100 % malt in the grist.
- a single hop addition at the beginning of the boil to increase the bitterness. For a more balanced bitterness, three different additions of hops are preferable: shortly after the beginning of the boil so that myrcene and other volatile aroma compounds can evaporate. The second hop addition should take place at the midpoint of the boil and the third at approximately 10 minutes before the end of the boil. The first addition can consist of bitter hops or high alpha hops, while the other two additions should only be aroma hops.
- Each hop addition can comprise 1/3 of the total quantity of alpha-acids, or 50 % of the amount can be given as the first addition and 25 % given in the two subsequent additions.

International Beers

- approximately 15 bittering units [IBU] for beers produced with a low hop rate, because the flavor profile of the beer is not as pronounced due to the use of adjuncts in the grain bill; consequently, bitterness is perceived more strongly in these beers
- conventional hop additions are possible (as described previously)
- employ isomerized pellets to increase hop utilization
- add isomerized kettle extract during wort boiling to increase hop utilization and to improve microbiological stability
- add iso-extract prior to filtration to correct bitterness
- add reduced iso-extract to stabilize foam (tetra-hydro-iso-α-acid or hexa-hydro-iso-α-acid products)
- employ a combination of the products listed above

2.1.2 BEER PRODUCED WITH A HIGH HOP RATE

A strongly hopped beer, that is, one produced with a high hop rate, typically contains around 20–40 bittering units and exhibits no perceptible hop aroma in the nose. The original gravity of this beer is close to 12 % (bottom-fermented). The beer should possess a readily perceptible bitterness. Descriptions, recommendations and brewing techniques for producing this type of beer are outlined below.

Provisional Beer Law

- approximately 30–40 bittering units, due to the very intense flavor profile of the beer from the use of 100 % malt in the grain bill
- strong beers should be more aggressively hopped in order to neutralize any sweet or heavy aftertaste; the correct balance between alcohol, residual extract and bitterness must be found to increase the drinkability and thus enjoyment of strong beers.
- the bitter hop addition should take the form of extract or pellets and be carried out shortly after the beginning of the boil to increase hop utilization and to avoid overloading the whirlpool. This

is followed by two further additions (aroma hops) at the midpoint of the boil and approx. 10 minutes prior to the end of the boil. Each hop addition can comprise 1/3 of the alpha-acid quantity, or 50 % of the amount can be given as the first addition and 25 % given in the two subsequent additions.
- select a suitable hop variety that has been tested in brewing trials with the beer in question; aroma hops may be used exclusively or in conjunction with bitter hops

International Beers
- approximately 20–27 bittering units, because the flavor profile of the beer matrix is not as pronounced due to the use of adjuncts; consequently, bitterness is perceived more strongly in these beers.
- conventional hop additions are possible (as described previously)
- employ isomerized pellets to increase hop utilization
- add isomerized kettle extract during wort boiling to increase hop utilization and to improve microbiological stability
- add iso-extract prior to filtration to correct bitterness; using a higher proportion of iso-extract can lead to a harsh, unbalanced, atypical bitterness
- add reduced iso-extract to stabilize foam (tetra-hydro-iso-α-acid or hexa-hydro-iso-α-acid products)
- employ a combination of the products listed above

2.1.3 BEER WITH A PRONOUNCED HOP AROMA

A beer with a pronounced hop aroma typically contains around 20–40 bittering units and exhibits a hop aroma that is readily perceptible in the nose. The original gravity of this beer is around 11–12 % (bottom fermentation). The beer should possess a distinct hop aroma and should be perceived both as a fine aroma and as a pleasant hop flavor [81]. Descriptions, recommendations and brewing techniques for producing this type of beer are outlined below.

Provisional Beer Law
- approximately 30–40 bittering units, due to the very intense flavor profile of the beer from the use of 100 % malt in the grain bill
- carry out the first hop addition at the beginning of the boil in order to increase hop utilization
- select a suitable hop variety for addition at the beginning of the boil; this variety should have already been tested in brewing trials with the beer in question; aroma hops may be used exclusively or in conjunction with bitter hops.
- introduce aroma with a second hop addition of aroma hops, make this addition shortly before the end of the boil or place the hops in the whirlpool before transferring the wort from the kettle. This serves to minimize losses through volatilization and evaporation and ensures maximum extraction of aroma compounds, from the hops through to the finished beer.
- introduce aroma with a third hop addition of aroma hops to beer. The later the hops are added in the process (fermentation-maturation-lagering/conditioning), the fewer aroma losses can be expected as a result of desorption and stripping by carbon dioxide generated during fermentation. This preserves more of the volatile aroma compounds commonly referred to as "top notes", which tend to be very pronounced in fresh, unprocessed hops.
- pre-cooling the wort before it enters the whirlpool can enhance the transfer of aroma. However, this diminishes subsequent isomerization in the whirlpool. A correlation exists between the drop in isomerization in the whirlpool and the extent to which the wort is pre-cooled [35, 83, 84].
- A hop addition performed at the midpoint of boiling has been proven to be equally beneficial for beers hopped at low rates and at high rates [96]. The harmonious character of the bitterness is improved. The EBC bittering units remain the same or are only slightly lower, as the continued exposure to heat in the whirlpool results in further isomerization. The difference between the EBC bittering units and the iso-α-acids increases.

- Decisions regarding which aroma hop variety is the most suitable should be made on the basis of results obtained from brewing trials. Each variety has a different influence on beer flavor and aroma. An overview of the various aroma characteristics for German hop varieties can be found in the CMA Hop Variety Portfolio [6].
- The intensity and quality of the hop aroma can be individually optimized based on the timing of the hop addition. The time interval for the last hop addition begins approximately 10 min before the end of the boil and extends to the placement of hops in the whirlpool before casting out the wort in the kettle. Hop extraction in beer is primarily influenced by the mixing of hops with the beer as well as by the contact time and the temperature.
- The larger the quantity of hops (e.g. type 90 pellets), the greater their contribution is to beer flavor. This applies to both aroma and bitter hops.
- The distribution of hop pellets in the whirlpool is not always homogeneous – especially with large hop additions, so the concentration of aroma compounds and bitter compounds may vary. In this situation, dissolving the pellets in water at 60 °C before adding them to the whirlpool is advantageous. The loss in hop oils while casting out the wort is lower and the distribution of the pellets is more uniform [84].
- For added assurance, an extraction device can be mounted between the whirlpool and the wort chiller to allow the hot wort to flow over the pellets en route to the heat exchanger during wort chilling. After the wort has passed through the extraction device, the remaining hop material is rinsed with hot water. This procedure also promotes proper whirlpool function by reducing the amount of hop material in the whirlpool.
- Introduce aroma by adding aroma hops directly to the beer. The later the hops are added in the process (fermentation-maturation-storage), the fewer aroma losses can be expected as a result of desorption and stripping by the carbon dioxide generated during fermentation. This preserves more of the volatile aroma compounds commonly referred to as "top notes", which tend to be very pronounced in fresh, unprocessed hops.

International Beers

- approximately 20–30 bittering units, because the flavor profile of the beer is not as pronounced due to the use of adjuncts in the grain bill; consequently, bitterness is perceived more strongly in these beers.
- conventional hop addition is possible (as described previously)
- add isomerized kettle extract during wort boiling to increase hop utilization and improve microbiological stability
- use iso-pellets made from aroma hops to improve aroma yields and hop utilization [85]
- add iso-extract prior to or after filtration to correct bitterness
- add hop oils or "hop oil fractions" prior to filtration to impart aroma
- add reduced iso-extract to stabilize foam (tetra-hydro-iso-α-acid or hexa-hydro-iso-α-acid products)
- employ a combination of the products listed above

Breweries generally calculate the amount of aroma hops needed based on the α-acid content of the hops. A more logical approach would be to make this decision according to the hop oil content, preferably the linalool content [86, 87, 88, 89] or the key compounds associated with hop aroma [43, 90]. This ensures that the same amount of aroma compounds is added each time. Fluctuations in oil content related to the crop year can be compensated with this approach. The α-acid content is relatively unimportant for late hop additions, since only a small portion of the α-acids are isomerized. The goal of these later hop additions is simply to transfer the maximum amount of aroma compounds possible. Linalool has been identified as the character impact compound for beers with a pronounced hop aroma [39, 91, 92]. Its sensory threshold in beer has been reported to be 5 µg/l [35, 93, 94], but may be as high as 80 µg/l [95] according to some sources. Depending on the beer matrix, hop aroma can be perceived sensorially starting at a linalool concentration of 20 µg/l. At concentrations around 50 µg/l, the hop aroma has been described as floral or perfume-like. Besides linalool, other compounds play a role in hop aroma. Some of these compounds occur in concentrations below their sensory threshold in beer, but still contribute to the flavor profile through synergistic and additive effects.

The flavor stability of beer can be impacted in a positive manner through late hop additions. Moreover, a subtle yet perceptible hop aroma has a masking effect on the aging compounds in beer [48, 96, 97, 98, 99, 100].

2.1.4 BEER ENRICHED WITH XANTHOHUMOL

As time progresses, the prenylflavonoid xanthohumol is attracting greater interest from the fields of medicine, physiology and brewing science (cf. Beer and Health). The xanthohumol content of beer can be increased through the use of specially enriched hop products and by altering the brewing process [54, 99, 100, 101, 102, 103, 104, 105, 106].

Xanthohumol can be found in whole hops, pellets and ethanol hop extract as well as in xanthohumol-enriched hop products. Xanthohumol is not present in CO_2 hop extract due to its solubility properties [107]. With the application of special brewing procedures, it is possible to produce beer containing higher concentrations of unisomerized xanthohumol while still complying with the German Provisional Beer Law [106].

The aforementioned purity law does not apply to beer-based beverages or non-alcoholic beverages. Therefore, the desired xanthohumol concentration in these beverages can be achieved through the addition of xanthohumol-enriched hop products as well as their addition to beer brewed in regions not subject to the German Purity Law. Stabilization with PVPP should be avoided with pale beer styles that have been enriched with xanthohumol, because PVPP removes most of the polyphenols from beer. With beer styles that are brewed with dark specialty malts in the grist, stabilization with PVPP can be carried out without losing very much xanthohumol (cf. Beer and Health).

2.1.5 ADDING HOPS ON THE COLD SIDE OF PRODUCTION

Influenced by craft beer, more and more breweries are electing to adopt the practice of dry hopping (adding hops on the cold side of production). Dry hopping is typical in the production of English ales (e.g. best bitter) but was also common in Germany around thirty years ago. A distinctive hop aroma was achieved by adding hops (dry hopping) to the barrels used for transporting beer (England) or to the maturation tank for certain beer styles, such as *Kölsch* or *Altbier* brewed in the west of Germany as well certain pilsners. Originally, dry hopping was carried out with whole hops pressed in bales. These were subsequently replaced by pellets, as whole hops were more likely to cause damage. Whole hops in wooden barrels or horizontal lager tanks could lodge in the bunging device, allowing excessive pressure to build, which would then cause deformation or rupture of the tank or barrel. Furthermore, whole hops are not suitable for storage, since they lose approximately 15 % of their α-acid content per year, even when they are stored under ideal conditions, e.g. at 0 °C in a room reserved exclusively for this purpose. Fluctuations also occurred in the bitterness and aroma of whole hops throughout the year. These problems were resolved with the introduction of hop pellets, a product that can easily be stored. Nevertheless, in Germany, dry hopping disappeared around the early 1980s. It was feared that the colloidal stability as well as the flavor stability could withstand neither the planned expansion into other distribution areas nor the (later) introduction of a best-before date. This was a legitimate concern because the procedure of involving the addition of hops to a lager tank prior to transferring the beer allows a significant amount of oxygen uptake. When late hop additions, either in the kettle or the whirlpool, are employed in place of dry hopping, other aroma notes are imparted to the beer. However, these aromas are partially lost through evaporation or during fermentation. Only through dry hopping techniques is it possible to transfer a significant amount of the aromas normally associated with fresh hops to beer. Today, dry hopping is widely established once again, especially in craft breweries. In part, the additions are significantly higher than they were in the past, typically ranging from 50 to 500 g/hl or even higher depending on the beer style.

The effects of dry hopping in the lager cellar (measured over a four-week period of maturation and cold storage) were monitored by measuring the transfer rates for each group of substances. A quantity of hops containing 1.5 ml hop oil/hl was added. This corresponded to 80–130 g/hl of hop pellets. The hopping rate for the whirlpool addition was 2.5 ml hop oil/hl. Compared to the control beer with 23 IBUs, the late-hopped beer showed an increase of 1–4 bittering units and the dry-hopped beer, 4–7 bittering units. The iso-α-acid concentration was less, 4.5 mg/l and 3.5 mg/l, respectively. The concentration of non-isomerized α-acids exhibited an increase of 3.0 and 4.5 mg/l, respectively. Total polyphenols were 17 % higher and the hop polyphenols were 62 % higher in the dry-hopped beer. Dry hopping imparted 50 % more terpenes to the beer, which was also the case for linalool and esters. Compared to beer produced with whirlpool hopping, dry hopping enhanced not only hoppy but also fruity notes in particular. However, it was obvious that dry hopping produced a distinctly different aroma impression, especially due to the terpenes. The aroma impression of dry-hopped beer was far closer to that of freshly harvested hops than the impression that results from hot-side addition [108, 109].

Dry hopping techniques: The inherent problem associated with dry hopping procedures using pellets is oxygen uptake in the beer. Therefore, pellets should be mixed with de-aerated, slightly carbonated water. Stripping with CO_2 is also possible. The loss of hop oils is almost negligible. This pellet solution can then be transferred to a lager tank that has been purged with CO_2. At this point, green beer or fully matured beer can be transferred to the tank containing the pellet solution. Depending on the tank design, special measures must be taken to ensure thorough mixing. Hop pellets prepared in this way can be transferred to the lager tank from the top through suitable CIP fittings. In horizontal lager tanks, the pellets remain at the bottom of the tank. In this situation, the last of the beer must be drawn from the tank over a standpipe. The volume of beer retained in the tank bottoms can be considerable. In cylindroconical tanks, the pellets can slowly be removed from the cone along with the yeast sediment. Pellet residues in beer can lead to filtration problems, even with a centrifuge installed upstream from the filter.

To avoid this problem, a sack can be filled with pellets or whole hops and placed in a (horizontal) lager tank through the manway. Of course, issues arise concerning, for instance, whether the contents of the sack are uniformly wetted and how the hop compounds are transferred to the beer. This can only be achieved by recirculating the beer in the tank with a pump. By doing so, the contents of the sack must come into contact with the beer in a consistent manner.

In dry hopping, the contact time between the beer and hops cannot be determined analytically. The beers must be evaluated sensorially at set intervals. The intervals become increasingly shorter as the dry hopping procedure nears completion, that is, when the optimal sensory profile has been achieved.

Once the appropriate amount of exposure time between the pellets and beer has been reached, the beer must be transferred to another tank, leaving the solid hop material behind. This should be carried out quantitatively under a CO_2 atmosphere. In breweries equipped with a CO_2 recovery system, the CO_2 can be collected and reused; however, the problem of extract loss still remains.

The least expensive solution – especially if larger tanks are involved – is to install extraction vessels to homogenize the quantity of pellets required. The beer is pumped through the vessel, which is outfitted with a screen to retain the pellets. The beer can be pumped again through the extraction vessel at specific time intervals. The majority of the pellet material is retained in the extraction vessel except for small amounts of fine particles. Potential extract loss through retention in the "spent pellets" is kept low by rinsing them with deaerated water. Extraction vessels are outfitted with an agitator. The system is fully automated and mobile, allowing it to be moved from tank to tank [110, 111, 112].

The polyphenols in the hops lower the colloidal stability of dry-hopped beers. "Post-filtration" treatment with PVPP or carbohydrate polymers offers a remedy. Nevertheless, measurements to determine a best-before date should be approached with caution. Since the majority of craft beers are not filtered, consumers generally do not consider an opalescent appearance or slight turbidity in the beer to be a problem.

In conclusion, the introduction of nitrate and pesticide residues into the beer through dry hopping deserves mention. In dry hopping, up to 88 % of the nitrate content in hops (roughly comparable to hop additions on the hot side) is transferred to the beer. At very high hopping rates, this results in nitrate levels slightly below or even exceeding the concentrations permitted in drinking water. In contrast, due to their solubility in water and the cold temperatures in fermentation, maturation and storage, the values measured for pesticide residues are not significantly higher in dry-hopped beers. This was confirmed in the results from the analysis of six samples of India pale ales and pale ales [113, 114, 115].

Hazy beers or those containing yeast often have a more distinctive and enduring hop aroma, because the hydrophobic hop oils attach to the equally hydrophobic surfaces of the particles responsible for turbidity, therefore allowing them to remain in the beer.

2.1.6 LIGHTSTRUCK FLAVOR

Lightstruck flavor, as it is commonly known, is a sensory change in beverages that occurs through non-enzymatic, photochemical reactions. This off-flavor also occurs in milk, wine and beer, among others, and is caused by sulfur compounds. 3-methyl-2-butene-1-thiol (3-MBT) has been identified as the primary compound responsible for lightstruck flavor [117]. Cleavage of the isoprene side chain of isohumulone and its reaction with a sulfur source generates this off-flavor. Riboflavin (RF) acts as a photoinitiator in the reaction. Through excitation as a result of exposure to UV light (350–500 nm), it is elevated to the high-energy triplet state (3RF*). In this state, riboflavin serves as a strong electron acceptor. Electron transfer from isohumulone to 3RF* is a key reaction in the formation of 3-MBT [117, 118, 119, 120, 122]. Through intramolecular or intermolecular cleavage of H^+, the free electron on the C4 atom of the molecule is localized, causing cleavage of the 4-methylpent-3-enoyl radical, which is then stabilized in the subsequent reaction through decarboxylation to a 3-methyl-2-butenyl radical [118].

The formation of this radical is depicted in figure 10. In the presence of a suitable sulfur source, the 3-methyl-2-butenyl radical reacts with a sulfhydryl group, generating 3-MBT [121, 122, 123, 124]. Cysteine is considered to be a source of sulfur for this reaction [125]. The lightstruck flavor is readily recognizable even when no degradation of bitter compounds is detectable [126]. The photochemical degradation of only one isohumulone molecule in a million is sufficient for the lightstruck flavor to be noticeable [127]. 3-MBT is one of the most aroma-active compounds known, with an aroma threshold of 0.2–0.4 ng/l in water [128] and 4.4–35 ng/l in beer [129, 130].

If light-stable or reduced hop products are used exclusively, the development of a lightstruck flavor can be prevented to a large degree. Any transfer of iso-α-acids to the beer must be avoided. Sources of contamination include residues from previous (conventional) hop additions, e.g. in the pipes, on cropped yeast used to ferment a previous batch of wort brewed using conventional hops or hop products or through the reuse of hot break material.

When using reduced hop products, the sensory bitterness, i.e. the level of bitterness perceived by the consumer, must be carefully considered. If the same amount of reduced iso-α-acids and regular iso-α-acids are added to two otherwise identical beers, there will be a perceptible difference in bitterness. For this reason, it is recommended that breweries produce pilot batches to determine the correct dosage of reduced products. The potential bitterness levels of various hop products are provided in table 2.

Figure 10: Reaction pathway for the generation of 3-methyl-2-butenyl radicals [118]

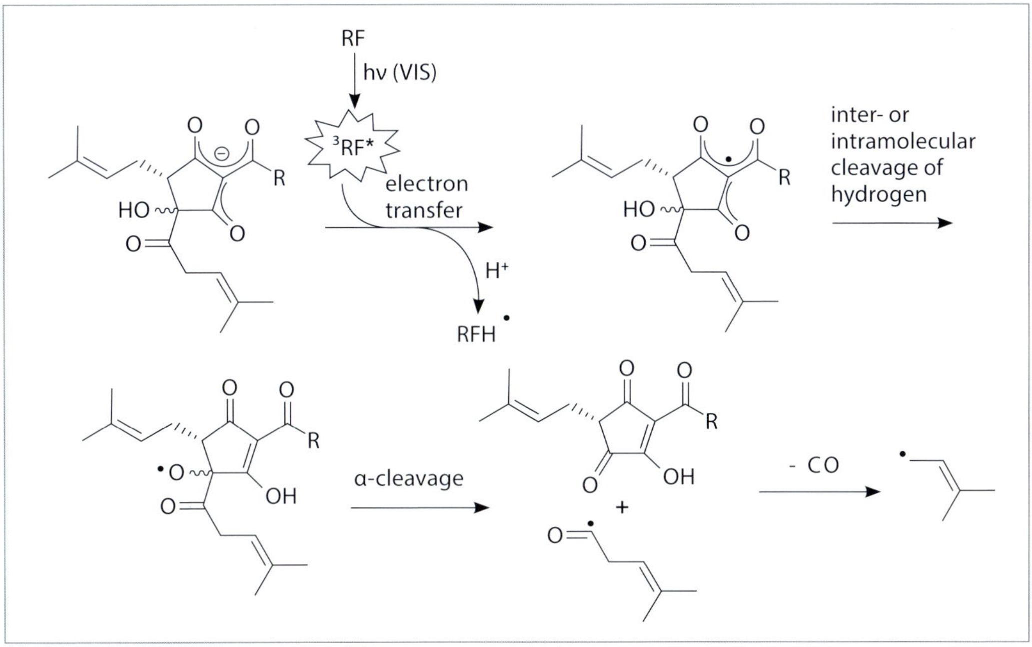

Table 2: Relative bitterness of various hop products compared to the bitterness of iso-α-acids [15]

	protection against UV light	foam improvement	bitterness
iso-a-acids	no	normal	1.0 (reference)
rho-iso-a-acids	yes	normal	0.6–0.8
tetrahydro-iso-a-acids	yes	improved	1.0–1.4
hexahydro-iso-a-acids	yes	improved	1.0–1.3

Although the bitterness impression is stronger with higher doses of reduced iso-α-acids, the quality of the bitterness decreases proportionally. Furthermore, a saturation point is reached with high doses of tetra-hydro-iso-α-acids with regard to the sensory impression of bitterness and improvement of foam stability. If the flavor profile of the beer is intense or the beer is consumed at temperatures near freezing, the bitterness can be masked, thus lowering the bitterness potential.

Among other things, this also explains why tetra-hydro-iso-α-acids exhibit a potential bitterness of 1.7 in water, but may drop to a value as low as 1.0 in certain beer styles.

2.2 HOP UTILIZATION

The key criteria determining the overall yield or utilization of the substances in hops (e.g. bitter substances, aroma, polyphenols) are numerous. They include the hop variety, the type of hop product, the quantity of hops added, the original gravity, the intensity of wort boiling, the wort boiling system, wort pH, contact time between the hops and the hot wort, the degree of break separation (hot break, cold break), reuse of hot break in the brewing process, yeast management, fermentation temperature, fermentation vessel design and the kind of filtration and stabilization measures employed [62, 131, 132, 133].

Figure 11 shows the progression of the isomerization process for standard pellets compared to iso-pellets for a hop addition at the beginning of the boil [1, 34, 131].

Hop utilization is generally calculated using one of two methods. Utilization is understood as the ratio of the number of bittering units in beer (EBC units) to the quantity of hops added. For pellets, the conductometer value (CV, EBC analysis method 7.5) serves as the basis for this calculation. For ethanol extract, the conductometer value (CV, EBC analysis method 7.6) plus half of the iso-α-acid content (EBC 7.8) is used as the basis. To determine hop utilization for CO_2 extract, the α-acid content is determined either conductometrically (CV, EBC 7.6) or by means of HPLC (EBC 7.7). However, it is also possible to calculate the ratio of iso-α-acids in beer to the quantity of α-acid in the hops. With this approach, the concentration of α-acids and iso-α-acids are determined by HPLC.

Figure 11: Change in hop utilization over the course of wort boiling

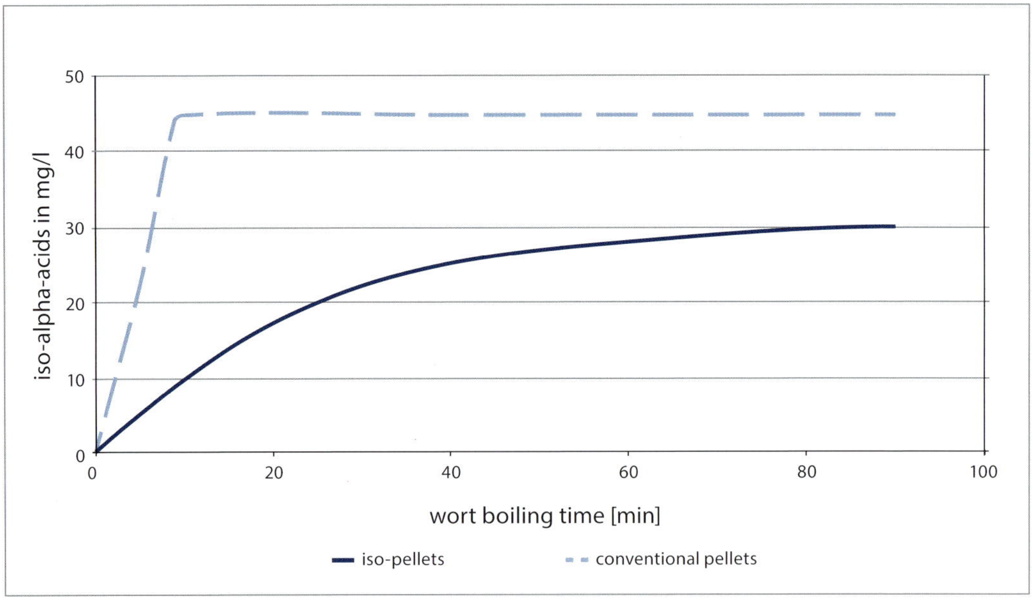

In order to carry out isomerization separate from the actual boiling process, the hops are boiled in a dedicated vessel to isomerize the bitter acids. This should be performed at a higher pressure and thus at higher temperatures. The parameters of temperature and time must be perfectly coordinated to optimize hop utilization [134]. Under excessive reaction conditions, the iso-α-acids react further, lowering the utilization, as depicted in figure 3.

The economic aspect of using isomerized hop products (in countries not subject to the *Reinheitsgebot*) in a brewery should always be carefully evaluated. An equation for calculating the break-even point is provided below.

$$\text{EUR/kg } \alpha\text{-acids} = \text{EUR/kg iso-}\alpha\text{-acids} \times \frac{IR_{conv.}}{IR_{isom.}}$$

$IR_{conv.}$ = isomerization rate of conventional hop product

$$IR_{conv.} = \frac{\text{iso}-\alpha-\text{acid content in beer}}{\alpha-\text{acid content added}}$$

$IR_{isom.}$ = isomerization rate of isomerized hop product

$$IR_{isom.} = \frac{iso-\alpha-acid\ content\ in\ beer}{iso-\alpha-acid\ content\ added}$$

If the hop utilization and the cost of the isomerized product are entered into the equation, the price limit for the conventional hop product can be determined. For instance, if the current market price for the conventional hop product is below that of the price limit calculated for the isomerized hop product, then the conventional product is less expensive.

2.3 FOAM

Hops have a positive effect on the foam stability of beer. Also within the realm of the *Reinheitsgebot*, the concentration of unisomerized α-acids can be increased in beer by a late hop addition. Although the unisomerized α-acids are only soluble in beer to small degree (approx. 5 mg/l at most) [16] and possess only a limited sensory effect, they serve to improve the foam stability.

Reduced hop products are permitted in the production of beers not subject to the *Reinheitsgebot*. Tetra-hydro-iso-α-acids and hexa-hydro-iso-α-acids can be dosed at a rate of 2 to 5 mg per liter of beer, greatly enhancing foam stability. Excessive amounts can result in beer foam with an unnatural appearance and structure. Furthermore, how this bitterness is perceived sensorially must also be taken into account, as shown in table 2 [1, 133]. The hydrophobic character of iso-α-acids increases with the degree of hydration, thus increasing foam stability. At the same time, however, their solubility in beer decreases, which can lead to technical problems when dosing these products (separation from the beer).

2.4 MICROBIOLOGY

Compounds present in hops increase the microbiological stability of beer. Humulone and isohumulone have an antibacterial effect on Gram-positive bacteria but not on Gram-negative bacteria [136, 137]. Unisomerized α-acids have a stronger antimicrobial effect on beer-spoiling microbes of the genera *Lactobacillus* and *Pediococcus* than iso-α-acids. The concentration of α-acids in beer can be increased through late hop additions. Similar to the effect encountered with foam, increasing hydration of the iso-α-acids impacts the microbiological stability of the beer in a positive manner.

The bitter substances in hops are ranked below according to their impact on microbiological stability, from the weakest effect (1) to the strongest effect (6) [138]:

1. unhopped beer
2. iso-α-acids
3. rho-iso-α-acids
4. tetra-hydro-iso-α-acids
5. 50 % tetra-hydro-iso-α-acids + 50 % hexa-hydro-iso-α-acids
6. α-acids

Table 3 provides an overview of compounds found in hops and beer that are relevant for microbiological stability (including dry-hopped beers).

Table 3: Substances in hops and beer relevant for microbiological stability (including dry-hopped beers)

	bitter hops (%)	aroma hops (%)
alpha-acids (alpha)	10-20	2-10
beta-acids (beta)	4-8	4-7
iso-alpha-acids (iso-alpha)	< 0.1	< 0.1
humulinone (HUM)	0.3-0.5	0.1-0.3
polyphenols (POLY)	3-4	4-6
xanthohumol (XN)	0.5-1.0	0.3-0.5
isoxanthohumol (IX)	< 0.1	< 0.1
ethereal hop oil (EHO)	1.5-3	0.6-1.5

Table 4: Substances in beer relevant for microbiological stability and their capacity for microbial inhibition

	beer (mg/l)	average conc. in standard-strength, pale lager beer (mg/l)
alpha	0.2-20	1
beta	n.d. -1	n.d.
iso-alpha	15-50	20
HUM	n.d. -20	1
POLY	150-300	150
XN	n.d. -6	0,2
IX	n.d. -2	1
EHO	n.d. -2	0.02

n.d. = not detectable

Table 5: Composition of various beers with relevant substances for microbiological stability, produced by adding hops at different points in the brewing process

Beer style	EBC	Iso-alpha	Alpha	XN	IX	HUM	microbial inhibition
	BU	mg/l	mg/l	mg/l	mg/l	mg/l	
trial beer 1	25.4	22.1	1.1	< 0.1	0.9	1.3	-
trial beer 2 (WP)	16.3	10.2	2.1	< 0.1	0.6	1.7	-
trial beer 3 (DH)	12.2	1.7	5.1	0.1	0.1	2.7	-
pilsner 1	26.4	23.3	0.2	n.n.	0.3	1.9	-
pilsner 2	33.3	29.4	0.7	n.n.	0.4	2.5	+
pilsner 3 (WP)	34.4	29.3	4.7	0.2	0.9	1.3	++
pilsner 4 (DH)	41.7	25.9	5.6	0.2	0.5	7.6	++
pilsner 5 (WP, DH)	53.3	34.9	5.5	0.3	0.9	10.3	++
non-alcoholic beer 1	27.6	27.7	1.1	< 0.01	0.1	n.n.	-
non-alcoholic beer 2 (DH)	30.7	25.4	3.6	< 0.1	0.2	2.7	++
wheat beer 1 (WP)	24.2	15.8	2.3	0.1	0.5	3	-
wheat beer 2 (DH)	22.8	9.2	8.9	0.6	0.3	2.9	+
wheat beer 3 (DH)	36.8	17.2	11.6	0.7	0.8	4.6	++
IPA (WP, DH)	60.7	36.3	14.1	0.5	1.1	11	++
stout (WP, DH)	45.8	12.5	24	6.1	1.1	7.8	++

WP = whirlpool addition, DH = dry hopping, XN = xanthohumol, IX = iso-xanthohumol, HUM = humulinones
– = no inhibition, + = inhibition, ++ = strong inhibition
(test organisms: *Lactobacillus brevis, L. lindneri, Pediococcus damnosus*)

When operating equipment used in the production of natural lactic acid in a brewery, care should be taken that no hop bitter compounds come into contact with the lactic acid bacteria culture (*Lactobacillus amylolyticus*), as this microorganism cannot tolerate hops. Only after the first wort has been transferred to the lactic acid plant should hot break material be added to the lauter tun (cf. Acidification with Natural Lactic Acid).

2.5 ADDITION OF DOWNSTREAM HOP PRODUCTS

Downstream hop products are generally added to the beer immediately prior to filtration. The solubility of certain downstream hop products is heavily dependent on pH and temperature. The higher the degree of hydration of the iso-α-acids, the better the stability of the beer foam is and its adhesion to the beer glass. However, their solubility in beer is normally reduced. For this reason, special units have been developed with specific instructions for dosing hop products. A widely established approach for dosing downstream products is to first dissolve the products in beer and then add the mixture proportionally to the remaining beer [1].

The quantity of iso-extract to be added to the beer can be calculated as follows:

$$E\left[\frac{mg}{l}\right] = \frac{BU[IBU] \cdot 10{,}000}{C[\%] \cdot Y[\%] \cdot F}$$

E: quantity of iso-extract to be added
BU: desired bitterness of beer expressed in EBC bittering units [IBU]
C: concentration of the iso-extract
Y: utilization factor
F: factor for the sensory bitterness of the product compared to the bitterness of the iso-α-acids

The following equation can be used to calculate the quantity of iso-extract required to correct the bitterness of the beer:

$$E[g] = \frac{(BU - AB) \cdot 1{,}000 \cdot V[HL]}{C[\%] \cdot Y[\%] \cdot F}$$

E: quantity of iso-extract to be added
BU: desired bitterness of beer expressed in EBC bittering units [IBU]
AB: actual bitterness of the beer prior to adding the iso-extract
V: the volume of beer to be corrected
C: concentration of the iso-extract
Y: utilization factor
F: factor for the sensory bitterness of the product compared to the bitterness of the iso-α-acids

2.6 STORING HOPS

Hops are susceptible to oxidation and other chemical reactions that alter their composition. If stored under unsuitable conditions, α-acids can become oxidized, lowering hop bitterness. The degree of α-acid degradation is dependent on the individual hop variety, storage conditions and how long the hops are stored.

The aged condition of bitter substances in hops can be determined by means of the Hop Storage Index (HSI). The HSI is the quotient of the absorption measured for an alkaline solution of methanol at a wavelength of 275 nm and 325 nm. The absorption maximum for oxidation products lies in the range of 255–280 nm. The maximum for humulone and lupulone is at 325 nm and 355 nm, respectively, while the absorption minimum is 275 nm for both.

$$HSI = \frac{A275nm}{A325nm}$$

If hops have been stored at an optimal temperature, then it can be assumed that they can be added at the previously determined rate. Hops stored under unsuitable conditions, both prior to and after processing, possess an unpleasant, cheesy aroma. This is especially applicable to whole hops and pellets. Therefore, hops should be stored at cold temperatures and protected against exposure to light and other external influences (oxygen, odors, moisture, etc.). α-Acids oxidize in the presence of oxygen and isomerize when placed in an inert atmosphere. This explains the loss in α-acids during storage. Hard resin compounds already exist in whole hops, and the proportion of these hard resin compounds is a measure of oxidation. Hard resin compounds are the oxidation products of bitter acids, which are formed through exposure to oxygen. They are also referred to as polar bitter substances.

Brewing with old hops (HSI > 0.5–0.6) produces beers with an unpleasant, cheesy aroma and a noticeably harsh, lingering bitterness [24]. A significant preference is shown for beers brewed with fresh hops compared to those brewed with old hops. This is especially noticeable in beer approaching the end of its shelf life. Consequently, cold storage of whole hops and pellets is strongly recommended. Pure resin extracts are much more stable than whole hops and pellets. The recommended storage temperatures for hops and hop products are listed in table 6.

Table 6: Recommended storage temperatures for hops and hop products

Stability of hops and hop products during storage

	1 year	2 years	3 years	5 years	8 years
whole (cone) hops	< 5 °C	< -20 °C	impractical		
pellets	< 5 °C			< -20 °C	impractical
CO_2 extract	< 25 °C		< 5 °C		
ethanol extract	< 10 °C				< 5 °C

Figure 12 traces the course of the degradation of α-acids stored at various temperatures. When measured with HPLC, the degradation of the α-acids was greater than the decline in the conductometer value. The explanation behind this difference is that the conductometer method measures the entire quantity of bitter substances present, including the non-specific substances. Therefore, especially when using old hop pellets, the quantity of the hop addition should be determined based on the conductometer value.

Figure 12: Relative decrease per month in the α-acid content of pellets (without oxygen exposure) as a function of storage temperature [1]

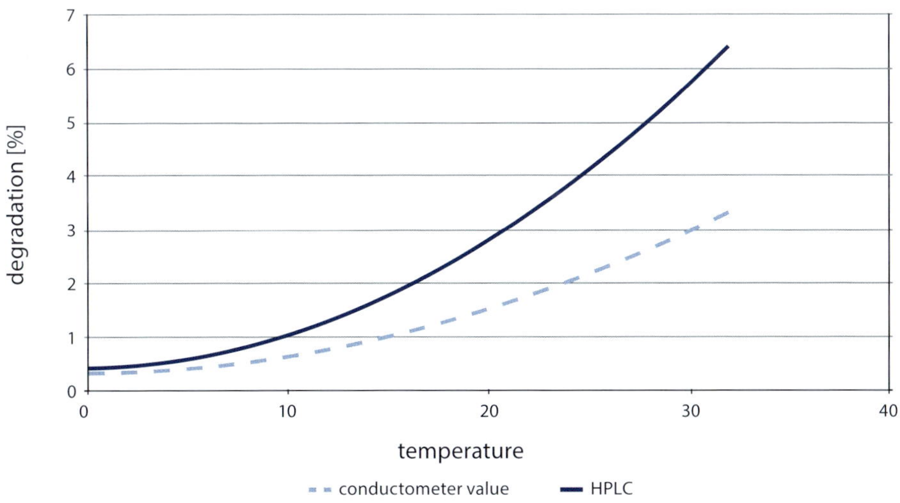

3 SUMMARY

Through modifying the techniques brewers normally employ to utilize the substances in hops, they can influence the character and sensory attributes of their beers. This enables them to brew beers that help set them apart from their competitors, giving them a market advantage.

This chapter is designed to provide practice-oriented brewers with a concise overview of the complex topic of hops and hop products. A range of topics concerning hops are discussed, including those

4 OVERVIEW

Problem	Cause	Corrective measures
poor foam quality	may occur for many reasons	add 20 % of the total quantity of α-acids at the end of the boil; add reduced hop products (in areas not subject to the German purity law); check yeast quality; analyze wort composition to evaluate the degree of proteolysis
cheesy aroma	late addition of oxidized hops combined with low evaporative efficiency; yeast autolysis	check the quality of the hops (HSI) and the storage conditions yeast vitality
harsh, astringent bitterness	may occur for many reasons, e.g. pellets stored under unsuitable conditions, oxidized hop oils, poor fermentation process, yeast coated with break material; excessive leaching of tannins (e.g. incorrect reuse of break material and last runnings) protein bitterness; yeast autolysis	utilize hops which have been stored under suitable conditions; improve yeast vitality; improve hot break separation; the lautering procedure was too intense, causing over-extraction; taste the brewing liquor for off-flavors; add $CaSO_4$ or $CaCl_2$
lightstruck flavor	3-methyl-2-butene-1-thiol	avoid filling beer in green or clear bottles or protect the beer from exposure to light (packaging); add reduced hop extract products (in areas not subject to the German purity law)
low hop utilization	pH of the wort is too low; wort was either not boiled or held at a high temperature for a sufficient period of time	acidify the wort starting at the end of the boil; check for proper automatic dosing of hops; add hops earlier; use extract instead of pellets; consider reintroducing hot break into the brewing process; pre-isomerize hops in an external vessel; correct bitterness level with iso-extracts (in areas not subject to the German purity law)

aspects particular to the German *Reinheitsgebot*. A review of the substances found in hops, information about hop products and methods of hop analysis are also provided. Furthermore, special topics are addressed, such as hop additions for different beer styles, hop utilization, beer foam, microbiology, hop storage and dosing downstream products.

Problem	**Cause**	**Corrective measures**
nitrate content of beer is too high	nitrate content of the pellets	use type 45 pellets instead of type 90 pellets; increase the proportion of hop extract
weakly hopped beers are too bitter	carryover or introduction of bitterness, e.g. with yeast used to ferment highly hopped beers, by yeast autolysis or when using a large quantity of high α bitter hops	wash the yeast; select appropriate cropped yeast to pitch
weak hop aroma	aroma hop addition is insufficient or the hops were added at the wrong time; loss of turbidity particles through filtration; strong CO_2 stripping effect in tall fermentation tanks	check the oil/linalool content of the aroma hops; add the hops at the end of the boil, in the whirlpool or to the finished beer based on the oil/linalool concentration; produce naturally cloudy beers; optimize the fermentation process (by regulating the stripping effect of CO_2)
whirlpool is overloaded	too much hot break; whirlpool is not functioning correctly	increase the enrichment level (use type 45 pellets instead of type 90 pellets or hop extract); employ an extraction vessel for whirlpool hop additions
low microbiological stability	may occur for many reasons; an insufficient quantity of hops	increase the hop rate; perform late hop addition to bring α-acids into the beer

5 REFERENCES

[1] Krottenthaler, M.: Hopfen. In: Heyse, K.-U.: Praxishandbuch der Brauerei. Behr´s Verlag, Hamburg, 2003.

[2] Barth-Haas Group (ed.). The Barth Report Hops 2015/2016. Nuremberg 2016.

[3] Simon H. Steiner Hopfen GmbH (Hrsg.). Entscheidungsdaten für den Hopfeneinkauf. 2016.

[4] Krottenthaler, M., Back, W., Engelhard, B. Großversuche mit der neuen Hüller Zuchtsorte "Hallertauer Merkur" (MR) – Large-Scale Trials with the new Hüll Variety "Hallertauer Merkur" (MR). Hopfenrundschau International (2000), pp. 43–46.

[5] Krottenthaler, M., Zürcher, A., Zarnkow, M., Back , W. Großversuche mit der neuen Hüller Zuchtsorte "Saphir" – Large-Scale trials with the new Hüll variety "Saphir". Hopfenrundschau International (2001/2002), pp. 60–67.

[6] CMA Centrale Marketing-Gesellschaft der deutschen Agrarwirtschaft mbH (Hrsg). Die Seele des Bieres, Hopfen aus Deutschland, Bonn, 2005. www.hops-and-malt.de.

[7] www.barthhaasgroup.com.

[8] www.hopsteiner.com.

[9] www.hvg-germany.de.

[10] www.yakimachief.com.

[11] Forster, A., Biendl, M., Schmidt, R.: Wie bestimmt sich der durchschnittliche alpha-Säuregehalt einer Hopfensorte? Brauwelt 144 (2004), No. 22, pp. 654–660.

[12] Moir, M.: Hops – A Millennium Review. Journal of the American Society of Brewing Chemists 58 (2000), No. 4, pp. 131–146.

[13] Back, W.: Colour Atlas and Handbook of Beverage Biology. Fachverlag Hans Carl, Nürnberg 2006.

[14] Kondo, K.: Preventive effects of dietary beer on lifestyle-related diseases. Proceedings of the 29th European Brewery Convention Congress Dublin, Fachverlag Hans Carl, Nürnberg 2003, pp. 1339–1350.

[15] Schönberger, C.: Bitter is better. BrewingScience 59 (2006), no. 3/4, pp. 56–66.

[16] Spetsig, L. O.: Electrolytic constants and solubilities of humulinic acid, humulone, and lupulone. Acta Chemica Scandinavica 9 (1955), no. 9, pp. 1421–1424.

[17] Fritsch, A., Shellhammer. T. H.: Alpha-acids do not contribute bitterness to lager beer. Journal of the American Society of Brewing Chemists 65 (2007), no. 1, pp. 26–28.

[18] Intelmann, D.: Molekulare, psychophysikalische und rezeptorbasierte Studien zum Bittergeschmack von Bier. Dissertation, Technical University of Munich (Weihenstephan), 2010.

[19] Haseleu, G.: Sensorische und quantitative Studien zu Bitterstoffen aus Hopfen (*Humulus lupulus*) und deren Beitrag zum Bittergeschmack von Bier. Dissertation, Technical University of Munich (Weihenstephan), 2010.

[20] Forster A., Schüll, F., Gahr, A.: Geringe Bittere und volles Hopfenaroma – ein Widerspruch? Brauindustrie 102 (2017), No. 1, pp. 24–27.

[21] De Taeye, L., De Keukeleire, D., Verzele, M.: The Anti-Iso-Alpha-Acids: Concept and Review of Current Research. Journal of the American Society of Brewing Chemists 39 (1981), no. 1, pp. 24–26.

[22] Aitken, R. A. The bitterness of hop-derived materials in beer. Journal of the Institute of Brewing 76 (1970), no. 1, pp. 29–36.

[23] Hartl, A. R.: Über die Umwandlung von Lupulonen in bierlösliche Bitterstoffe. Proceedings oft he 15th European Brewery Convention Congress Nizza. Elsevier 1975, pp. 141–152.

[24] Peacock, V. Fundamentals of Hop Chemistry. Master Brewers Association of the Americas Technical Quarterly 35 (1998), no. 1, pp. 4–8.

[25] Verzele, M., De Keukeleire, D.: Chemistry and Analysis of Hop and Beer Bitter Acids. Elsevier, Amsterdam 1991.

[26] Hughes, P.: Differences in the UV-Spectra of the Hop-Derived *cis*- and *trans*-Iso-alpha-Acids. Journal of the American Society of Brewing Chemists 64 (2006), no. 1, pp. 47–51.

[27] De Cooman, L., Aerts, G., Overmeire, H.: Alterations of the Profiles of Iso-alpha-Acids During Beer Ageing, Marked Instability of *Trans*-Iso-alpha-Acids and Implications for Beer Bitterness Consistency in Relation to Tetrahydroiso-alpha-Acids. Journal of the Institute of Breweing 106 (2000), no. 3, pp. 169–178.

[28] Techakriengkrai, I., Paterson, A., Taidi, B., Pigott, J. R.: Relationships of sensory bitterness in lager beers to iso-alpha-acid contents. Journal of the Institute of Brewing 110 (2004), no. 1, pp. 51–56.

[29] Verzele, M. Jansen, H. E., Ferdinandus, A.: Organoleptic Trials with Hop Bitter Substances. Journal of the Institute of Brewing 76(1970), no. 1, pp. 25–28.

[30] Bamforth, C. W.: Beer flavour: malt and hop character. Brewer's Guardian 130 (2001), no. 11, pp. 16–18.

[31] Hughes, P., Simpson. W. J.: Sensory impact of hop-derived compounds. EBC Monograph 22 Symposium on Hops, Fachverlag Hans Carl, Nürnberg 1994, pp. 128–140.

[32] Hughes, P., Simpson. W. J.: Bitterness of Congeners and Stereoisomers of Hop-Derived Bitter Acids Found in Beer. Journal of the American Society of Brewing Chemists 54 (1996), no. 4, pp. 234–237.

[33] Biendl M., Engelhard, B., Forster, A., Gahr, A., Lutz, A., Mitter, W., Schmidt, R., Schönberger, C.: Hops – Their Cultivation, Composition and Usage. Fachverlag Hans Carl, Nürnberg 2014.

[34] Benitez, J. L., Forster, A., De Keukeleire, D., Moir, M., Sharpe, F. R., Verhagen, L. C., Westwood, K. T.: Hops and Hop Products. EBC Manual of Good Practice. Fachverlag Hans Carl, Nürnberg 1997.

[35] Kaltner, D.: Untersuchungen zur Ausbildung des Hopfenaromas und technologische Maßnahmen zur Erzeugung hopfenaromatischer Biere. Dissertation, Technical University of Munich (Weihenstephan), 2000.

[36] Steinhaus, M., Fritsch, H. T., Schieberle, P.: Quantitation of (R)- and (S)-Linalool in Beer Using Solid Phase Microextraction (SPME) in Combination with a Stable Isotope Dilution Assay (SIDA). Journal of Agricultural and Food Chemistry 51 (2003), no. 24, pp. 7100–7105.

[37] Steinhaus, M., Schieberle, P.: Comparison of the Most Odor-Active Compounds in Fresh and Dried Hop Cones (*Humulus lupulus* L. Variety Spalter Select) Based on GC-Olfactometry and Odor Dilution Techniques. Journal of Agricultural and Food Chemistry 48 (2000), no. 5, pp. 1776–1783.

[38] Gresser, A.: Beiträge zur Kenntnis der Hopfenaromastoffe unter besonderer Berücksichtigung technologischer Verfahren. Dissertation, Technical University of Munich (Weihenstephan), 1985.

[39] Kaltner, D. Steinhaus, M., Mitter, W. Biendl, M., Schieberle, P.: (R)-Linalool als Schlüsselaromastoff für das Hopfenaroma in Bier und sein Verhalten während der Bieralterung. Monatsschrift für Brauwissenschaft 56 (2003), no. 11/12, pp. 192–196.

[40] Forster, A., Gahr, A., Schmidt, R., Faltermeier, A.: The behaviour of the Stereoisomers of Linalool during beer Ageing. Brewing Science Yearbook 2015, pp. 146–152.

[41] Praet T., van Opstaele, F., de Causmaecker, B., Bellaio, G., de Rouck, G., Aerts, G., de Cooman, G.: De novo Formation of Sesquiterpene Oxidation Products during Wort Boiling and Impact of the Kettle Hopping Regime on Sensory Characteristics of Pilot-Scale Lager. BrewingScience, 68 (2015), no. 11/12, pp. 130–145.

[42] Lermusieau, G., Bulens, M., Collins, S.: Use of GC-Olfactometry to Identify the Hop Aromatic Compounds in Beer. Journal of Agricultural and Food Chemistry 49 (2001), no. 8, pp. 3867–3874.

[43] Steinhaus, M.: Wichtige Aromastoffe im Hopfen (*Humulus Lupulus* L.). Verlag Dr. Hut, München 2000.

[44] Steinhaus, M., Schieberle, P.: Transfer of potent hop odorants linalool, geraniol and 4-methyl- 4-sulfanyl-2-pentanone from hops into beer. Proceedings of the 31st European Brewery Convention Congress Venice. Fachverlag Hans Carl, Nürnberg 2007, pp. 1004–1011.

[45] Kaltner, D. Moderne Hopfenprodukte – Aus vielen guten Zutaten kann der Brauer ein spitzen Bier-Image brauen. Getränkeherstellung, (2005), August, pp. 25–27.

[46] Forster, A., Beck, B., Massinger, S., Schmidt, R.: Niedermolekulare Polyphenole beim Hopfenwachstum. Brauwelt 144 (2004), no. 46/47, pp. 1556–1561.

[47] Forster, A., Beck, B., Schmidt, R.: Untersuchungen zu Hopfenpolyphenolen. Proceedings of the 25th European Brewery Convention Congress Brussels. Oxford University Press 1995, pp. 143–150.

[48] Forster, A., Beck, B., Schmidt, R.: Hopfenpolyphenole – mehr als nur Trübungsbildner – Hop Polyphenols do more just cause turbidity. Hopfenrundschau International, (1999), pp. 68–74.

[49] Forster, A., Gahr, A., Ketterer, M., Beck, B., Massinger, S.: Xanthohumol in Bier – Möglichkeiten und Grenzen einer Anreicherung. Monatsschrift für Brauwissenschaft 55 (2002), no. 9/10, pp. 184–194.

[50] Bellmer, H. G.: Hopfenpolyphenole im Brauprozeß. Brauwelt 121 (1981), no. 8, pp. 240–245.

[51] Pöschl, M., Geiger, E., Biendl, M.: The effect of using polyphenolic extracts during mashing in the colloidal stability of beer. Proceedings of the 31st European Brewery Convention Congress Venice. Fachverlag Hans Carl, Nürnberg 2007, pp. 101–111.

[52] RÖMPP Online. www.roempp.com. Georg Thieme Verlag KG, Stuttgart 2008.

[53] Biendl, M.: Grünes Gold. Anticancerogene Aktivität – Ein neuer Aspekt bei Hopfeninhaltsstoffen. Brauindustrie 84 (1999), no. 9, pp. 502–507.

[54] Biendl, M., Methner, F.-J., Stettner, G., Walker, C. J.: Brauversuche mit einem xanthohumolreichen Hopfenprodukt. Brauwelt 144 (2004), no. 9/10, pp. 236–241.

[55] Biendl, M., Mitter, W., Peters, U., Methner, F.-J.: Einsatz eines xanthohumolreichen Hopfenproduktes bei der Bierherstellung. Brauwelt 140 (2000), no.46/47, pp. 2006–2011.

[56] Forster, A.: Hopfen – mehr als nur ein alpha-Säureträger. 48. Arbeitstagung der österreichischen Braumeister und Brautechniker. Innsbruck 1998.

[57] Kammhuber, K. Stand der Erkenntnisse zum Hopfeninhaltsstoff Xanthohumol. Brauwelt 138 (1998), no. 36, pp. 1633–1636.

[58] Forster, A., Beck, B., Gehrig, M.: Der Verbleib von Pestizigrückständen des Hopfens bei der Herstellung von Hopfenprodukten. Proceedings of the 23rd European Brewery Convention Congress Lisbon. Oxford University Press 1991, pp. 193–200.

[59] Hartl, A. (1992). Kaum Pestizide in Hopfenextrakten. Beitrag zur Frage der Schadstoffreduzierung mit Hopfenprodukten. Brauwelt 132 (1999), no 7/8, pp. 262–266.

[60] Schmidt, R., Anerdegg, P., Biendl, M.: Umweltkontaminanten in Hopfen. Brauwelt 143 (2003), no. 45, pp. 1493–1497.

[61] Bradley, L.: Extracting success. Brewer's Guardian 130 (2001), no. 10, pp. 24–26.

[62] Kaltner, D.: Moderne Aspekte zur Hopfung des Bieres. Brauindustrie 90 (2005), no. 2, S. 22–25.

[63] Mitter, W. (1989). Analytik und Bewertung von Hopfen und Hopfenprodukten im Hinblick auf die Bierbittere. Schweizer Brauerei-Rundschau 100(11), pp. 279–283.

[64] European Brewery Convention: Analytica-EBC. https://analytica-ebc.com. Fachverlag Hans Carl, Nürnberg 1998 and 2017.

[65] Jacob, F. (ed.): Collection of Brewing Analysis Methods. Wort, Beer and Beer-based Beverages. Selbstverlag der MEBAK, Freising, 2013.

[66] Bayerische Landesanstalt für Landwirtschaft. Jahresbericht 2004 Sonderkultur Hopfen. Freising 2005.

[67] Ashurst, P. R.: Chemistry of the hop resins. Fortschritte der Chemie Organischer Naturstoffe 25 (1967), pp. 63–89.

[68] Meilgaard, M.: Hop Analysis, Cohumulone Factor and the Bitterness of Beer: Review and Critical Evaluation. Journal of the Institute of Brewing 66 (1960), no. 1, pp. 35–50.

[69] Rigby, F. L.: A Theory of the Hop Flavor of Beer. American Society of Brewing Chemists. Proceedings 30 (1972), pp. 46–50.

[70] Hughes, P.: The Significance of Iso-alpha-Acids for Beer Quality. Journal of the Institute of Brewing 106 (2000), no. 5, pp. 271–276.

[71] Kusche, M., Stettner, G., Stephan, A., Mitter, W., Kaltner, D.: Influence of the new high alpha hop variety Herkules on beer quality. Proceedings of the 31st European Brewery Convention Congress Venice. Fachverlag Hans Carl, Nürnberg 2007, pp. 220–225.

[72] Lindner, S.: Praxisversuche mit verschiedenen Hopfenprovenienzen. Schweizer Brauerei-Rundschau 73 (1962) no. 8, pp. 135–138.

[73] Wackerbauer, K., Balzer, U.: Hopfenbitterstoffe in Bier Teil 2: Der Einfluß des Cohumulons auf die Bierqualität. Brauwelt 122 (1992), no. 10/11, pp. 396–398.

[74] Mitter, W., Kessler, H., Biendl, M.: Bitterhopfengabe in Form von Pellets und Extrakt und deren Einfluss auf Würze und Bier. Brauindustrie 84 (1999), no. 10, pp. 560–563.

[75] Forster, A. Zur Nitratdosage durch Hopfen und Hopfenprodukte. Brauwelt 128 (1988), no. 6, pp. 188–191.

[76] Ertlmaier, S.: Reduzierung der Nitratdosage bei Einsatz von Ethanolextrakten. Brauwelt 131 (1991), no. 42, pp. 1864–1865.

[77] Ketterer, M., Forster, A., Gahr, A., Beck, B., Massinger, S., Schmidt, R.: The influence of isomerized hop pellets on beer quality. Proceedings of the 29th European Brewery Convention Congress Dublin. Fachverlag Hans Carl, Nürnberg 2003, pp. 268–275.

[78] Guzinski, J. A.: Practical considerations of reduced hop extracts. European Brewery Convention Monograph 22 Symposium on Hops Zoeterwoude. Fachverlag Hans Carl 1994, pp. 105–113.

[79] Seldeslachts, D., De Winter, D., Melotte, L., De Bock, A., Haerinck, R.: The use of high tech hopping in practice. Proceedings of the 27th European Brewery Convention Congress Cannes. Oxford University Press 1999, pp. 291–297.

[80] Weiss, A., Schönberger, C., Mitter, W., Biendl, M., Back, W., Krottenthaler, M.: Sensory and analytical characterisation of reduced, isomerised hop extracts and their influence and use in beer. Journal of the Institute of Brewing 108 (2002), no. 2, pp. 236–242.

[81] Kaltner, D., Mitter, W.: The Influence of Hops upon the Aroma and Taste of Beer. Scandinavian Brewers' Review 60 (2003), no. 3, pp. 8–16.

[82] Coors, G., Krottenthaler, M.: Auswirkungen einer Würzevorkühlung beim Ausschlagen. Brauwelt 140 (2000), no. 42/43, pp. 1696–1699.

[83] Coors, G., Krottenthaler, M., Back, W.: Pre-Cooling and its Influence on Casting. Brauwelt International 21 (2003), no. 1, pp. 40–41.

[84] Yamaguchi Y., Gastl, M., Becker, T., Narziß, L.: Pre-treatment of hop pellets before hop dosage into the brewing process, MBAA Annual Conference, Florida 2015.

[85] Wilson, R. J. H., Roberts, T. Smith, R. J., Biendl, M.: Improving hop utilization and flavor control through the use of pre-isomerized products in the brewery kettle. Master Brewers Association of the Americas Technical Quarterly 38 (2001), no. 1, pp. 11–21.

[86] Kaltner, D., Thum, B., Forster, C., Back, W.: Untersuchungen zum Hopfenaroma in Pilsner Bieren bei Variation technologischer Parameter. Monatsschrift für Brauwissenschaft 54 (2001), no. 9/10, pp. 199–205.

[87] Mitter, W., Biendl, M., Kaltner, D.: Behaviour of hop derived aroma substances during wort boiling. European Brewery Convention, Monograph 31 Symposium on Flavour and Flavour Stability, Nancy 2001, pp. 125–136.

[88] Herrmann, M., Hanke, S., Kaltner, D., Back, W.: Hop Volatile Compounds (Part I): Analysis of Hop Pellets and Seasonal Variations. BrewingScience 61 (2008), no. 7/8, pp. 135–139.

[89] Hanke, S., Herrmann, M., Rückerl, J., Schönberger, C., Back, W.: Hop Volatile Compounds (Part II): Transfer Rates of Hop Compounds from Hop Pellets to Wort and Beer. Brewing-Science 61 (2008), no. 7/8, pp. 140–147.

[90] Hanke, S.: Untersuchungen zum Einfluss der Hopfungstechnologie auf die Geschmacks-stabilität und Harmonie untergäriger Biere. Dissertation, Technical University of Munich (Weihenstephan), 2009.

[91] Fritsch, H., Kaltner, D., Schieberle, P., Back, W.: Entschlüsselung des Hopfenaromas in Bier Ein Paradebeispiel anwendungsorientierter Forschung. Brauwelt 144 (2004), no. 39/40, pp. 1206–1207.

[92] Fritsch, H., Schieberle, P.: Changes in key aroma compounds during boiling of unhopped and hopped wort. Proceedings of the 29th European Brewery Convention Congress Dublin. Fachverlag Hans Carl, Nürnberg 2003, pp. 259–267.

[93] Fritsch, H. S. (2005). Bewertung von Hopfenaromastoffen in Pilsener Bieren. Brauerei-Forum (2005), no. 3, pp. 66–67.

[94] Fritsch, H. Einfluss des Hopfens auf wertgebende Aromastoffe in Pilsener Bieren sowie in Zwischenstufen des Brauprozesses. Dissertation, Technical University of Munich (Weihenstephan), 2001.

[95] Guadagni, D. G., Buttery, R. G., Harris, J.: Odour intensities of hop oil components. Journal of the science of food and agriculture 17 (1966), no. 3, pp. 142–144.

[96] Forster, A., Gahr, A., Schmidt, R., Faltermeier, A.: The Behaviour of the Stereoisomers of linalool during beer Aging. BrewingScience Yearbook 2015, pp. 146–152.

[97] Moir, M.: Hop aromatic compounds. European Brewery Convention Monograph 22 Symposium on Hops, Zoeterwoude. Fachverlag Hans Carl, Nürnberg 1994, pp. 165–180.

[98] Back, W. Forster, C., Krottenthaler, M., Lehmann, J., Sacher, B., Thum, B.: Neue Forschungserkenntnisse zur Verbesserung der Geschmacksstabilität. Brauwelt 137 (1997), no. 38, pp. 1677–1692.

[99] Kaltner, D., Forster, C., Thum, B., Back, W.: Untersuchungen zur Ausbildung des Hopfenaromas während des Brauprozesses. Proceedings of the 27th European Brewery Convention Congress Cannes. Fachverlag Hans Carl, Nürnberg 1999, pp. 63–70.

[100] Back, W. Z. (2004). Patentnr. DE 10256166.

[101] Biendl, M.: Untersuchungen zum Xanthohumol-Gehalt in Hopfen – Research on the Xanthohumol content in hops. Hopfenrundschau International, (2002/2003), pp. 71–75.

[102] Forster, A., Beck, B., Schmidt, R., Jansen, C., Mellenthin, A.: Über die Zusammensetzung von niedermolekularen Polyphenolen in verschiedenen Hopfensorten und zwei Anbaugebieten. Monatsschrift für Brauwissenschaft 55 (2002), no. 5/6, pp. 98–108.

[103] Forster, A., Ketterer, M., Gahr, A.: Ein neuartiger Xanthohumol-angereicherter Hopfenextrakt und sein Einsatz für unfiltrierte Biere – A nex xanthohumol-enriched hop-extract and its use for unfiltered beers. Hopfenrundschau International (2003/2004), pp. 65–71.

[104] Mitter, W., Biendl, M., Kaltner, D.: Possible effects of prenylated flavonoids and their presence in beer. 28th Convention of the Institute & Guilt of Brewing – Asia Pacific Section. Hanoi 2004.

[105] Walker, C. J., Biendl, M., Fernández Lence, C..: Investigation into the high levels of xanthohumol found in stout and porter style beers. Brauwelt International 22 (2004), no. 2, pp. 100–103.

[106] Wunderlich, S., Zürcher, A., Back, W.: Mehr Xanthohumol im Bier für mehr Gesundheit? Gesellschaft Deutscher Chemiker, Lebensmittelchemische Gesellschaft, Bonn 2004.

[107] Wunderlich, S., Zürcher, A., Back, W.: Enrichment of xanthohumol in the brewing process. Molecular Nutrition & Food Research 49 (2005), no. 9, pp. 874–881.

[108] Forster, A. und Köberlein, A.: Der Verbleib von Xanthohumol aus Hopfen während der Bierbereitung. Brauwelt 138 (1998), no. 37, pp. 1677–1679.

[109] Fritsch, H., Biendl, M., Stephan, A., Stettner, G., Methner, F.-J.: Sensory characterization of xanthohumol and isoxanthohumol. Proceedings of the 30th European Brewery Convention Congress Prague. Fachverlag Hans Carl, Nürnberg 2005, pp. 947–953.

[110] Forster, A., Gahr, A.: A Comparison oft the Analytical and Brewing Characteristics of Cascade and Comet Hop Varieties as Grown in Yakima (USA) and Hallertau (Germany). BrewingScience 67 (2014), no. 7/8, pp. 137–148.

[111] Forster, A., Gahr, A.: On the Fate of Certain Hop Substances during Dry Hopping. BrewingScience 66 (2013), no. 7/8, pp. 93–103.

[112] Mitter, W. Cocuza, S.: Die Kalthopfung – Untersuchung verschiedener Parameter, Auswirkungen durch die Art der Einbringung. Brauindustrie 97 (2012), no. 11, pp. 74–78.

[113] Mitter, W., Cocuzza, S.: Wiederbelebtes Verfahren, die Kalthopfung, Grundsätzliches und Techniken. Brauindustrie 97 (2012), no. 4, pp. 10–12.

[114] Schönberger, C., Gahr, A., Wiesen, E.: Hopfenstopfen – Gut gestopft ist halb gewonnen. Brauwelt 152 (2012), no. 9/10, pp. 251–254.

[115] Kippenberger, M., Hanke, S., Biendl, M., Stettner, G., Lagemann, A.: Transfer of Nitrate and various Pesticides into Beer during Dry Hopping. BrewingScience 67 (2014), no. 1/2, pp. 1–9.

[116] Charpentier, N., Maujean, A.: Sunlight flavors in champagne wines. Flavour' 81 3rd Weurman Symposium, München. De Gruyter 1981, pp. 609–615.

[117] Patton, S.: The Mechanism of Sunlight Flavor Formation in Milk with Special Reference to Methionine and Riboflavin. Journal of Dairy Science 37 (1954), no. 4, pp. 446–452.

[118] Kuroiwa, Y., Hashimoto, N.: Composition of Sunstruck Flavor Substance and Mechanism of its Evolution. American Society of Brewing Chemists Proceedings 19 (1961), pp. 28–36.

[119] Kuroiwa, Y. Hashimoto, N., Hashimoto, H., Kokubo, E., Nakagawa, K.: Factors Essential for the Evolution of Sunstruck Flavor. American Society of Brewing Chemists Proceedings 21 (1963), pp. 181–193.

[120] Huvaere, K., Andersen, M. L., Skibsted, L. H., Heyerick, A., De Keukeleire, D.: Photooxidative degradation of beer bittering principles: A key step on the route to lightstruck flavor formation in beer. Journal of Agricultural and Food Chemistry 53 (2005), no. 5, pp. 1489–1494.

[121] Huvaere, K., Olsen, K., Andersen, M. L., Skibsted, L. H., Heyerick, A., De Keukeleire, D.: Ribo- flavin-sensitized photooxidation of isohumulones and derivatives. Photochemical & Photobiological Sciences 3 (2004), no. 4, pp. 337–340.

[122] Huvaere, K., Sinnaeve, B., Van Boxclaer, J., De Keukeleire, D.: Photooxidative degradation of beer bittering principles: product analysis with respect to lightstruck flavour formation. Photochemical & Photobiological Sciences 3 (2004), no. 9, pp. 854–858.

[123] Irwin, A. J., Bordeleau, L., Barker, R. L.: Model Studies and Flavor Threshold Determination of 3-Methyl-2-Butene-1-Thiol in Beer. Journal of the American Society of Brewing Chemists 51 (1993), no. 1, pp. 1–3.

[124] Sakuma, S., Rikimaru, Y., Kobayashi, K., Kowaka, M.: Sunstruck Flavor Formation in Beer. Journal of the American Society of Brewing Chemists 49 (1991), no. 4, pp. 162–165.

[125] Gray, P. P., Stone, I., Rothchild, H.: The Action of Sunlight on Beer. Wallerstein Laboratories Communications 4 (1941), no. 11, pp. 29–40.

[126] Gunst, F., Verzele, M.: On the Sunstruck Flavor of Beer. Journal of the Institute of Brewing 84 (1978), no. 5, p. 291–292.

[127] Templar, J., Arrigan, K., Simpson, W. J.: Formation, Measurement and Significance of Lightstruck Flavor in Beer: A Review. Brewers Digest 70 (1995), no. 5, pp. 18–25.

[128] Heyerick, A., Zhao, Y. N., Sandra, P., Huvaere, K., Roelens, F., De Keukeleire, D.: Photolysis of hop-derived *trans*-iso-alpha-acids and *trans*-tetrahydro-iso-alpha-acids: product identification in relation to the lightstruck flavour of beer. Photochemica & Photobiological Sciences 2 (2003), no. 3, pp. 306–314.

[129] Hughes, P.: The lightstruck flavor problem. Cerevisia 24 (1999), no. 2, pp. 21–25.

[130] Holscher, W., V. Vitzthum, O. G., Steinhart, H.: Prenyl alcohol – source for odorants in roasted coffee. Journal of Agricultural and Food Chemistry 40 (1992), no. 4, pp. 655–658.

[131] Hanke, S., Back, W., Tauscher, F.: Einflüsse auf die Hopfenausbeute und Trubbildung bei der Würzekochung. Brauindustrie 93 (2008), no. 2, pp. 34–37.

[132] Kaltner, D., Mitter, W., Binkert, J., Preis, F., Zimmermann, R., Biendl, M.: Würzekochsystem SCHOKO und Hopfenkomponenten . Brauwelt 145 (2005), no. 19/20, pp. 582–588.

[133] Mitter, W., Kessler, H., Biendl, M.: Beobachtung einer Verwandlung. Isomerisierungs verlauf der Alphasäuren im großtechnischen Maßstab. Brauindustrie 85 (2000), no. 3, pp. 142–145.

[134] Malowicki, M. G., Shellhammer, T. H: Isomerization and Degradation Kinetics of Hop (*Humulus lupulus*) Acids in a Model Wort-Boiling System. Journal of Agricultual and Food Chemistry, 53 (2005), pp. 4434–4439.

[135] Smith, R. J., Davidson, D., Wilson, R. J. J.: Natural foam stabilizing and bittering compounds derived from hops. Journal of the American Society of Brewing Chemists 56 (1998), no. 2, pp. 52–57.

[136] Sakamoto, K., Konings, W. N.: Beer spoilage bacteria and hop resistance. International Journal of Food Microbiology 89 (2003), no. 2–3, pp. 105–124.

[137] Teuber, M.: Low antibiotic Potency of Isohumulon. Applied Microbiology 19 (1970), no. 5, p. 871.

[138] Kaltner, D., Bohak, I., Forster, A., Gahr, A., Back, W.: Investigations of the Influence of Hop Products on the Microbiological Stability of Beer. Proceedings of the 28th European Brewery Convention Congress Budapest. Fachverlag Hans Carl, Nürnberg 2001.

MASHING

1 INTRODUCTION

The brewhouse operations encompassing mashing must not only be performed in a cost-effective manner but must also meet certain standards of quality. Given the quality of the malt and the available brewhouse equipment, the regime selected for the mashing process should be chosen according to its capacity to yield wort of a composition that will produce beer of the highest quality possible. The proteolytic degradation should be carried out in such a way that abundant assimilable nitrogen is present in the wort in order to ensure optimal conditions during fermentation (cf. Yeast Technology and Fermentation) while sufficient amounts of the higher molecular weight nitrogen fractions are nevertheless retained for promoting head retention (cf. Beer Foam). The cytolytic processes should occur so that they ensure trouble-free lautering and filtration (cf. Filterability – Issues with Turbidity). The amylolytic processes are responsible for producing the desired level of fermentable extract. Furthermore, the brewer is also able to influence the aroma and flavor profile of the beer through the various degradation products produced during mashing.

2 THE TECHNOLOGY OF MASHING

Mashing essentially entails bringing the compounds in the malt into solution through enzymatic, physical and chemical processes. Starch, protein and the molecules comprising the cell walls are the most important groups of substances in the malt, which are brought into solution through hydrolysis. Additionally, organic phosphate is transformed into primary phosphates through the action of phosphatases during the mashing process (phosphorolysis). These increase the buffer capacity of the mash and the wort and can thus affect the drop in pH during fermentation. Lipids are degraded to produce a wide variety of products through both auto-oxidative processes and enzymatic activity. Polyphenols are subject to oxidation and polymerization reactions, which lead to a decrease in the concentration of antioxidants, thus undermining flavor stability (cf. Flavor Stability) [1].

For the reasons outlined above, a number of factors must be taken into account when planning and designing mashing equipment and technology. In choosing the most expedient mash program, the quality of the raw materials (malt and brewing liquor) and the available brewhouse equipment have to be considered in order to produce the desired beer style. The malt principally determines the color and aroma of the finished beer. Through the accumulation of low molecular weight nitrogenous compounds, it is possible to increase the concentration of reactants necessary for the Maillard reactions in the mash. At higher temperatures, for example, during the wort boiling process, these compounds can be transformed into substances which impart color and contribute to the aroma of the finished beer. Not only should the individual vessels in the brewhouse be taken into consideration but also important aspects of their design, such as how rapidly they can heat the mash and the rotation speed of the agitators. Vessels, in which the mash can be heated rapidly, are especially critical in mashing procedures requiring the suppression of certain enzymatic processes, for example, during the formation of maltose in the production of non-alcoholic beer. Adjusting the speed of the agitator is important for minimizing oxygen uptake during mashing and ensuring homogenization, e. g. as the temperature of the mash is raised during mashing in or when returning the decoction mash to the main mash.

Figure 1: Mash parameters

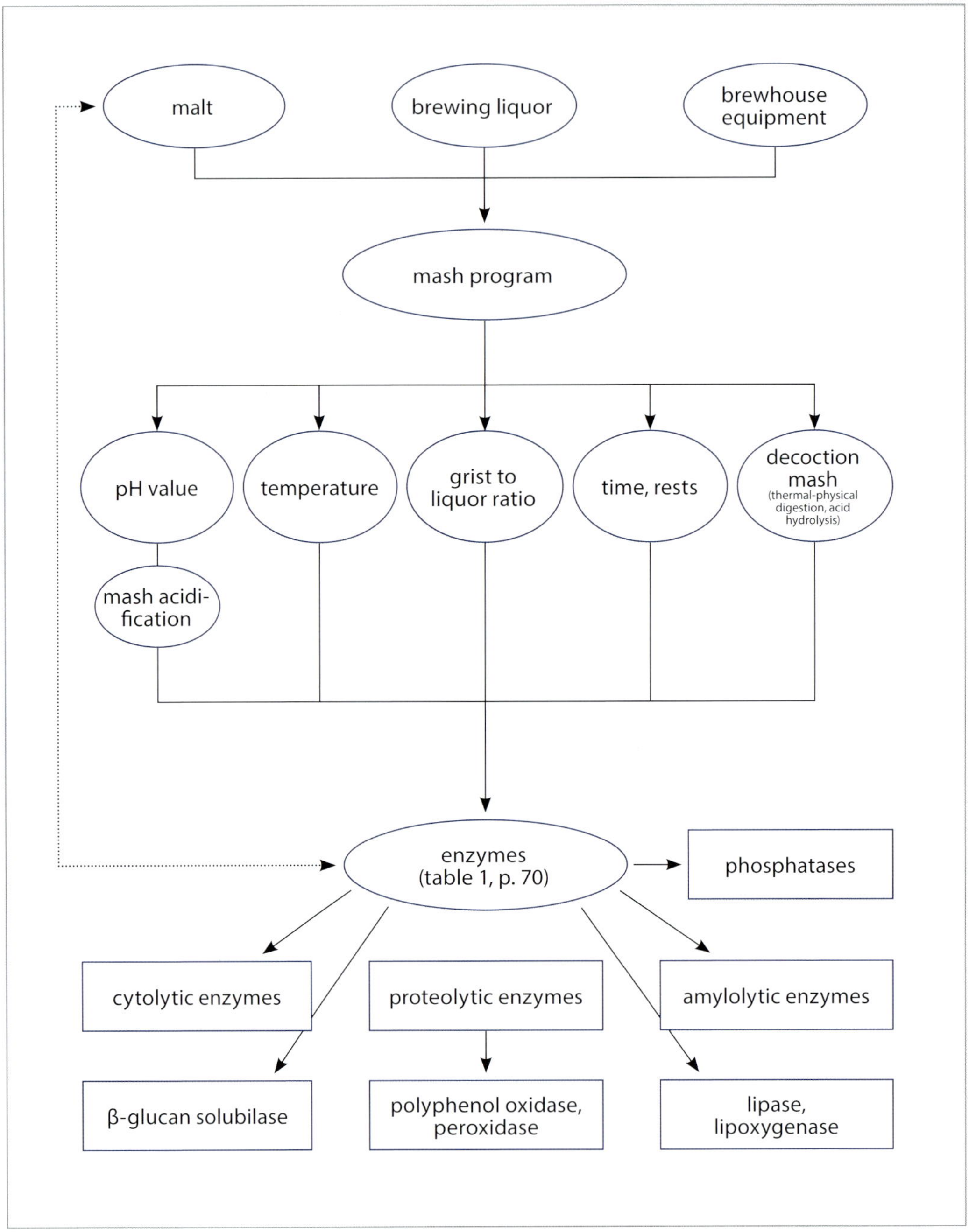

2.1 MASH PARAMETERS

The enzymatic degradation processes, which occur during mashing, can be regulated through the parameters of temperature, pH, time and the volumes of mash and sparge liquor relative to the amount of grist (fig. 1, tab. 1). Over time, a wide range of different mash methods have been developed, and on the whole they can be divided into two categories: infusion and decoction. The latter reinforces the enzymatic degradation of the endosperm through a form of starch digestion, a process that is both physical and thermal. This is achieved by boiling a portion of the mash, which brings about greater thermal stress and more pronounced protein precipitation during the mashing process. Also of consequence for the quality of the beer is the act of pumping a portion of the mash into and out of the mash kettle, which can be problematic with regard to hot-side aeration (cf. Flavor Stability) and the mechanical stress imposed on the husks and the mash colloids.

Due to advances in breeding, modern malting barley varieties achieve satisfactory cytolytic and proteolytic modification during the malting process. This allows brewers producing pale beers to concentrate solely on the amylolytic processes during mashing and to alter their procedures accordingly. For this reason, mashing in at higher temperature is possible and is now common in the industry. An example is the *hoch-kurz* (German term meaning "high-short") mash program, modified to become an infusion procedure as depicted in figure 2 and utilizing a high mash-in temperature between 60 and 64 °C. Mashing in at such temperatures requires a certain level of malt quality, which is described in the chapter Malt in this book.

Aside from saving energy, a necessity in the modern world, the *hoch-kurz* mash program also offers a number of technical advantages. Through suppression of the activity of phosphatase enzymes, whose temperature optimum lies between 50 and 53 °C (tab. 1), the buffering capacity will increase to a much lesser degree, resulting in a more dramatic drop in the pH during fermentation. The finished beer will be lighter in color and possess a higher colloidal stability. The chapter Filterability – Issues with Turbidity includes more information on these processes. Furthermore, the proteolytic activity is reduced, improving head retention since the high molecular weight proteins are not degraded to as great an extent; however, one must ensure that the malt can provide the necessary amount of FAN to the wort. An especially noteworthy version of the *hoch-kurz* mash program incorporates the addition of *Sauergut* or natural lactic acid to adjust the pH to 5.4–5.5 (cf. Flavor Stability).

Figure 2: The *hoch-kurz* mash program modified for infusion mashing

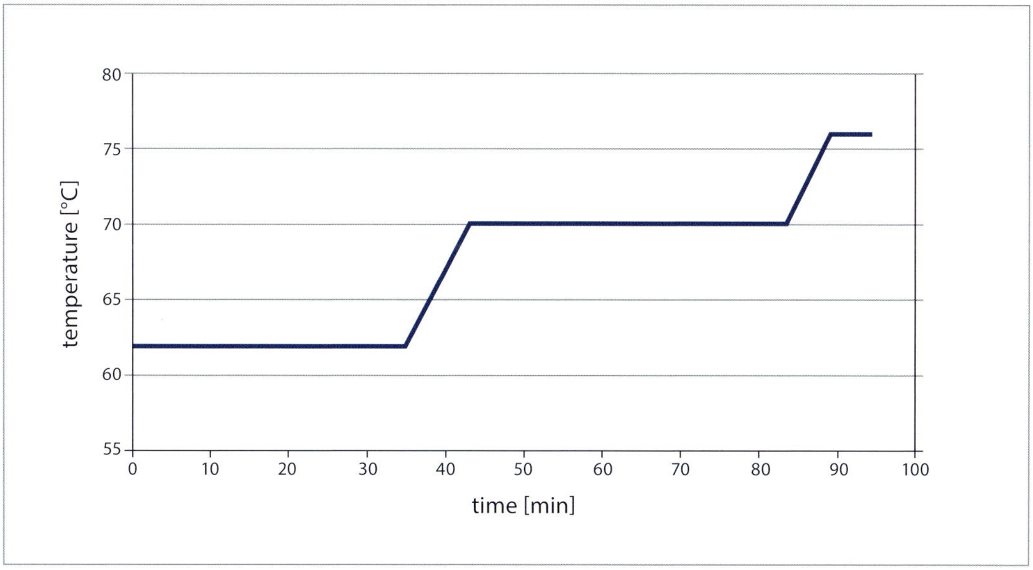

Table 1: pH and temperature optima of selected barley malt enzymes in the mash

	enzyme	temperature optimum in the mash [°C]	pH optimum in the mash	substrate	product
cytolysis	β-glucan solubilase	62–65	6.8	β-glucan bound to the matrix	soluble, high molecular weight β-glucan
	endo-1,3-β-glucanase	< 60	4.6	soluble, high molecular weight β-glucan	low molecular weight β-glucan, cellobiose, laminaribiose
	endo-1,4-β-glucanase	40–45	4.5–4.8	soluble, high molecular weight β-glucan	low molecular weight β-glucan, cellobiose, laminaribiose
	exo-β-glucanase	< 40	4.5	cellobiose, laminaribiose	glucose
proteolysis	endo-peptidase	45–50	3.9–5.5	proteins	peptides, free amino acids
	carboxy-peptidase	50	4.8–5.6	proteins, peptides	free amino acids
	emino-peptidase	45	7.0–7.2	proteins, peptides	free amino acids
	dipeptidase	45	8.8	dipeptides	free amino acids
amylolysis	α-amylase	65–75	5.6–5.8	high and low molecular weight α-glucans	megalosaccharides, oligosaccharides
	β-amylase	60–65	5.4–5.6	α-glucans	maltose
	maltase	35–40	6.0	maltose	glucose
	limit dextrinase	55–60	5.1	limit dextrins	dextrins
additional enzymes	lipase	55–65	6.8–7.0	lipids, lipid hydroperoxides	glycerin + free, long chain fatty acids, fatty acid hydroperoxides
	lipoxygenase (LOX)	45–55	6.5–7.0	free, long chain fatty acids	fatty acid hydroperoxides
	polyphenol oxidase	60–65	6.5–7.0	polyphenols	oxidized polyphenols
	peroxidase	> 60	6.2	organic and inorganic substrates	free radicals
	phosphatases	50–53	5.0	organically bound phosphates	inorganic phosphate

4-vinyl guaiacol is the substance responsible for the clove-like aroma present in Southern German wheat beer. Ferulic acid is its precursor and it is a constituent of the pentosans. During mashing, ferulic acid is liberated from its ester-like bond with arabinoxylans and subsequently decarboxylated by weissbier (wheat beer) yeast to 4-vinyl guaiacol over the course of fermentation. In order to increase the ferulic acid content of the mash, certain variables can be adjusted during mashing; for instance, a rest at 45 °C for 15–30 minutes at a pH of 5.7 results in an increase in the ferulic acid content. Accordingly, mashing in above 50 °C in a mash acidified to a pH of 5.4–5.5 suppresses the activity of ferulic acid hydrolase [2]. The glycoproteins in the malt are among those that have a positive influence on beer foam. They are brought into solution in the mash at low mash temperatures to a certain extent, but are subsequently degraded to low molecular weight substances. The glycoproteins more readily go into solution with a rest at 70 °C. Moreover, they are no longer degraded at this temperature resulting in a positive contribution to head retention (cf. Beer Foam) [3].

The primary objective of mashing is to degrade the starch to such an extent that it is broken down into fermentable sugars. Starch comprises 55 to 65 % of barley (dry basis) and is thus its principal constituent. Starch granules appear in two different sizes: large (type A, 10–25 µm) and small (type B, 1–5 µm). The larger ones make up a mere 10 % of the total number of starch granules but comprise 90 % of the mass. Normal starch granules consist of 20–30 % amylose and 70–80 % amylopectin. From the standpoint of the brewing process, the most important attribute of starch is its propensity to gelatinize, which refers to the swelling of the starch granules caused by an uptake of liquid. As a result of gelatinization, the crystalline structure of the starch undergoes irreversible deterioration [4, 5]. The temperature at which this process begins to take effect is known as the gelatinization temperature. For large starch granules, the gelatinization temperature lies around 61–62 °C, whereas small granules gelatinize at approximately 75–80 °C [6]. A rotation viscometer can be employed to determine the gelatinization temperature as well as the gelatinization behavior of barley and barley malt [4]. It should be noted that when determining the gelatinization temperature, the results might vary from method to method by a few degrees [7]. The gelatinization temperature of barley malt lies between 58 and 65 °C and thus differs slightly from that of the barley from which it was produced [8]. The gelatinization temperature itself is influenced by the barley variety and its provenance as well as the conditions over the growth season of that particular year (fig. 3), especially if the temperatures are high when the barley is ripening [1, 4, 22]. For this reason, the 2010 harvest exhibited a gelatinization temperature 2–4 °C higher compared to that of the 2011 harvest. The elevated gelatinization temperature for the 2010 barley crop was manifested in an overall loss in final attenuation of approximately 1 % in beers produced in that year compared to those produced in 2004.

Below the gelatinization temperature, the starch granules that had been enzymatically degraded during the malting process [9] or had been mechanically damaged during milling are hydrolyzed [10]. Only such starch granules will swell in contact with water below the gelatinization temperature and are therefore able to be hydrolyzed enzymatically [5, 8, 10, 22].

Native, i.e. unmodified, starch can only be degraded once the gelatinization temperature has been exceeded in the mash. Starch degradation during mashing is subject to the accessibility of the starch and the activity of the amylolytic enzymes. Rapid gelatinization and high amylolytic activity ensure a timely conversion and a high final attenuation.

Research to date has indicated that there is little that maltsters can do to influence the gelatinization temperature. Thus, for all barley harvests exhibiting higher gelatinization temperatures (e.g. 2006 and 2010), practical technological approaches to increasing the α-amylase activity have to be taken into consideration. If the unfortunate combination of high gelatinization temperatures and low α-amylase activity occur in one harvest, the mash pH and the temperature of the individual mash rests must be precisely controlled. High gelatinization temperatures and low α-amylase activity lead to difficulties in hydrolyzing the starch. The low α-amylase activity already aggravates a difficult situation with the β-amylase activity being compromised by limited access to the starch due to a lack of gelatinization. Figure 3 indicates the gelatinization temperature above which the starch will become accessible to the amylolytic enzymes. However, at the same temperature, the initial stages of enzyme denaturation are already occurring, bringing about a decrease in the residual activity of the β-amylase. As a result of hot, dry growing seasons, the gelatinization temperature of starch will often be higher than, for example, the optimal temperature for β-amylase activity, which can result in a low final attenuation in the finished beer. It is not infrequent after a year of that kind to obtain barley with a high gelatinization temperature coupled with low α-amylase activity. This, in turn, may increase the amount of time required for saccharification, i.e. until the iodine test is negative. This combination can even result in an incomplete iodine reaction (in the mash, in the first wort or later in the sparged lauter wort). The impact of a high gelatinization temperature can be reduced by increasing the concentration of calcium and chloride ions, i.e. the "hardness", of the brewing liquor, by blending the malt with other malt exhibiting a lower gelatinization temperature (with the gelatinization temperature of the resulting blend below the calculated mean value) or through decoction [12].

In developing a mash procedure, it is therefore necessary to obtain information concerning the gelatinization temperature and the α-amylase activity in the malt. Even during the β-amylase rest, the conditions present in the mash should be adjusted to the optima for α-amylase (pH 5.6–5.8). Mash rests at the gelatinization temperature promote the accessibility of the starch and thus improve final attenuation in the finished beer.

The two enzymes essential for the mashing process, α-amylase and β-amylase, both cleave the α-(1,4)-D-glycosidic bonds found in starch [13]. The enzyme β-amylase cleaves maltose from the non-reducing end of the starch molecule and is thus crucial in establishing the degree of attenuation possible for the wort [14]. β-Amylase is heat-sensitive and can lose up to 60 % of its activity after only 10 min. at 65 °C [15]. The maltose concentration increases only slightly during a mash rest at 62 °C lasting more than 20 min. during a mash rest at 62 °C. Therefore, in the end, lengthening the duration of the maltose rest in order to raise the degree of attenuation accomplishes very little. By mashing in at a lower temperature, the inactivation of β-amylase can be slowed, because the enzyme is more thermostable in the presence of proteins (fig. 4) [1]. This is of particular interest in high gravity brewing operations. In contrast to α-amylase, β-amylase is incapable of directly accessing starch granules. For this reason, as already explained in the chapter on malt, α-amylase is indispensable because it sets the pace of starch conversion [1, 16].

Figure 3: Gelatinization temperature

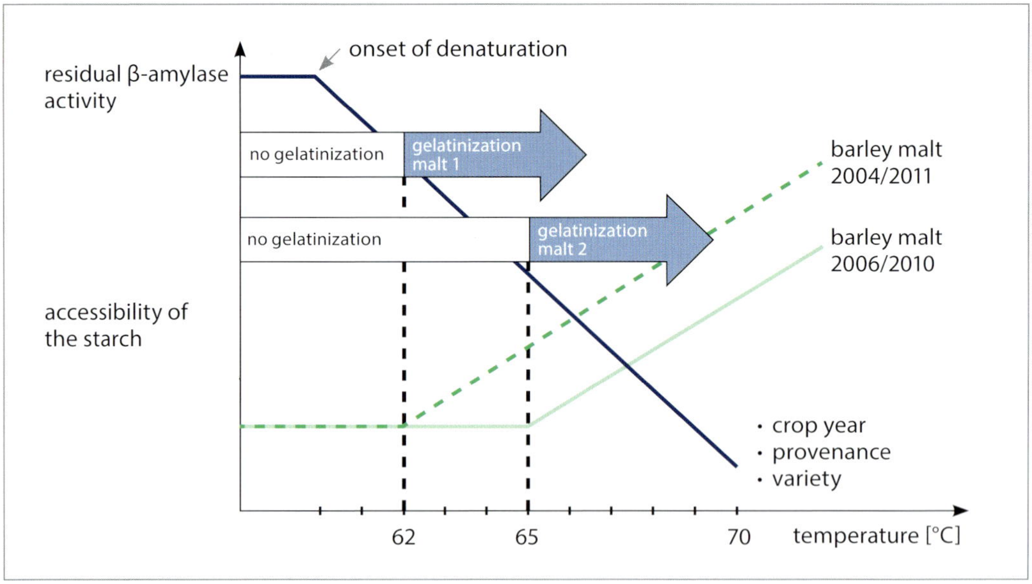

By comparison, α-amylase is much more thermostable, still exhibiting a residual activity of 30 % of its original activity after 10 min at 80 °C. Since α-amylase is an endoenzyme, the number of starch fragments with non-reducing ends increases while it is active, steadily releasing substrate for β-amylase. This synergistic effect between the two enzymes can only be fully exploited if gelatinization is far enough advanced that the starch granules become accessible to the α-amylase while the β-amylase enzymes remain active. This effect is only possible above the gelatinization temperature and is thus dependent upon the specific malt (60–65 °C) utilized by the brewer.

The enzyme limit dextrinase cleaves the α-(1,6)-D-glycosidic bonds present in amylopectin and is crucial for thorough starch conversion [13, 14, 17]. However, limit dextrinase only exhibits marginal activity with complete inactivation at 65 °C, so its contribution to starch degradation in barley malt is minor.
Incorrect amylolysis procedures can result in a degree of attenuation that is either excessive or lacking. Moreover, should the results of the iodine test be positive in the wort, turbidity generated by α-glucans (starch fragments) can occur later in the beer. These substances can also interfere with filtration due to their high viscosity. Even if the iodine test performed on the mash is shows conversion has occurred, the results of the iodine test performed on wort samples taken further downstream can still yield a different result. This is due to the starch granules not undergoing gelatinization and then finding their way into the lauter wort, where they rupture during the wort boiling process, releasing high molecular weight α-glucans (cf. Filterability – Issues with Turbidity).

Figure 4: Residual activity of β-amylase at 65 °C as a function of the grist to liquor ratio

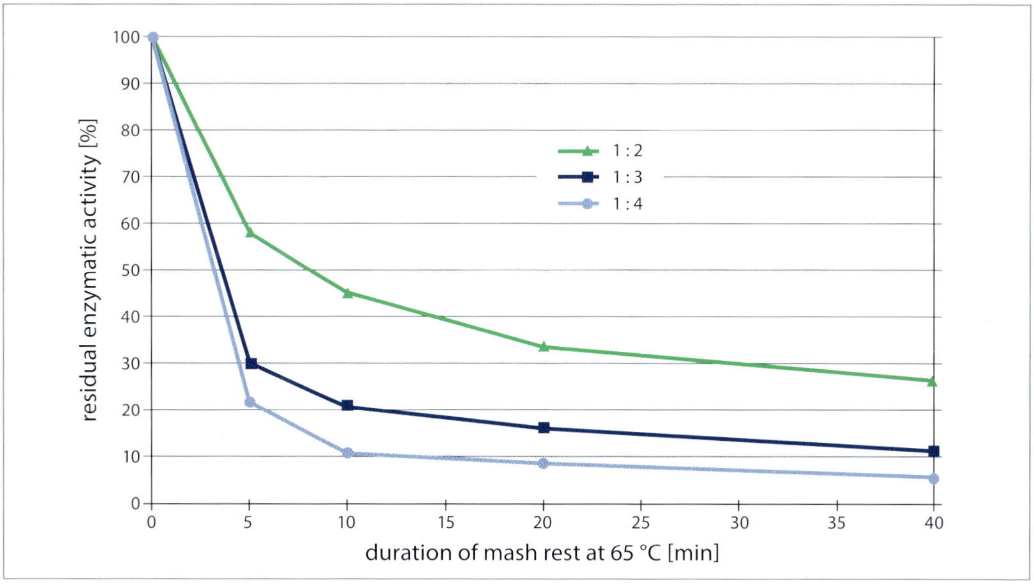

Unnecessarily excessive degradation during mashing can be countered by mashing in at a higher temperature and performing abbreviated rests. Decoction procedures can also be altered correspondingly to prevent this from occurring.

Compared to the malts of the past, those produced today are more potent enzymatically and exhibit considerably higher degrees of modification. For these reasons and bearing the purpose of brewhouse automation in mind, it is a reasonable choice to replace decoction with infusion mashing in many cases. By doing so, the beer becomes somewhat more neutral in flavor. Precisely due to the high enzymatic content, which intensifies "over-conversion" during mashing, or a high gelatinization temperature of a particular type of malt, decoction can at times make sense. The mash does not necessarily need to be boiled, simply heated to just below the boiling point and held there for 5–15 min. By doing so, Maillard products will also form, providing the beer with some character. The Strecker aldehydes evaporate downstream in the wort kettle, and consequently the flavor stability is not compromised by their presence. On the contrary, the flavor stability is improved since the oxidases in the decoction mash – the lipoxygenases, peroxidases and polyphenol oxidases – are inactivated, so that the reactions they catalyze are less pronounced overall. The excessive amounts of proteolytic enzymes are also reduced, which leads to a greater percentage of proteins belonging to the higher molecular weight fraction. The gelatinized starch re-entering the mash tun improves the gelatinization of the entire mash at a lower temperature than could be achieved using an infusion procedure. In view of these factors, decoction still offers an attractive alternative. Additionally, desmoenzymes are liberated through heating the decoction mash. These findings have been confirmed in practice on an industrial scale.

Concern about wasting energy is unwarranted, since the mash is not boiled and thus very little vapor rises from the brewhouse. In centuries past, the decoction mash employed in the production of pale beer was usually boiled for 15–20 min and up to 30 min for dark beer, even though the heating surfaces in the mash kettles were relatively large and resulted in a significant amount of evaporation. In modern breweries, a minuscule percentage of additional energy – ultimately lost as radiant heat – is required to complete a decoction procedure. Only when using hot water from the hot liquor reservoir for the purpose of heating the mash is it necessary to achieve somewhat higher temperatures, which can nonetheless be generated by supplementary heat sources, e.g. a cogeneration unit. Modern brewhouse vessels with internal heating surfaces accommodate these needs.

Decoction mashing is a simple means for enhancing the diversity among individual beer styles. Decoction does not lengthen the duration of the mash schedule, since the enzyme rests can be curtailed to some extent.

With the standard types of malt currently available on the market, the degree of attenuation in the beers one finds today is often too high. For those looking to counteract this tendency, one alternative would be to fairly rapidly heat the mash to beyond the temperature optimum for β-amylase or to acidify the mash to lower the pH. The procedure depicted in figure 5 provides a mash program designed to hasten past the β-amylase temperature optimum when heating the mash. When carrying out this procedure, it is important that the decoction mash be heated rapidly to a temperature over 70 °C, so that β-amylase does not have the opportunity to remain active for very long. The temperature of the main mash should not rise above 52 °C. The decoction mash should be returned back to the main mash as quickly as possible, so that the β-amylase in the main mash is promptly deactivated. The same objective is achieved through the spring mash program (cf. section 2.2.1). Lowering the pH to at least 5.4 by means of mash acidification can also reduce the degree of attenuation. Though this is actually advantageous for β-amylase, acidifying the mash shifts the pH away from the optimum for α-amylase (5.6–5.8). Gelatinization transpires more slowly at lower pH values [4], restricting the enzymatic degradation of the starch by α-amylase and limiting substrate availability for β-amylase.

Figure 5: Mash program for reducing the degree of attenuation

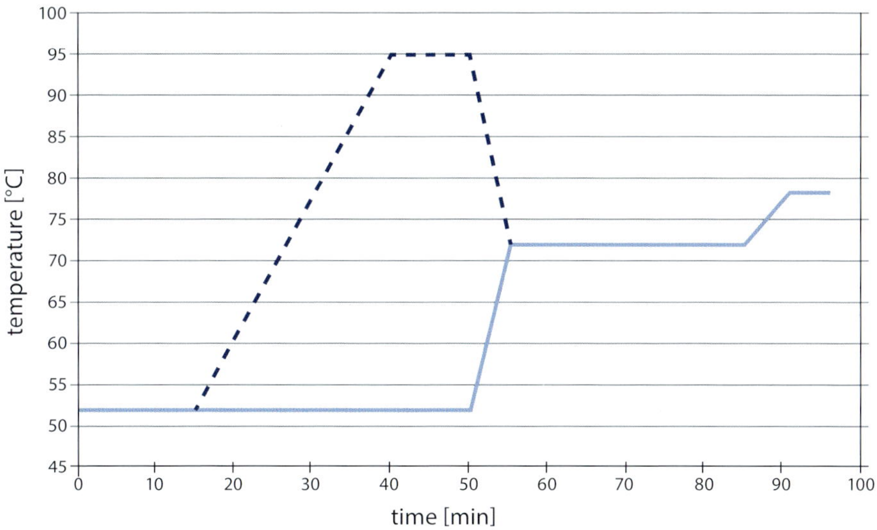

Modern brewhouses made of stainless steel are designed to avoid oxygen uptake, which over the course of a 2.5–3.5 hour mash program can amount to 50–100 mg of oxygen per liter [1]. Fuelled by the catalytic effect of the copper or black steel material typical of earlier brewhouse vessels, groups of substances in the mash interacted with the oxygen. This had a pronounced influence on protein degradation, since the sulfhydryl groups from cysteine formed through oxidation of the disulfide bridges, leading to an increase in the size of the protein molecules. Forming aggregates like this is evidenced by the "dough" as it is known, which causes difficulties during lautering but also impedes the hydrolases from engaging their respective substrates. This affects both protein degradation and starch conversion, and with regard to the latter, gives rise to lengthier saccharification rests and a lower degree of attenuation. Furthermore, oxidation processes take place causing the peroxidases and polyphenol oxidases to react with polyphenols resulting in color increases in the mash, wort and the finished beer. Of significance is also the oxidation of long-chain fatty acids stemming from lipid degradation, particularly unsaturated linoleic acid and linolenic acid. These two fatty acids are oxidized to carbonyls, i.e. forming aging compounds later in the finished beer.

"Inert" brewhouses reduce the amount of oxygen uptake by allowing mashing in to occur from below, i.e. through the floor of the vessel. This is an uncomplicated process with wet-milled grist, while there are special grist hydrators or pre-mashers able to perform this task for dry-milled or finely milled grist. The mash (plus any decoction mashes) is also pumped into the respective vessels of the brewhouse from below (mash tun, mash kettle, lauter tun, mash filter). Variable-speed agitators help prevent the uptake of air and reduce the effects of shearing. Further measures can be taken as well: de-aerated brewing liquor or even a layer of nitrogen gas above the mash can be employed. In implementing these measures, the oxygen uptake begins to approach zero.

The effect is as follows: Improvements in the activity of the hydrolytic enzymes means that starch degradation is more intense, resulting in a higher limit of attenuation. More proteins go into solution, and there is less high molecular weight nitrogen and an increase in the soluble nitrogen, low molecular weight fractions and FAN, creating a tendency toward more neutral, empty tasting beers with poor head retention.

The effect on the substances providing structure and support in the endosperm is minimal. Reducing the oxidation increases the concentration of polyphenols with a lower polymerization index. This improves the reducing capacity, promoting better flavor stability, but also consequently brings about more protein precipitation (foam!) during the wort boiling process. Moreover, the flavor stability is enhanced by suppressing the activity of the lipoxygenases (prevent contact with air, higher mash-in temperatures, lower pH through mash acidification).

The formation of "dough" is halted to some degree, and the lautering process occurs much more rapidly. The mash in the individual chambers of the mash filter is also more homogeneous, so that the filtration process takes place more quickly and efficiently (better yield).

2.2 SELECTED MASH PROGRAMS

Depending on the malt quality and taking the desired beer style into consideration, various specialized mash programs are conceivable. Several examples are provided below.

2.2.1 SPRING MASH PROGRAM

The spring (German for "jump") mash program (fig. 6) is especially well suited for non-alcoholic beers [1, 18]. This procedure is based upon the principle of rapidly pumping boiling water into the mash within a very short time, causing the temperature to abruptly rise from 50 °C to over 70 °C. This causes β-amylase to be quickly inactivated, halting maltose formation for the most part. Wort with a low limit of attenuation can be produced in this way. The enzyme α-amylase is also damaged, and for this reason, the iodine test performed on the wort is often positive [19]. It is imperative that the temperature not fall below 73 °C. The mash-in temperature and the rests should be adjusted so that the wort contains ample assimilable nitrogen (22 % of the total nitrogen) (cf. Yeast Technology and Fermentation). For this reason, malt should be employed that is well modified, both proteolytically and cytolytically, and exhibits a low limit of attenuation (cf. Malt).

Figure 6: Spring mash program

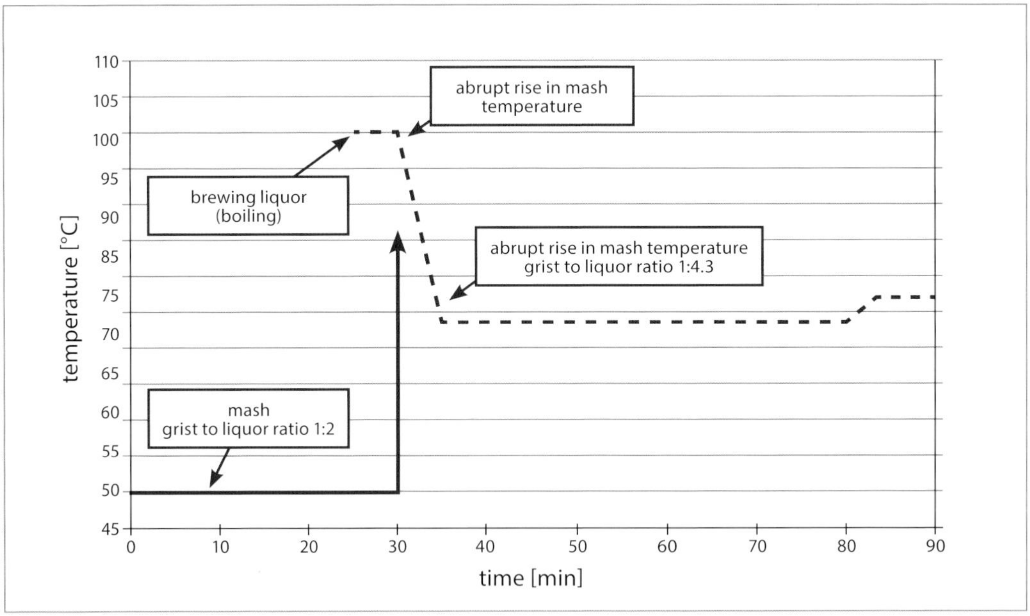

2.2.2 SEPARATING THE MASH TO ACHIEVE β-Glucan DEGRADATION

If only extremely under-modified malt is available, a mash program, in which a thick mash is pulled at 35 °C and heated to 65 °C, can provide relief (fig. 7). At this temperature, high molecular weight β-glucan can be dissolved from its location in the cell wall matrix by the enzyme β-glucan solubilase. Through the addition of cold brewing liquor, the separate mash can be cooled, so that when the two are combined, the total mash reaches a temperature of 45 °C. This facilitates the subsequent degradation of high molecular weight β-glucans released in the separate mash by the endo-β-1,4 glucanases in the total mash. Afterwards, an infusion mash with a maltose rest and starch conversion rest can be carried out.

Figure 7: Separating the mash to achieve greater β-glucan degradation

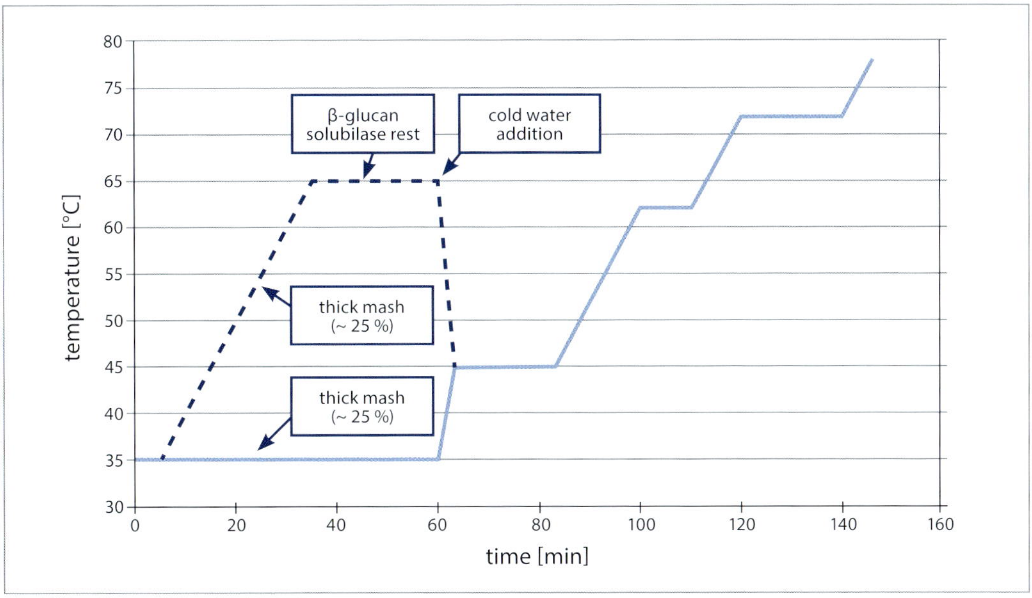

The maltase mash program illustrated in figure 8 offers another method for processing cytolytically under-modified malt [20].

2.2.3 THE MALTASE MASHING METHOD

A high ester content may be desirable in certain styles of beer, but it is especially important in weissbier (cf. Southern German Wheat Beer). Among other things, the ester content can be adjusted through the glucose concentration. A high glucose concentration can be obtained through favoring maltase activity. At a temperature optimum of 45 °C, this enzyme cleaves maltose into two glucose units. With a typical infusion mash, there is very little maltose available in the mash at that temperature. However, a mash procedure would rectify this, in which there is a maltase rest [21] consisting of two parts (fig. 8). In the first part, 60 % of the grist is mashed-in at 62 °C with 45 % of the mash liquor. After a 40 min rest, the mash is heated up to 70 °C for starch conversion. After the starch conversion rest, cold liquor is added to cool the mash, at which point the remainder of the grist is added. The maltase in the grist just added to the mash hydrolyzes the maltose released during the starch conversion rest carried out earlier. This procedure can also be employed to reduce high β-glucan levels in extremely under-modified malt (cf. section 2.2.2).

Figure 8: Maltase mash program

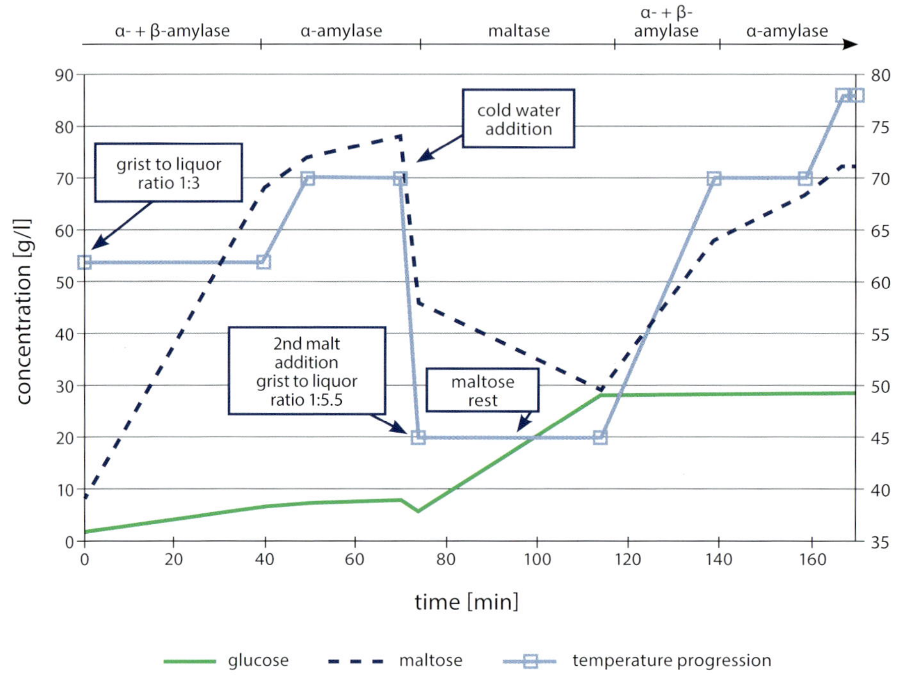

2.2.4 DARK BEER STYLES

The most recognizable attributes of dark beer styles are their color (50–80 EBC units) and malty aroma. Dark malts, dark caramel malts (3–5 %), coloring malt (usually 0.5–2 %, max. 5 %) or a liquid malt coloring agent are utilized to acquire these characteristics in beer. As little coloring malt as possible should be used, since substances with a burnt, tar-like aroma, e.g. pyrazines (cf. Flavor Stability) are found in these products. The addition of 10–20 % pale malt to the grist is recommended in order to ensure that an adequate supply of enzymes is present. Hard brewing liquor with a higher residual alkalinity will produce beer with a darker color, whereas soft water increases the aromatic malty character of dark beer [1].

In the production of dark malt, FAN becomes depleted in the production of melanoidins as part of the Maillard reactions. For this reason, proteolysis must be more extensive during malting, in order to prevent scarcity later during fermentation when the yeast requires amino acids (cf. Yeast Technology and Fermentation).

Figure 9: Decoction procedure for the production of dark beer styles

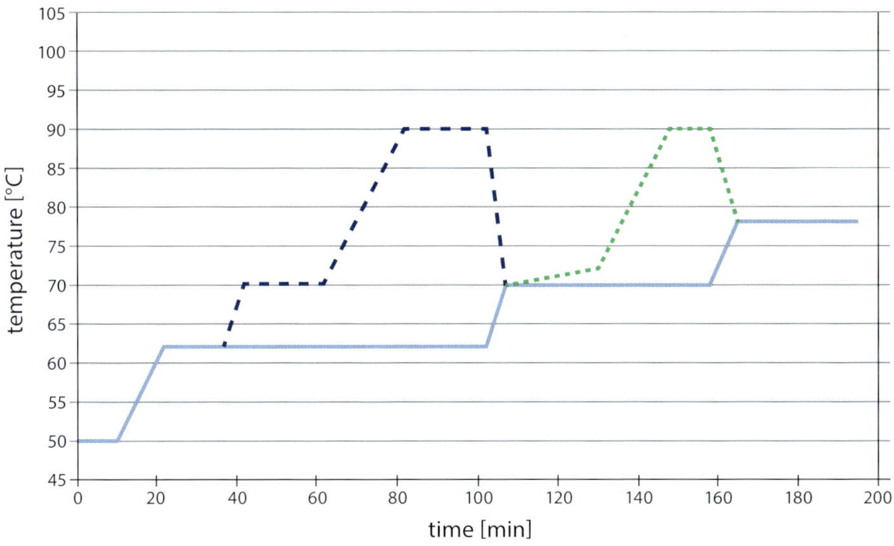

The intense decoction mash program employed for the production of dark beers is inevitably accompanied by greater energy usage. A double decoction mash program with a mash-in temperature of 50 °C expends 35 % more energy than a comparable infusion program mashed in at the same temperature (basis for the calculation: 300 hl total mash, 5 % heat loss, mash program according to fig. 9). The energy necessary for boiling both decoction mashes is evident in the level of evaporation resulting from those processes. In modern mash programs, the decoction mashes are often not boiled but rather held at temperatures of 96–97 °C for 10–20 min. This also reduces the energy losses due to thermal radiation.

3 SUMMARY

Due to the highly advanced proteolytic and cytolytic modification typical in modern malt, mashing procedures can largely focus on amylolytic processes. Should the mash-in temperature be raised to 62 °C, a careful review of the quality of the malt is essential; otherwise, problems downstream can arise during lautering, fermentation and filtration. These difficulties in the brewing process can also negatively affect the quality of the finished beer (e.g. poor head retention and flavor stability). Differences in quality owing to the barley variety, provenance or a particular harvest year necessitate that the mash parameters be modified on a regular basis. Not only is mashing employed to maintain a consistent level of quality, the mash parameters can also be modified to create a broad spectrum of beer styles.

4 OVERVIEW

Attribute	Cause	Measures/Suggested Solution	Chapter
inefficient filter operations	high β-glucan content due to undermodified malt, β-glucan gel formation due to the emergence of shearing forces, homogeneity < 70 %	use highly modified malt, increase intensity of cytolytic degradation during mashing; use homogenous malt	malt, filtration
sluggish fermentation	FAN content is too low	use malt with more FAN; more intense proteolytic rests	malt
low final attenuation	fermentable sugar content is too low	intensify β-amylase rest; set mash-in temperatures to β-amylase optimum	
low final attenuation	high gelatinization temperature; above all accompanied by low α-amylase activity	optimal conditions for α-amylase (pH 5.6–5.8); rests within the range of gelatinization temperature	
high final attenuation	excessive amounts of fermentable sugars	reduce the maltose rest; mash acidification	filtration
turbidity	insufficient saccharification through α-amylase activity	intensify the amylolytic rests	
high iodine values	mash-out temperatures are too low; sparge liquor temperature is too high; mill gap settings are not optimal	maintain a maximum temperature of 78 °C; monitor grist composition	
foam issues	proteolysis is too intensive	select a higher mash-in temperature	foam
diminished flavor stability	oxygen uptake is too high	avoid oxygen uptake; do not pre-mill the malt; review the mill settings, the agitator settings, and the pump operations; employ higher mash-in temperatures; acidify the mash	flavor stability
protracted lautering time	excessive β-glucan values are excessively high; milling is too fine	use more highly modified malt; increase intensity of cytolytic degradation during mashing; check grist composition, use more homogenous malt	malt
color increase	oxygen uptake is too high; thermal stress is too high	use infusion mash program; reduce thermal stress	wort boiling
insufficient pH decline	buffering effect is too strong; mash pH is too high	mash-in temperature is above the phosphatase optimum (> 50 °C); if necessary forego mash acidification	acidification with natural lactic acid
brewhouse yield is too low	malt is undermodified; available residual extract is too high; saccharification by α-amylase is insufficient; milling is too coarse	lower mash-in temperature; use more highly modified malt; intensify the amylolytic rests; check grist composition	malt

5 REFERENCES

[1] Narziß, L.; Back, W.: Die Bierbrauerei, Band 2: Die Technologie der Würzebereitung, 8th ed. Wiley-VCH, 2009.

[2] Nitzsche, W. F.: Untersuchungen zur Bildung phenolischer Komponenten bei der Herstellung von bayerischem Weizenbier. Dissertation, Technical University of Munich (Weihenstephan), 1992.

[3] Narziß, L., Reicheneder, E., and Barth, D.: Untersuchungen über den Einfluss der Glykoproteide auf die Schaumeigenschaften des Bieres. Brauwissenschaft 35 (1982), no. 11, pp. 275–283.

[4] Tegge, G.: Stärke und Stärkederivate. 3rd ed. B. Behr's Verlag GmbH & Co. KG, Hamburg 2004.

[5] French, D.: Organization of Starch Granules. In: Paschall, E. F.: Starch: Chemistry and Technology. Academic Press, London 1984, pp. 184–247.

[6] Palmer, G. H.: Cereals in Malting and Brewing. In: Palmer, G. H.: Cereal Science and Technology. Aberdeen University Press, Aberdeen 1989, pp. 61–242.

[7] Qian, J. Y., and Kuhn, M.: Evaluation on Gelatinization of Buckwheat Starch: a Comparative Study of Brabender Viscoamylography, Rapid Visco-Analysis, and Differential Scanning Calorimetry. In: European Food Research and Technology 209 (1999), 3–4, pp. 277–280.

[8] Keßler, M., Kreisz, S., Zarnkow, M., and Back, W.: Investigations about the relative Hartong Extract at 45 °C. BrewingScience 58 (2005), no. 7/8, pp. 56–62.

[9] Bertoft, E., and Henriksnas, H.: Starch Hydrolysis in Malting and Mashing. In: Journal of the Institute of Brewing 89 (1983), no. 4, pp. 279–282.

[10] Gibson, T. S., Alqalla, H., and McCleary, B. V.: An Improved Enzymatic Method for the Measurement of Starch Damage in Wheat-Flour. Journal of Cereal Science 15 (1992), no. 1, pp. 15–27.

[11] Bathgate, G. N., and Palmer, G. H.: The In-Vivo and In-Vitro Degradation of Barley and Malt Starch Granules. Journal of the Institute of Brewing 79 (1973), no. 5, pp. 402–406.

[12] Schüll, F. A., Einfluss spezifischer Eigenschaften der Stärke auf den Brauprozess. Dissertation, Technical University of Munich (Weihenstephan), 2012.

[13] Kruger, J. E., and Lineback, D. R.: Carbohydrate-Degrading Enzymes in Cereals. In: Stauffler, C. E.: Enzymes and their Role in Cereal Technology. St. Paul. American Association of Cereal Chemists, 1987, pp. 117–135.

[14] MacGregor, A.W., Bazin, S. L., Macri, L. J., and Babb, J. C.: Modelling the Contribution of Alpha- Amylase, Beta-Amylase and Limit Dextrinase to Starch Degradation During Mashing. Journal of Cereal Science 29 (1999), no. 2, pp. 161–169.

[15] Eglinton, J. K., Langridge, P., and Evans, D. E.:Thermostability Variation in Alleles of Barley Beta- Amylase. Journal of Cereal Science 28 (1998), no. 3, pp. 301–309.

[16] Ziegler, P.: Cereal Beta-Amylases. Journal of Cereal Science 29 (1999), no. 3, pp. 195–204.

[17] Stenholm, K., and Home, S.: A new Approach to Limit Dextrinase and its Role in Mashing. In: Journal of the Institute of Brewing 105 (1999), no. 4, pp. 205–210.

[18] Narziß, L., Miedaner, H., and Krottenthaler, M.: Development of a Selection Method Suitable for Serial Application for Improving the Quality of the Beer and the Quality Parameters of Barley which Influence the Taste of Beer. Monatsschrift für Brauwissenschaft 48 (1995), no. 3/4, pp. 108–111.

[19] Schur, F., Anderegg, P., Pfenninger, H.: Charakterisierung filtrationshemmender Stoffe. In: Brauerei Rundschau 89 (1978), no. 8, pp. 129–144.

[20] Hermann, M.: Technologische Möglichkeiten zu Beeinflussung des Aromaprofils von Weizenbieren. Dissertation, Technical University of Munich (Weihenstephan), 2005.

[21] Herrmann, M., Back, W., Sacher, B., and Krottenthaler, M.: Options to Technologically Influence Flavour Compounds in Beer. Monatsschrift für Brauwissenschaft 56 (2003), no. 5/6, pp. 99–106.

[22] Keßler, M. T.: Analytische Erfassung und Interpretation der Stärkedegradation im Gersten- und Malzkorn und die Aussagekraft für den Brauprozess. Dissertation, Technical University of Munich (Weihenstephan), 2006.

WORT BOILING AND WORT BOILING SYSTEMS

1 INTRODUCTION

Around 15 years ago, the topic of wort boiling seemed to have been researched from every angle. High quality wort was being produced with a total evaporation of 7 to 8 % in vessels equipped with internal or external calandrias. However, problems often associated with a particular harvest or a barley's provenance arose, impacting attributes such as foam, flavor dimethyl sulfide (DMS) and flavor stability (thermal stress) in the finished beer. Due to the frequency of these phenomena, the general consensus was reached that this topic required further investigation. By controlling and adjusting the individual parameters – including the temperature – of the heat transfer medium, evaporative efficiency, wort temperature, wort circulation as well as the duration of the boil or the amount of time the wort remains at a constant temperature, it is now feasible to produce wort of a specific composition, thus tailoring it to the requirements of an individual beer style. In doing so, it is also possible to reduce the total evaporation to a mere 4 to 5 % with no detrimental effects.

2 THE TECHNOLOGICAL BASIS FOR WORT BOILING

The processes necessary for wort production, which occur during the boiling process, can be divided roughly into two main categories: the first involves maintaining the heated wort at a constant temperature and the second, wort evaporation.

2.1 MAINTAINING THE HEATED WORT AT A CONSTANT TEMPERATURE

A number of processes and reactions take place while hot wort is held at a constant temperature. These include hop isomerization, formation of color and aroma compounds, increased solubility of certain substances, enzyme inactivation and elimination of microbes. The inactivation of the malt enzymes is crucial, since unusual flavor profiles may otherwise result, e.g. overly attenuated beer. Furthermore, protein molecules and protein-tannin complexes (hot break) are removed during wort boiling through precipitation, facilitating the production of bright and delicately luminous wort. Through excessive protein coagulation, the high molecular weight proteins instrumental for foam formation (10 and 40 kDa) are also precipitated and the foam stability of the finished beer suffers as a result. This is of particular significance if no additives for promoting foam are employed to enhance foam formation. On the other hand, if protein coagulation is unsatisfactory, this can cause colloidal instability in the finished beer. In many cases, consumers consider hazy beer to be unacceptable and as grounds for complaint. However, the substances generating turbidity can be partially removed by adsorptive means through the use of filter aids, either prior to or during beer filtration (cf. Filterability – Issues with Turbidity).

The Maillard reactions commence at temperatures around approximately 80 °C and bring about the formation of aroma compounds. Of special significance are the Strecker aldehydes, compounds formed from amino acids, which can negatively affect the flavor stability of beer. Strecker aldehydes are the primary and secondary products of the Maillard reactions and can be reduced through vaporization during wort boiling, depending upon their volatility, the kind of wort boiling system in use and the amount of evaporation. This means that boiled wort contains fewer of these aroma compounds than unboiled wort [1, 2].

2.2 EVAPORATION

Evaporation serves to drive out undesirable aroma compounds, such as myrcene derived from hops, various carbonyls as well as some sulfur compounds, especially dimethyl sulfide.

Aroma compounds produced as part of metabolic processes involving lipids are eliminated through vaporization during the wort boiling process. Some of these substances can be employed as analytical indicators for the efficiency of evaporation occurring during the boiling process, since they are not produced when the wort is boiling.

Additionally, boiling the wort sets the original gravity. This is essential in order to ensure a consistent product, to fulfill legal requirements and to guarantee a safe and marketable product. A high level of evaporation allows for the use of more sparge liquor during lautering and thus increases extract yield, thereby conserving malt. At the same time, more energy has to be expended for evaporation. In doing so, the wort is subjected to greater thermal stress, negatively affecting the flavor stability of the beer.

2.2.1 EVAPORATIVE EFFICIENCY [3]

Experimental trials have shown that in practice, different levels of reduction in the concentrations of indicators for evaporation can result, despite comparable values for total evaporation. The decline in the concentrations of these indicators can be described through a dimensionless number, the evaporative efficiency (EE), which is characterized by a corresponding indicator for evaporation (e.g. hexanal), and is dependent upon the total evaporation and the wort boiling system. The evaporative efficiency is defined as follows:

$$\frac{\textit{decrease in concentration evaporation indicator } [\%]}{\textit{total evaporation } [\%]} = \textit{evaporative effeciency}$$

In turn, it follows that:

$$EE = \frac{(c_0 - c) \times 100\,\%}{c_0 \times TE}$$

Equation 1: Evaporative efficiency
c_0 = original concentration; c = concentration at the time of measurement; TE = total evaporation [%]

The key metric as defined in equation 1 specifies the extent of the percental loss of a substance for each percent of evaporation. This means that the greater the evaporative efficiency, the more a particular substance dissipates based on the degree of evaporation – regardless of whether this is considered a positive or negative attribute of the wort boiling system. The evaporative efficiency must be separately determined for each substance chosen as an indicator for evaporation. The maximum amount able to be volatilized depends on the level of evaporation. From a purely physical standpoint, evaporation is not dependent on the wort boiling system. However, the matrix of beer wort is very complex, and the way currents flow through the numerous types of wort boiling systems can vary immensely. Thus, the level of mixing differs considerably from system to system. If these conditions are not favorable, the evaporative efficiency may suffer since some of the wort may not be adequately boiled.

Using this key metric, the conditions in a wort boiling system can be evaluated based on the data acquired during the boiling process itself. The evaporation of certain substances can therefore be extrapolated from these data, meaning that in practice, the duration of wort boiling can be curtailed to a time sufficient to remove these unwanted aroma compounds. Likewise, the energy consumption (lower total evaporation) and the thermal stress on the wort (flavor stability) are reduced under these conditions, while retaining some of the residual coagulable nitrogen (better foam, more full-bodied).

An error can result in calculating the evaporation or total evaporation when using this key metric to monitor and evaluate a wort boiling system or the boiling process. The increase in the extract concentration, the volume of lauter wort present in the kettle prior to boiling, the volume of cast-out wort or the consumption of heat energy can be used in the calculation.

An assessment of the boiling process on the basis of evaporative efficiency must be carried out for each aroma compound or indicator for evaporation, since the volatility of the individual compounds in wort – a matrix composed of a wide range of constituents – will be different. This key metric can only be applied to substances that are not produced over the course of wort boiling, such as the indicator substances hexanal, 1-hexanol or 1-pentanol.

The values listed in table 1 form the mathematical basis for determining the evaporative efficiency. Figure 1 was generated from this table and serves as a tool for graphically determining a desired value when two values of three are known (e.g. total evaporation, the original and final concentrations of the indicator substances).

Table 1: Data table for determining evaporative efficiency

	total evaporation [%]												
$(c_0 - c)/c_0$ [%]	**2.0**	**2.5**	**3.0**	**3.5**	**4.0**	**4.5**	**5.0**	**5.5**	**6.0**	**6.5**	**7.0**	**7.5**	**8.0**
100	50.0	40.0	33.3	28.6	25.0	22.2	20.0	18.2	16.7	15.4	14.3	13.3	12.5
95	47.5	38.0	31.7	27.1	23.8	21.1	19.0	17.3	15.8	14.6	13.6	12.7	11.9
90	45.0	36.0	30.0	25.7	22.5	20.0	18.0	16.4	15.0	13.8	12.9	12.0	11.3
85	42.5	34.0	28.3	24.3	21.3	18.9	17.0	15.5	14.2	13.1	12.1	11.3	10.6
80	40.0	32.0	26.7	22.9	20.0	17.8	16.0	14.5	13.3	12.3	11.4	10.7	10.0
75	37.5	30.0	25.0	21.4	18.8	16.7	15.0	13.6	12.5	11.5	10.7	10.0	9.4
70	35.0	28.0	23.3	20.0	17.5	15.6	14.0	12.7	11.7	10.8	10.0	9.3	8.8
65	32.5	26.0	21.7	18.6	16.3	14.4	13.0	11.8	10.8	10.0	9.3	8.7	8.1
60	30.0	24.0	20.0	17.1	15.0	13.3	12.0	10.9	10.0	9.2	8.6	8.0	7.5
55	27.5	22.0	18.3	15.7	13.8	12.2	11.0	10.0	9.2	8.5	7.9	7.3	6.9
50	25.0	20.0	16.7	14.3	12.5	11.1	10.0	9.1	8.3	7.7	7.1	6.7	6.3
45	22.5	18.0	15.0	12.9	11.3	10.0	9.0	8.2	7.5	6.9	6.4	6.0	5.6
40	20.0	16.0	13.3	11.4	10.0	8.9	8.0	7.3	6.7	6.2	5.7	5.3	5.0
35	17.5	14.0	11.7	10.0	8.8	7.8	7.0	6.4	5.8	5.4	5.0	4.7	4.4
30	15.0	12.0	10.0	8.6	7.5	6.7	6.0	5.5	5.0	4.6	4.3	4.0	3.8
25	12.5	10.0	8.3	7.1	6.3	5.6	5.0	4.5	4.2	3.8	3.6	3.3	3.1
20	10.0	8.0	6.8	5.7	5.0	4.4	4.0	3.6	3.3	3.1	2.9	2.7	2.5
15	7.5	6.0	5.0	4.3	3.8	3.3	3.0	2.7	2.5	2.3	2.1	2.0	1.9
10	5.0	4.0	3.3	2.9	2.5	2.2	2.0	1.8	1.7	1.5	1.4	1.3	1.3
5	2.5	2.0	1.7	1.4	1.3	1.1	1.0	0.9	0.8	0.8	0.7	0.7	0.6
0	0.0	0.0	0.0	0.0	0.0	0.0	0.0	0.0	0.0	0.0	0.0	0.0	0.0

Additionally, characteristic values for existing wort boiling systems can be developed. Results from practical trials for each substance can be delineated in order to generate a graph of the evaporative efficiency (fig. 2). This allows wort boiling systems to be evaluated with regard to whether they operate within known limits, i.e. those typical for the evaporation of volatile aroma compounds.

Figure 1: Determination of the decrease in the concentration of a substance [%] at a specific total evaporation

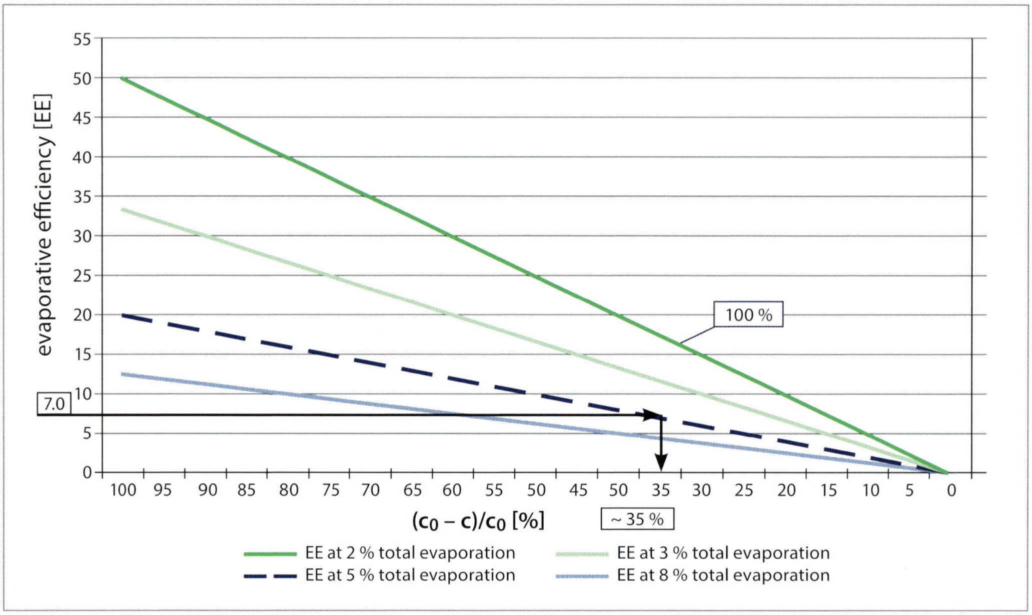

Figure 2: Using hexanal to determine the evaporative efficiency of an individual wort boiling system

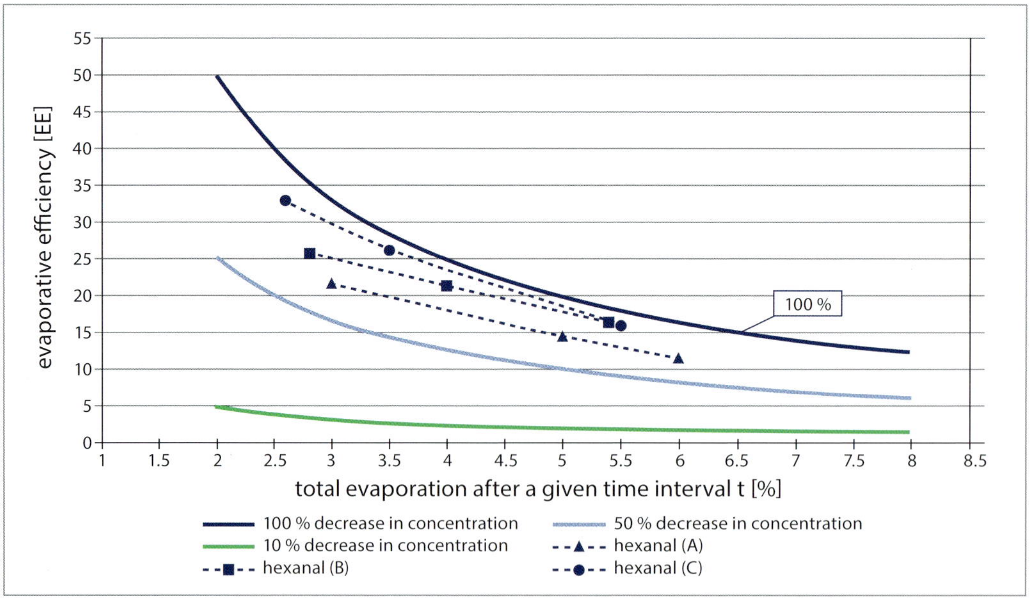

The solid lines shown in figure 2 correspond to the values for the evaporative efficiency in relation to the decreases in the hexanal concentration (100 %, 50 % and 10 %) with respect to the initial concentration. Of course, there can be no value above the 100 % line, as this curve merely defines the maximum values for evaporative efficiency as a function of evaporation. The points connected by dotted lines reflect the evaporative efficiency of the respective indicator for each system. By combining the values from figure 2 with those in figure 1, the percentage of hexanal dissipated at any given rate of evaporation can be found.

For purposes of comparison, data points for the evaporative efficiency of three different wort boiling systems (A, B, C) were entered. The curves for hexanal in figure 2 exemplify that each system has its own distinctive progression with regard to the evaporative efficiency at various degrees of evaporation.

In the case of Strecker aldehydes, an increase is most likely due to thermal reactions occurring in the wort boiling process; however, their evaporation far outweighs their generation. For this reason, boiled wort contains fewer Strecker aldehydes. The wort handling procedures downstream from the wort kettle (e.g. in the whirlpool) causes the concentrations of these substances to once again rise as a function of time spent at higher temperatures. Furfural, on the other hand, shows an increase during boiling and also in wort handling procedures downstream [11].

Sufficient levels of evaporation using a particular wort boiling system and thus a high evaporative efficiency result in good quality wort, since the fundamental analytical parameters of boiling (fig. 3) – free DMS, the thiobarbituric acid index and coagulable nitrogen – are able to ideally complement one another. Consequently, in a wort boiling system with a high evaporative efficiency but low total evaporation, undesirable aroma compounds will dissipate while the TBI increases only slightly. This also allows the concentration of residual coagulable nitrogen to be adjusted to the desired level.

Figure 3: Key metrics for wort boiling and their interdependence

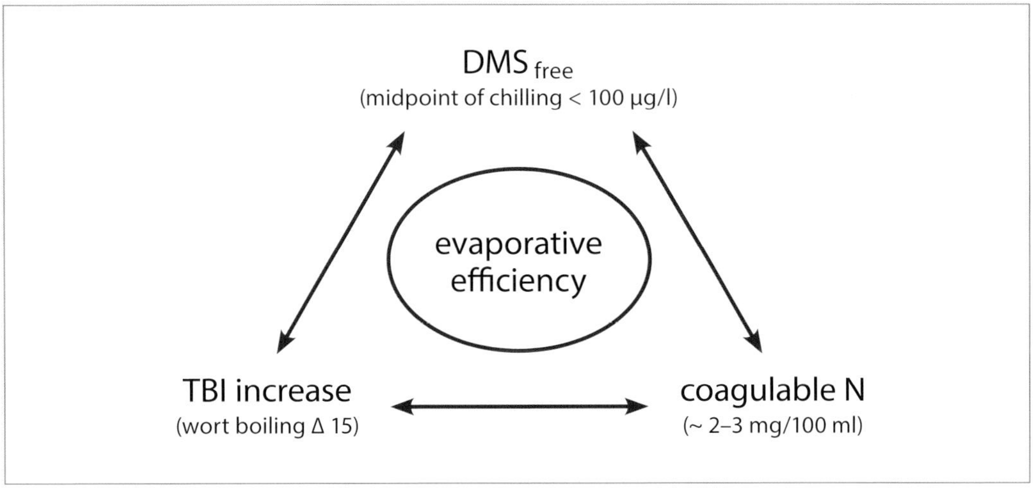

In conventional wort boiling systems (fig. 4), such as an internal or external calandria, the wort is heated at the same time evaporation takes place. Once boiling is completed, the hot break material is removed from the wort, usually in a whirlpool.

Figure 4: A conventional wort boiling process [4]

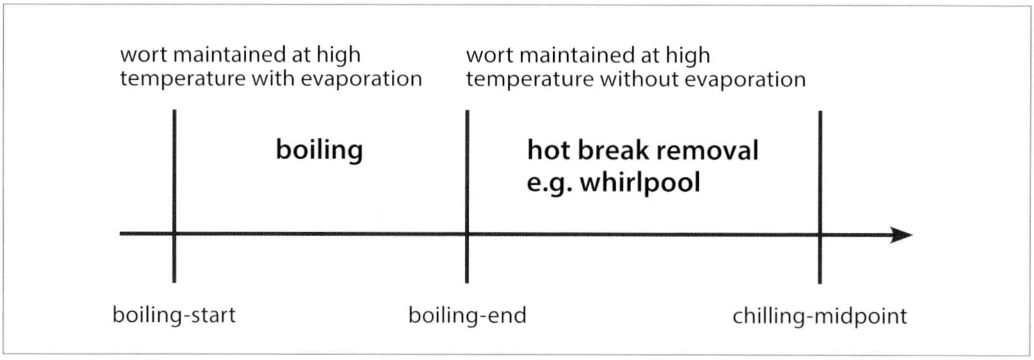

With modern wort boiling systems, the wort can either be heated with simultaneous evaporation or this can occur in two distinct steps. Modern wort boiling systems (fig. 5) offer more options and greater flexibility. In this case, flexibility refers to the fact that the timing of certain processes, the energy input and other attributes affecting quality can be deliberately controlled and coordinated.

Figure 5: Possible procedures for modern wort boiling [4]

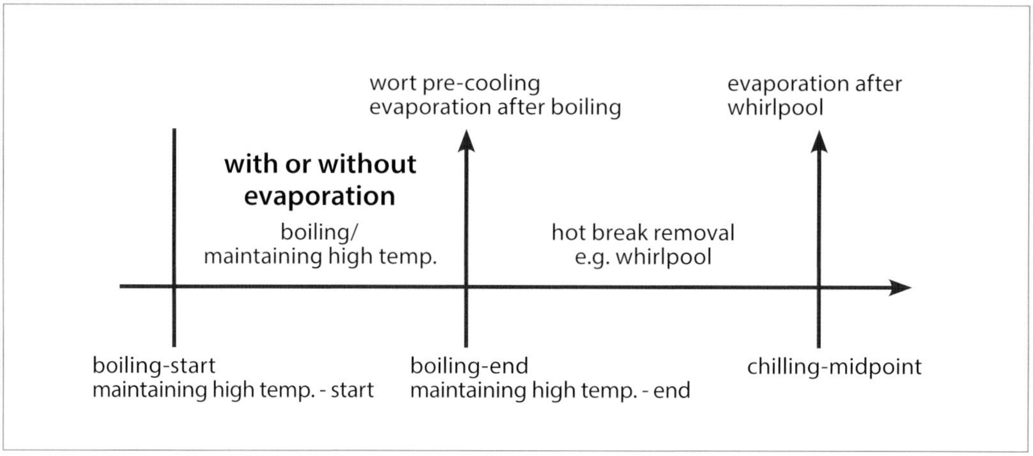

Evaporation can, for example, occur before or after hot break separation. Wort pre-cooling is also possible prior to hot break separation. Wort pre-cooling or downstream evaporation, either while casting out the wort or downstream from the whirlpool, allows brewers to shorten the time required for boiling the wort [5]. This is why the increase in TBI during the boiling process is lower. A significant decline in TBI occurs downstream due to the lower temperatures and/or vacuum evaporation.

2.2.2 PRE-COOLING THE WORT

Pre-cooling the wort prior to hot break separation entails lowering the temperature of the wort from 98–99 °C to below 90 °C. This can be achieved either by means of a heat exchanger (fig. 6) or a flash evaporator (fig. 18). When casting out, the wort can be run through the wort chiller in a bypass or alternatively through a separate plate heat exchanger specifically installed for this purpose.

After the boiling process is complete, the wort can also be recirculated through a flash evaporator until the desired temperature is reached. The temperature is reduced in the flash evaporator by setting the vacuum to the corresponding negative pressure. As a consequence, the thermal stress on the wort is lower [6]. Pre-cooling benefits flavor stability and results in less generation of free DMS from S-methylmethionine, a compound also known as DMS precursor [5]. However, due to the lower temperatures in the whirlpool, any remaining hop isomerization reactions will slow, thus reducing the hop yield.

Figure 6: Pre-cooling the cast-out wort (source: GEA Huppmann GmbH, Kitzingen, Germany)

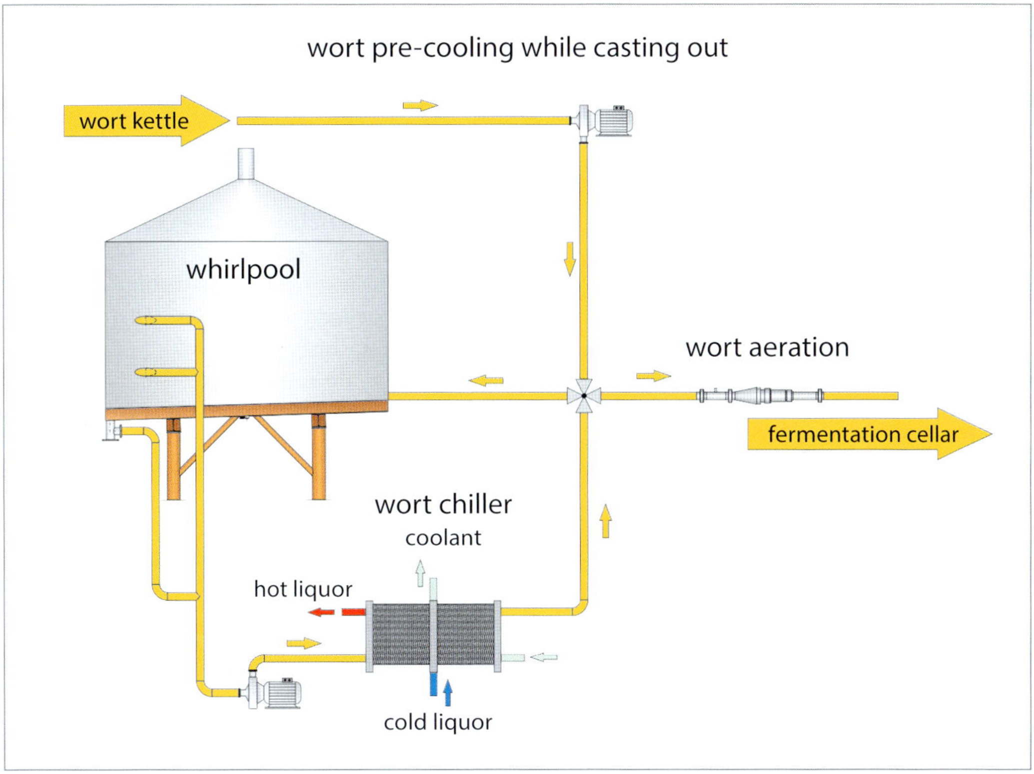

2.2.3 DOWNSTREAM EVAPORATION

Evaporation can be carried out either upstream from hot break separation or directly prior to wort chilling. It can be achieved by two separate methods: one, under atmospheric conditions by once again boiling the wort, or two, through vacuum evaporation.

The fundamental advantage of evaporation downstream from hot break separation is that it enables any unwanted aroma compounds generated from the end of boiling up to that point to be eliminated from the wort. These aroma compounds may otherwise produce off-flavors in the finished beer. However, subjecting the wort exiting the whirlpool to vacuum evaporation also results in the loss of hop aroma compounds. In this case, the wort can be diverted through an extraction vessel immediately upstream from the chiller. Alternatively, a whirlpool hop addition can be performed (cf. Hops).

2.3 INTERPLAY BETWEEN KEY ANALYTICAL METRICS FOR WORT BOILING

Table 2 provides the reference values for free dimethyl sulfide, the thiobarbituric acid index and coagulable nitrogen for wort employed in the production of wort that is pale in color.

Of course, other criteria beyond these three values also play a role in the flavor profile of beer. For example, boiling wort at a lower temperature or with very little total evaporation often produces more characterful and substantial beers.

Table 2: Reference values for wort boiling parameters [8]

wort	TBI	total for free DMS and DMS precursor [µg/l]	coagulable nitrogen [mg/l], adjusted to 12 % extract by weight
cast-out wort	< 45	< 100	15–25 (practical experience: 20–35)

The thiobarbituric acid index is a dimensionless key metric, which provides information about the amount of thermal stress to which the wort has been subjected. Maillard products formed from amino acids and reducing sugars during wort boiling are detected using this method. Values presented in the MEBAK [7] or DIN 8777 [8] guidelines are often used as a basis for commissioning and evaluating brewhouse equipment. Measuring the increase in the TBI during the wort boiling process and in the whirlpool provides clear evidence of the degree of thermal stress incurred during individual steps in the brewing process or during the occupancy of certain vessels or equipment. A low level of thermal stress is beneficial for the flavor stability of beer, since fewer carbonyl compounds (Strecker aldehydes) are formed [9]. This does not apply to the wort used to produce darker beer styles. Copious amounts of Maillard products and Strecker aldehydes are purposely formed in the malt, because they are desirable in the aroma profiles of these beer styles and are essential for their character.

The concentration of the amino acid L-glutamine decreases as a function of the temperature and the duration of the wort boiling process, that is, as a function of the increase in thermal stress [3]. A connection between total evaporation and the decrease in L-glutamine concentration has not been observed. Since L-glutamine is not a volatile compound, the reaction must solely involve chemical degradation. In order to obtain comparable values, the amounts of the amino acid measured in the wort – both prior to and after boiling – must be corrected to reflect the original gravity prior to boiling, because the wort becomes more concentrated through evaporation.

Based on these results, the change in the concentration of L-glutamine over time as a function of thermal stress was examined in more detail during the period that the hot wort is held at a constant temperature or is boiling in the wort kettle. An application of this degradation reaction for assessing thermal stress over the course of wort production was compared to the increase in the TBI (fig. 7) [3].

Figure 7: Relationship between the increase in TBI and the decrease in the concentration of L-glutamine

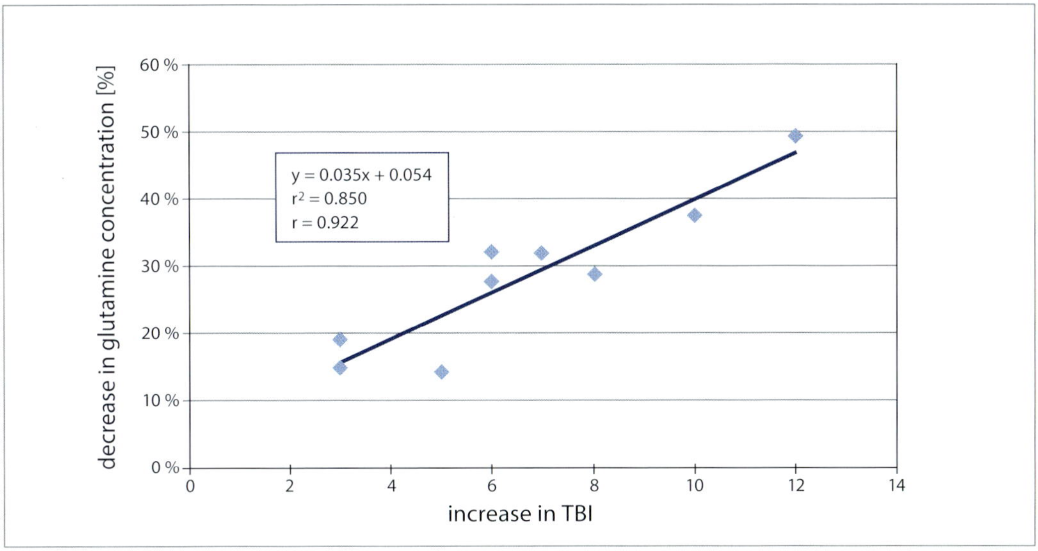

The half-life of L-glutamine degradation varies according to temperature and can be derived mathematically (tab. 3) [3].

Table 3: Half-life times for the degradation of L-glutamine as a function of reaction temperature

reaction temperature: 80 °C	reaction temperature: 90 °C	reaction temperature: 100 °C
192.3 min	**113.6 min**	**64.1 min**

On the basis of these results, it appears that the concentration of the amino acid L-glutamine can also serve as a thermal indicator along with the TBI value [3].

DMS-P is transformed into free DMS over the course of the wort boiling process and should be present in concentrations of < 100 µg/l (the flavor threshold in beer) [10]. Depending on the matrix, concentrations of more than 130 µg/l in the wort may result in a vegetal off-flavor in the beer.

One should note that during hot break material separation in the whirlpool, free DMS formation continues, but it will not evaporate at this stage in the process. This must be taken into account with brewhouse practices that involve conventional wort boiling. Instead, evaporating the volatiles in the wort after separation of the hot break material permits a final elimination of DMS to take place. Suitable concentrations can thus be reached in the pitching wort, as shown in figure 8 (the appropriate sample collection point for this analysis: wort at the mid point of chilling).

Figure 8: Minimum and maximum concentrations of free DMS and DMS-P over the course of wort boiling and subsequent evaporation [3]

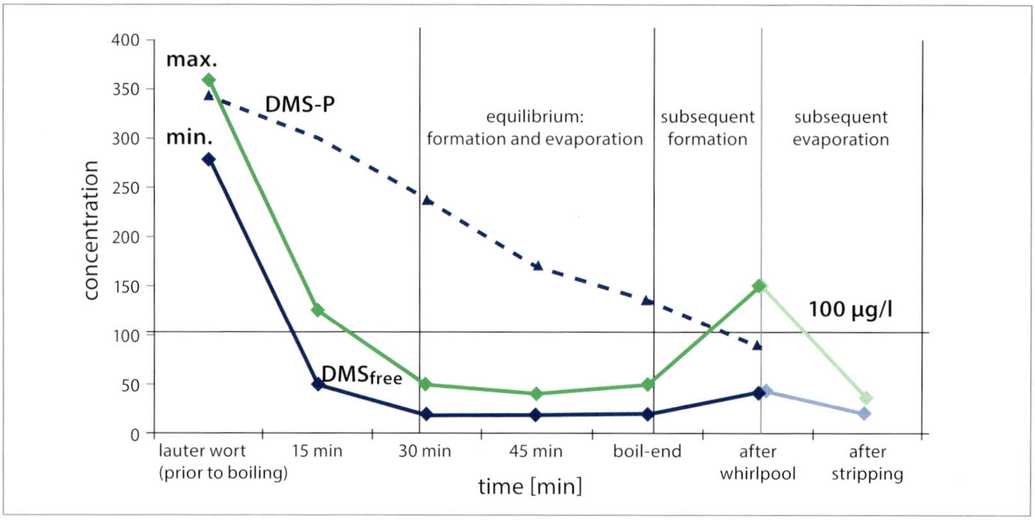

The equipment required for evaporation downstream from the whirlpool or for pre-chilling the wort can be retrofitted in any existing system, allowing brewers to achieve the desired wort composition in the brewhouse in an energy-saving manner. This equipment also makes it possible for brewers to react in a targeted manner to fluctuations in the quality of the raw materials.

Protein coagulation takes place in two stages during the wort boiling process. First, the proteins are denatured, and then afterwards the actual coagulation occurs. This process is dependent upon the temperature and duration of the wort boiling process, the degree to which the wort is mixed (homogeneity) and the wort pH. A pH value below 5.2 is propitious for protein coagulation [11] and can be reached through acidification of the wort. The concentration of the residual coagulable nitrogen in the chilled wort influences the colloidal stability, the beer foam, the flavor and the mouthfeel. Reference values for residual coagulable nitrogen [7] range from 15 to 25 mg/l. It has repeatedly been shown in practice that concentrations up to 35 mg/l should not pose a problem and can even be beneficial for flavor and foam stability.

2.4 MODERN WORT BOILING SYSTEMS

As a result of better mixing and excellent evaporative efficiency, modern wort boiling systems are routinely capable of achieving a total evaporation of 4 to 6 %. If the total evaporation is too low, the energy balance has to be taken into consideration when an energy storage system is in operation. Lower percentages of total evaporation means that less brewing liquor is required for sparging. This may result in some losses in the volume of extract obtained during lautering and also affects the maximum volume of last runnings that can be subsequently utilized as mash liquor. In order to positively impact efficiency and profitability, the volume of last runnings used for mashing in the following batch should not impart any perceptible negative effects on the flavor or flavor stability of the finished beer. Of additional concern at low rates of total evaporation are the atypical aroma notes that may arise in the finished beer if total evaporation levels fall below 4 %.

Table 4 provides a brief overview of the fundamental parameters for evaluating the wort boiling systems described below. The table presents information beginning at the point the lauter wort is in the kettle to the start of the wort chilling process.

Table 4: Fundamental process parameters for selected modern wort boiling systems [12]

	boiling		subsequent evaporation prior to separation of hot break material	subsequent evaporation after separation of hot break material
	Duration of the boil	Boiling temperature		
internal calandria/ external calandria	varies	atmospheric	–	–
high temperature wort boiling (HTW)	at T_{max} for 2–3 min	128–135 °C	X	–
dynamic wort boiling > p_{atm}	40–60 min	100–103 °C	–	–
SchoKo	40–60 min holding time at specified temperature	97–99 °C	–	X
wort stripping	approx. 40 min + stripping	atmospheric	–	X
vacuum evaporation	40–50 min + vacuum	atmospheric	–	X
Varioboil	40 min + vacuum	atmospheric	X	–
Merlin®	35–40 min + stripping	atmospheric	–	X
Stromboli	60 min	atmospheric	–	–
Ecotherm	70 min	atmospheric	–	–

Figure 9: Classification of wort boiling systems according to their mode of operation

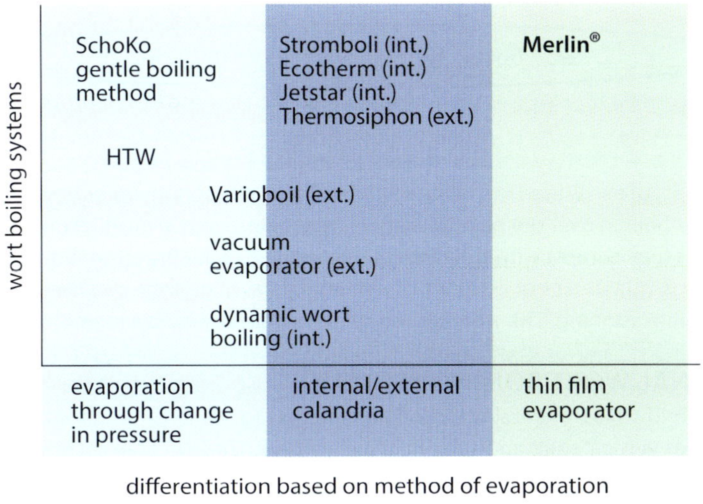

In table 4, one sees that in many modern wort boiling systems, stripping is carried out once hot break separation is complete. This allows the DMS formed in the whirlpool to be evaporated to the desired level.

An overview of wort boiling systems over the last hundred years is provided in figure 10.

Figure 10: Timeline of developments in wort boiling systems

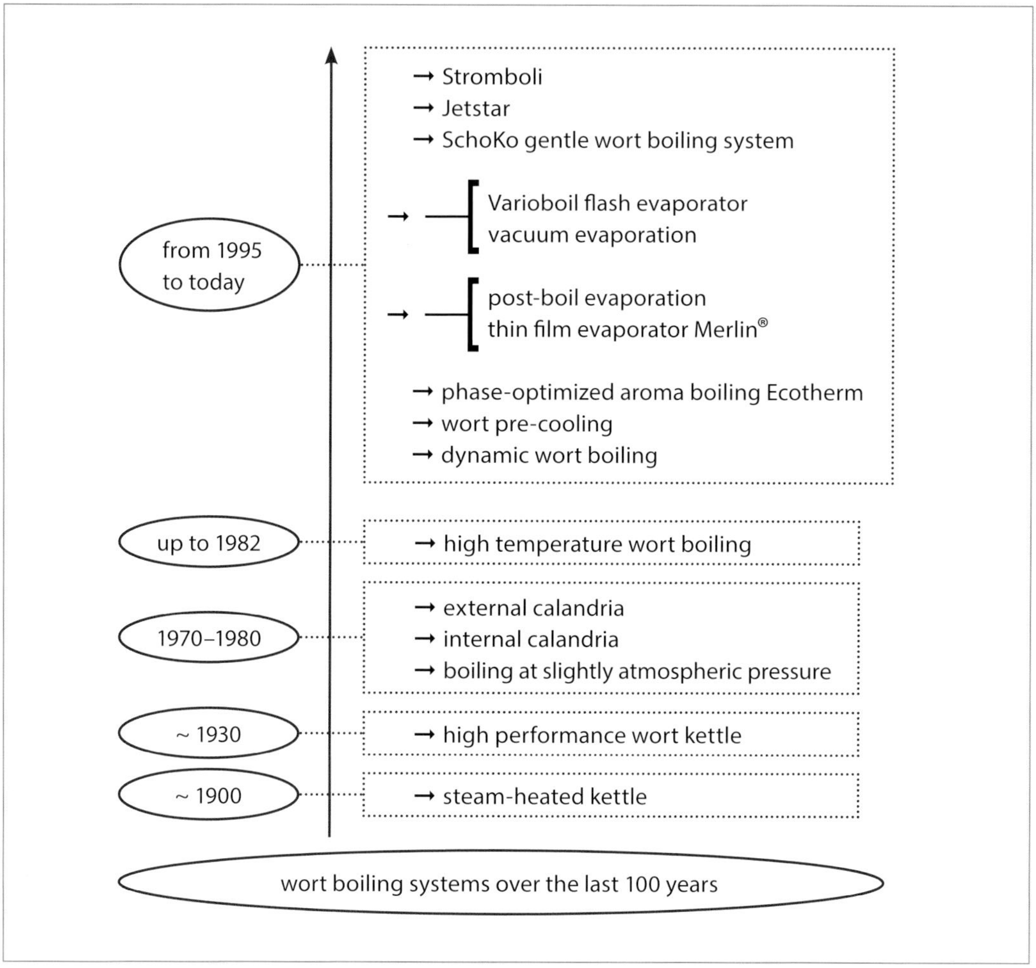

The following is a list of existing wort boiling systems. The list also enumerates important developments in the field of wort boiling over the past 10 years. Flawless wort can be produced with any of these wort boiling systems as they comply with the guidelines compiled by MEBAK. The requirements in DIN 8777 [8] are often cited in guarantees of contractual agreements as one of the conditions for assessing brewhouses during commissioning. The wort boiling systems itemized below meet these requirements.

2.4.1 INTERNAL WORT BOILING SYSTEMS

Beginning in the mid-1970s, internal wort boiling systems (fig. 11) were used for the first time. They are usually designed as vertical shell-and-tube heat exchangers. These heat exchangers operate according to the principle of natural circulation, meaning that the wort is heated in the bundle of tubes with steam or hot water, and the temperature difference between the inlet and the outlet causes the formation of wort vapor and subsequent evaporation. The formation of vapour results in a difference in density. The hot, less dense wort ascends the vertical tubes until it emerges at the top, thus ensuring natural circulation. The contents of the wort kettle circulate 20–30 times per hour through the heat exchanger. One issue inherent to internal wort boiling systems is their propensity to bring about recurring pulsation of the wort inside the tubes while the wort is being heated to boiling. This should normally cease once boiling commences. Nevertheless, this drawback can be partially or entirely eliminated through installation of a recirculation pump [13]. If the difference in the temperatures at the heat exchanger outlet and inlet (i.e. the ambient wort) is small, the tendency for the wort to pulsate and foam in the tubes will be negligible.

Figure 11: Internal calandria design (source: GEA Brewery Systems GmbH, Germany)

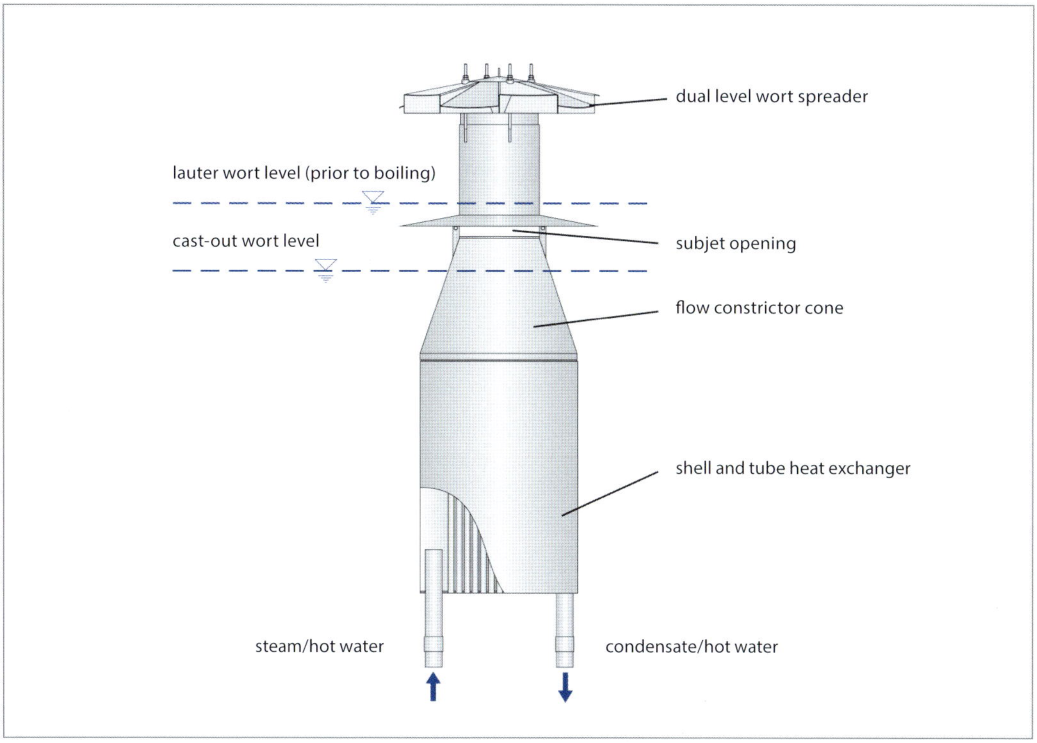

2.4.2 EXTERNAL WORT BOILING SYSTEMS

Most external wort boiling systems (fig. 12) are also designed as shell-and-tube heat exchangers. In this type of system, the wort is continuously pumped out of the wort kettle through the heat exchanger and back into the wort kettle six to eight times per hour. The wort temperature can be regulated by adjusting a throttle valve situated between the heat exchanger outlet and the wort kettle inlet. The wort enters the kettle and is then directed over the surface of a wort spreader. The hot wort cascades back into the kettle in a thin, conical layer, enlarging its surface area and thus intensifying the evaporation of both undesirable aroma compounds and water. Drawbacks of external wort boiling systems include substantial space requirements and a relatively high level of technical complexity. There are many practical advantages associated with external calandrias. These include uniform mixing of the contents of the wort kettle and an enormous versatility with regard to installation of heating surface area. In addition, external calandrias can maintain the wort at a certain temperature by varying the flow rate of wort through the heat exchanger by regulating the pump throughput. Each of these aspects allows for highly versatile and precise wort handling [13]. Furthermore, several wort kettles can be served by a single external wort boiling system. In brewing applications, a thermosiphon refers to a wort boiling system not equipped with a recirculation pump that relies solely on natural convection to move the wort.

Figure 12: Schematic diagram of a wort boiling system equipped with an external calandria (source: KRONES AG, Germany)

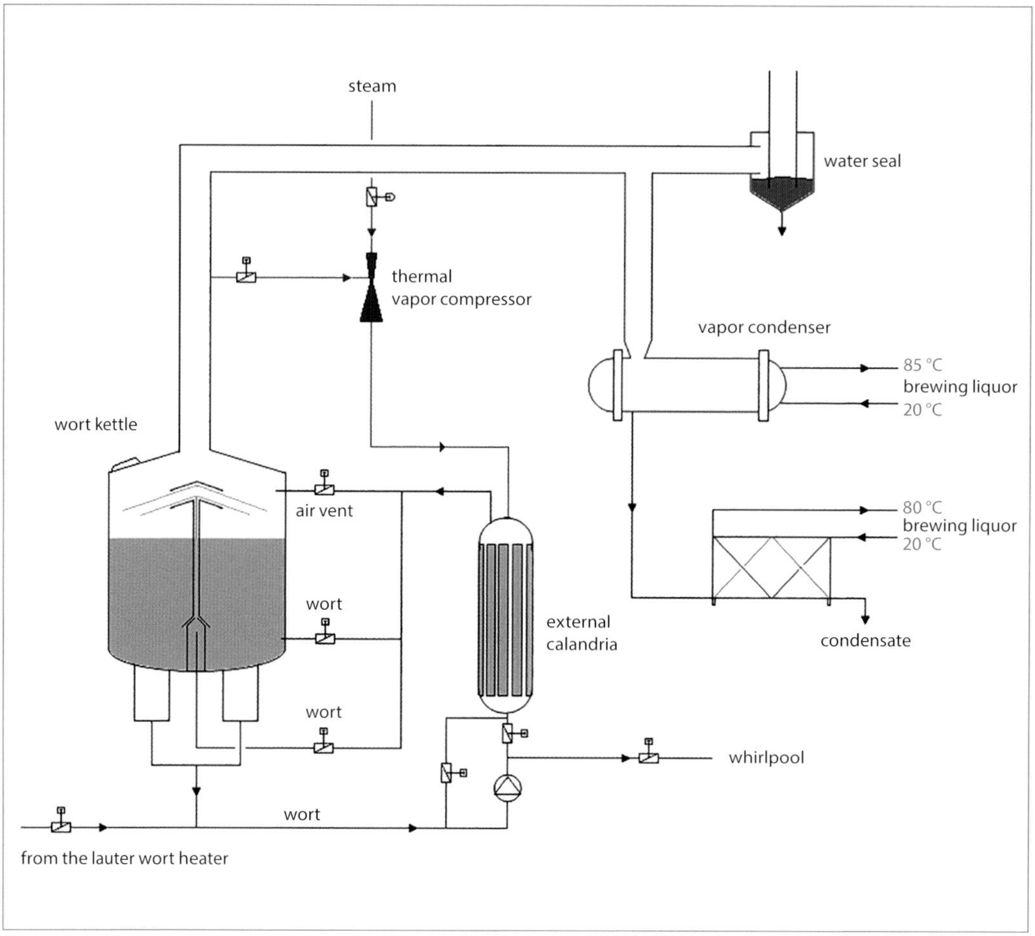

A calandria is an internal or external heat exchanger, either a thermosiphon or one equipped with a pump, in which the heated medium wells up and is recirculated through heated vertical tubes. The pump can be used to recirculate the wort until the boiling point is reached, after which the thermosiphon effect can assume that function. Having flowed through the calandria, the wort enters the kettle from below, ascending a relatively wide vertical pipe with a hydrodynamic, streamlined inlet, creating a "mild" Venturi effect. This draws the surrounding wort up into the vertical pipe as well, so that overall a greater quantity of wort is circulated through the calandria. The wort then wells up until it reaches a plate hydrodynamically designed to spread the wort stream optimally and evenly over the surface of the liquid wort in the kettle [14]. In Germany, it is now standard practice for an external calandria to be equipped with a recirculation pump. A throttle valve downstream from the calandria outlet allows the pressure to be increased slightly inside the calandria. Consequently, the wort temperature rises, accelerating the physical and chemical reactions in the wort, thus allowing the duration of the boil to be shortened. A wort temperature of 102–103 °C is generally preferred, in order to prevent excessive protein coagulation. In this case, the wort ordinarily enters the kettle from below as well, but upon flowing past the throttle valve, enters a narrower space at the base of the kettle. Therefore, less wort is drawn up into the vertical pipe with the ascending wort. The wort is then impelled at a higher velocity against the wort spreader, and despite smoothing and streamlining to foster a less turbulent downward flow, the hot break material can be sheared to such an extent that it does not settle into a sufficiently compact cone in the whirlpool. This design is not completely free of "short circuit flow", i.e. the wort is not boiled in a uniform manner, resulting in a somewhat flawed wort boiling process. According to a suggestion by Denk [15], the vertical pipe should be removed, and instead 2/3–3/4 of the wort volume should be returned to the wort kettle at an angle of 23 ° to the radius ("mixing angle"). This portion of the wort would enter at the approximate level of the surface of the cast-out wort in the kettle, while a smaller portion would be directed over the wort spreader from above. This prevents excessive foaming. In either an internal or external calandria, the temperature of the heat transfer medium should not be too high, e.g. 110–115 °C for saturated steam or 125–130 °C for hot water. Counter-pressure builds at the top of an internal calandria where the cone narrows and at the throttle valve of the external calandria. The free flow of the wort is thus obstructed and more of the thermal energy from the heat transfer medium enters the wort, causing certain reactions in the wort to be accelerated. These include thermal reactions that increase the color of the wort and precipitate high molecular weight, foam-positive proteins. This can lead to fouling, requiring that the brewhouse equipment be cleaned more frequently. These disadvantages can be avoided by maintaining the target temperatures and by improving the homogeneity of the wort.

2.4.3 HIGH-TEMPERATURE WORT BOILING

High-temperature wort boiling (fig. 13) frequently involves heating the wort in three stages to temperatures between 128 and 135 °C in separate wort heaters. Afterwards, the wort temperature is typically reduced in two expansion stages. Through releasing the pressure on the wort, the required evaporation takes place, including that of undesirable volatile wort compounds. The energy released through the phase change of the accumulated vapor is recovered and subsequently utilized to heat the wort incrementally.

The wort is heated to 87 °C and then to 107 °C in the first two wort heaters by utilizing the energy obtained from decompression of the accumulated wort vapor. Heating the wort to 130 °C takes place with primary energy. The wort is held at this temperature for 2.5–3.0 min and then cooled by releasing the overpressure on the wort in two evaporation vessels to 117 °C and finally to 100 °C. Separation of the hot break material can be carried out in a whirlpool. The evaporation rate is 6–8 % [11]. During this process, a large part of the heat energy is recovered, making this wort boiling system one of the most efficient systems on the market. According to Narziss [16], the best beers were obtained from wort held at 130 °C for 3 min, followed by those held at 140 °C for 2 min. On the other hand, with dark beers, an unpleasant overheated or burnt note can result due to the presence of manifold Maillard products and the high temperatures. In the early development of this technology, fouling occurred on the heating surfaces of the wort heater at stage 3 of the boiling process at the beginning of the workweek. This resulted in an inconsistent thermal treatment of the wort. The problem was remedied through installation of a second, parallel boiling system, which allows a cleaning cycle to be carried out every 3–6 h without shutting down the system.

Practical experience has shown that at stage 3, a shell-and-tube heat exchanger is preferable to a plate heat exchanger.

Figure 13: Schematic diagram of a high temperature wort boiling system

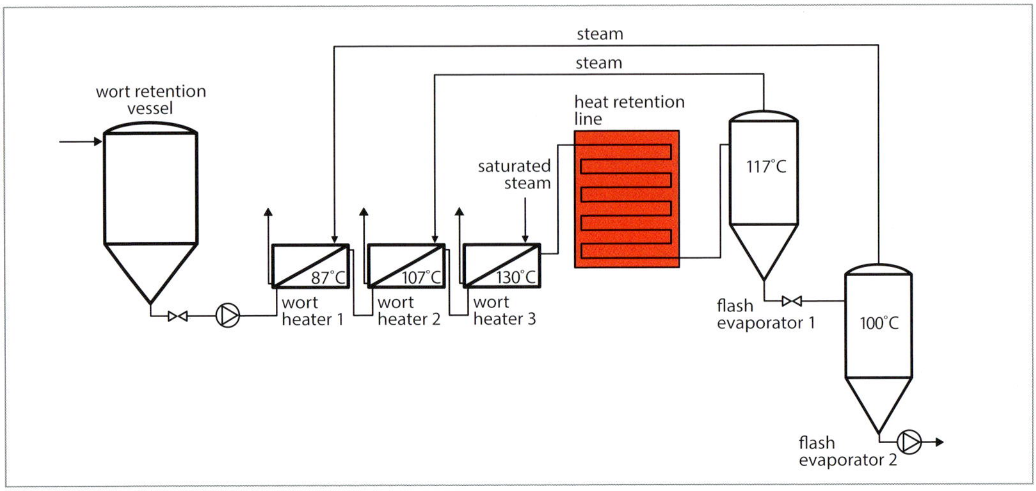

2.4.4 DYNAMIC WORT BOILING SLIGHTLY ABOVE ATMOSPHERIC PRESSURE

Conventional wort boiling slightly in excess of normal atmospheric pressure has been practiced worldwide since 1979. Using this method, the wort is boiled at temperatures between 102 °C and 105 °C. A dynamic iteration on this method of wort boiling involves increasing and then releasing the pressure on the wort several times over the course of the process (fig. 14). The system components include an internal calandria and a vapor condenser. The total evaporation can be reduced to 4.5 % (based on the volume of the cast-out wort) and the duration of boiling to 40–50 min [17, 18]. Repeatedly raising and lowering the pressure on the wort effectively volatilizes the aroma compounds. Normally the pressure only peaks three to four times during this process. Nevertheless, when the pressure is released causing the temperature to drop from 103 °C to 101 °C, for instance, an intense surging motion erupts across the surface of the wort, since bubbles of vapor form throughout the entire volume of the wort and then are suddenly liberated. The bubbles of vapor bring about desorption, which greatly increases the evaporative efficiency. In order to perform this type of wort boiling, the wort kettle must have the capacity to withstand an absolute pressure of 1.5 bar.

Figure 14: The sequence of operations for dynamic wort boiling slightly above atmospheric pressure (source: GEA Huppmann GmbH, Kitzingen, Germany).

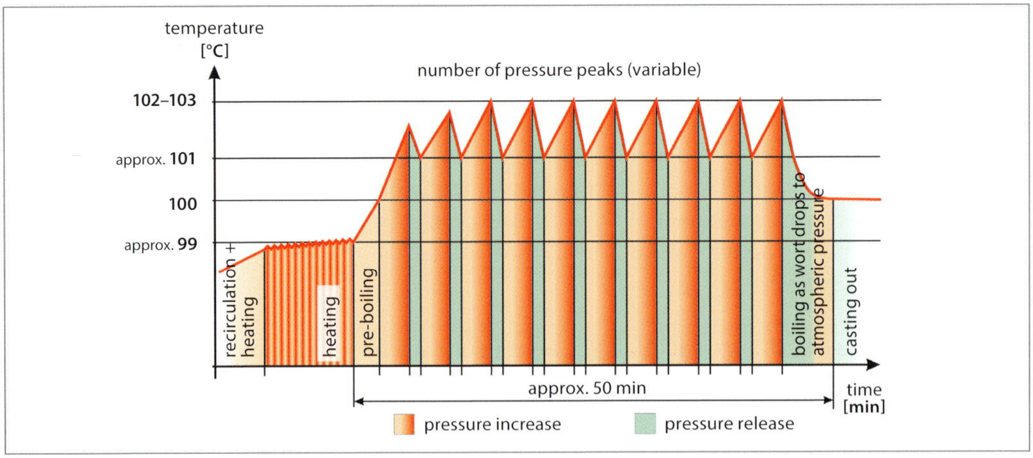

2.4.5 GENTLE BOILING WITH SCHOKO

With the SchoKo system, the wort boiling process occurs in two discrete stages: first, the wort is heated to a temperature slightly below the boiling point and remains there for a certain amount of time; second, evaporation is carried out between the whirlpool and the wort chiller (fig. 15). During the first stage, the wort is heated to between 97 and 99 °C and held there for 40–60 min, resulting in an evaporation of < 1 % in practice. If this vessel is equipped, for example, with an internal calandria, the wort could also be temporarily boiled in a conventional manner. During the second phase, about 5 % is evaporated in a vacuum evaporator at an absolute pressure of approximately 400 mbar. In order to be able to carry out vacuum evaporation, the wort is pumped tangentially into a specially designed evacuated chamber, so that the liquid wort flows around the internal wall of the chamber as a thin film. As a result, the surface area of the wort is increased greatly, inducing evaporation of aroma compounds and some of the water content. This unique system design allows operators a high degree of flexibility with regard to the parameterization and control of the boiling process. This vacuum evaporator can also be retrofitted quite easily into an existing brewhouse [19, 20, 21].

Figure 15: Process diagram for the SchoKo wort boiling system, (source: Kaspar Schulz Brauereimaschinenfabrik, Bamberg, Germany)

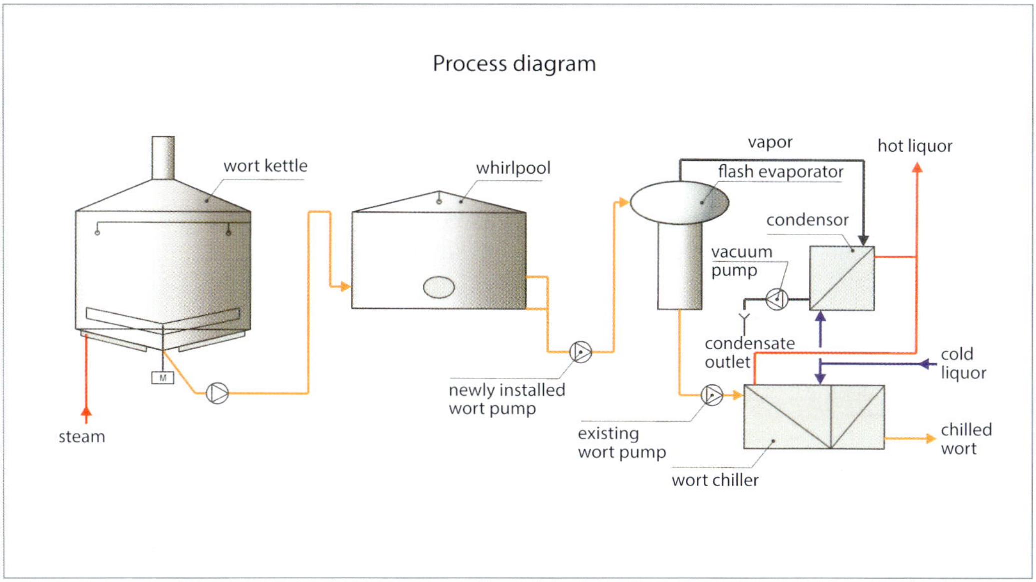

2.4.6 WORT STRIPPING

The equipment manufacturer Meura offers a system that allows evaporation downstream from the wort kettle in a stripping column, positioned at a point between hot break separation and wort chilling. The system consists of a wort kettle, a hot break sedimentation tank or whirlpool and a stripping column followed by a plate and frame chiller (fig. 16). Once the temperature reaches the boiling point, the wort is held for 30–50 min under gentle agitation. Subsequently, the wort is pumped into a sedimentation vessel or a whirlpool. Once hot break separation is complete, the stripping process can begin. The wort passes through the packed column from top to bottom as steam flows in the opposite direction. The wort must first be heated back up to boiling before stripping commences. The water and aroma compounds exiting the stripping column at the top are condensed and removed through an auxiliary device [22].

Figure 16: Process diagram for a wort stripping system (source: Meura, Tournai, Belgium)

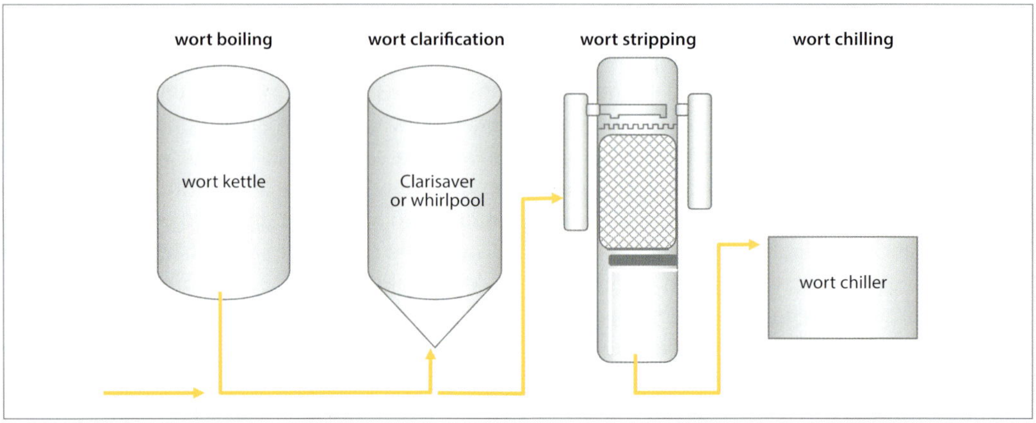

2.4.7 VACUUM EVAPORATION

The Ziemann vacuum evaporator (fig. 17) offers yet another method for evaporation downstream from the whirlpool. The wort is first boiled conventionally. After hot break separation in the whirlpool, the wort enters a flash evaporator and the volatile compounds are removed at an absolute pressure of approximately 600 mbar. Conventional boiling usually lasts 40–50 min with a total evaporation of about 4 %. A further 2 % is evaporated in the vacuum evaporator. This system also allows flexible adjustment of parameters and can easily be retrofitted into an existing brewhouse or combined with other equipment [19, 23].

Figure 17: Process diagram for a vacuum evaporation system (source: Ziemann Holvrieka GmbH, Ludwigsburg, Germany)

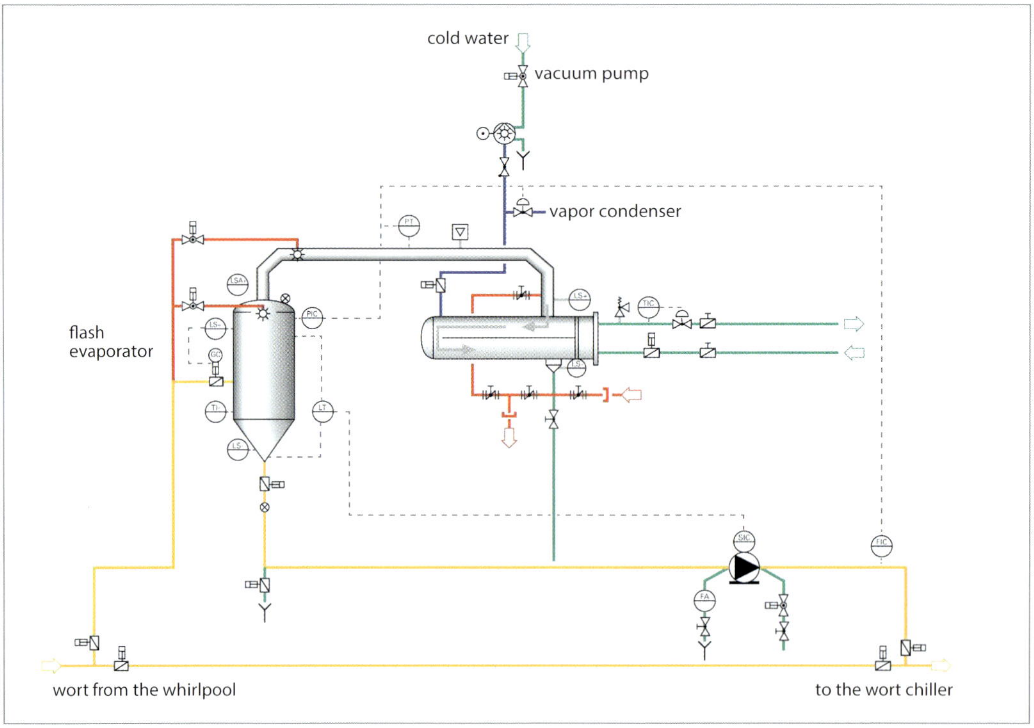

2.4.8 FLASH EVAPORATION WITH VARIOBOIL

The primary components of this system are an external calandria, a flash evaporator and a storage tank. The wort is boiled conventionally in the first stage of the process using the Varioboil system (fig. 18). The evaporation of water and aroma compounds takes place in the flash evaporator. The resulting vapor is liquefied in a condenser. In the second stage of the process, a vacuum is pulled on the flash evaporator, resulting in evaporation below 100 °C, which cools the wort. After these two stages of wort boiling are complete, hot break separation occurs in a whirlpool. In the case of vacuum evaporation, the wort should not be cooled below 80 °C, otherwise the cone of hot break may disintegrate in the whirlpool. It is also possible to heat brewing liquor up to a temperature of at least 80 °C as the wort is chilled.

Even with this system, retrofitting or upgrading is relatively simple, i.e. existing vessels can be repurposed to some extent. Furthermore, as with other systems, the parameters for boiling the wort can be adjusted with little to no difficulty [24].

Figure 18: Process diagram for the Varioboil system (source: Esau & Hueber GmbH, Germany)

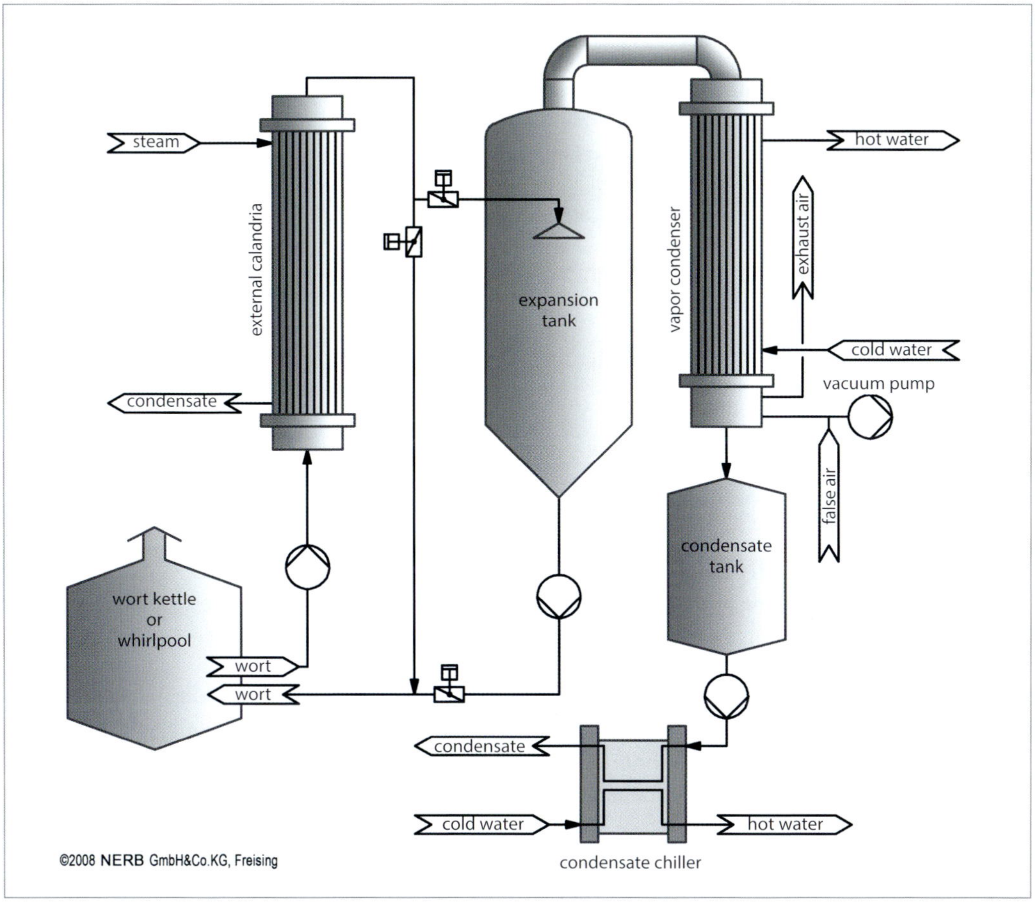

2.4.9 THE MERLIN® THIN FILM EVAPORATOR

The Merlin® wort boiling system (fig. 19) consists of two vessels: one with a heating surface in the form of an inverted cone and a whirlpool. In addition to hot break separation, the whirlpool serves as a collection vessel for the wort. First, the wort exits the lauter tun and enters the whirlpool. At the end of lautering, the wort is heated to boiling by recirculating the wort over the conical heating surface with a pump. The wort flows in a thin yet turbulent layer down the inverted cone, allowing for efficient heat transfer to take place. The entire volume of wort in the whirlpool is recirculated 4 to 6 times over the cone during the boiling process, which lasts for 35–40 min. Hot break separation takes place during the recirculation process, since the returning wort is pumped tangentially into the whirlpool. Because it begins so early, hot break separation can be reduced to around 10 min. The wort once again passes over the cone at atmospheric pressure immediately prior to entering the plate and frame chiller to evaporate any volatiles which may have formed during hot break separation [25, 26].

Figure 19: Process diagram for the Merlin® wort boiling system (source: KRONES AG Werk Steinecker, Freising, Germany)

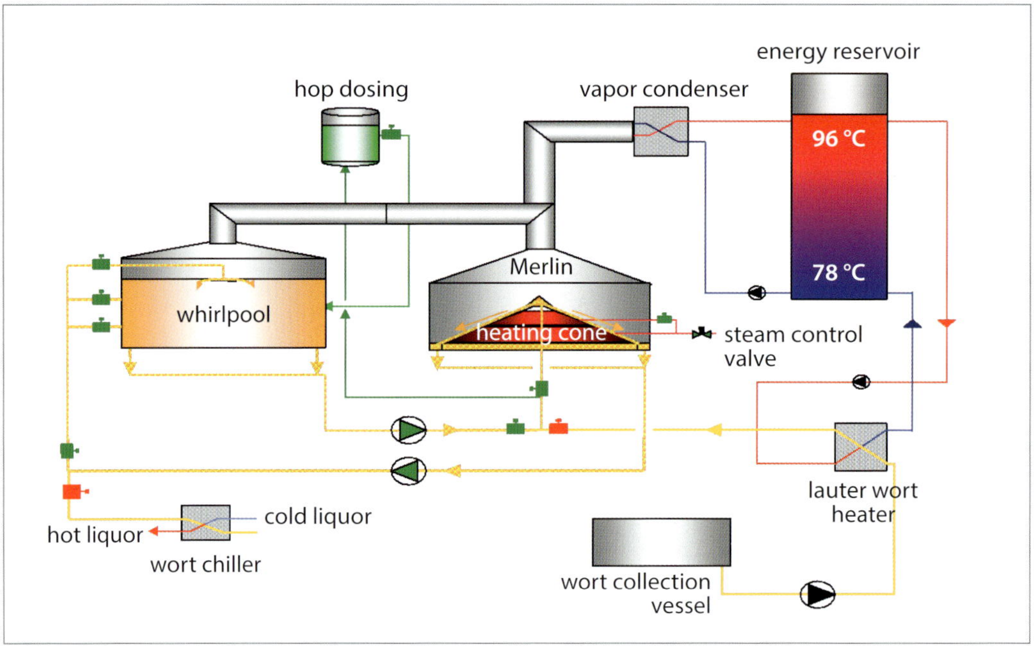

2.4.10 THE JETSTAR INTERNAL CALANDRIA

The Jetstar internal calandria (fig. 11, p. 95 and fig. 20, p. 103) has an adjustable aperture (subjet) below the surface of the wort.

Upon exiting the tubes of the heat exchanger, the wort from the internal calandria flows back into the kettle slightly below the surface if this aperture is open. This results in piston flow in the wort kettle and thus guarantees that the wort is homogenously mixed and the boiling process is uniform. Short circuit flow and dead zones can thus be avoided. No backpressure builds in the internal calandria since all movement takes place below the surface of the wort. This increases the circulation rate and facilitates homogenous mixing of the wort in the kettle. As a result, the thermal stress on the wort is reduced, particularly when heating the wort to boiling.

Preferably, after reaching the boiling point, the wort can be directed over the wort spreader by closing the subjet. This increases the surface area of the wort and allows more evaporation. The wort cascading over the surface also hinders foam formation in the wort kettle.

Figure 20: Jetstar internal calandria (source: GEA Huppmann GmbH, Kitzingen, Germany)

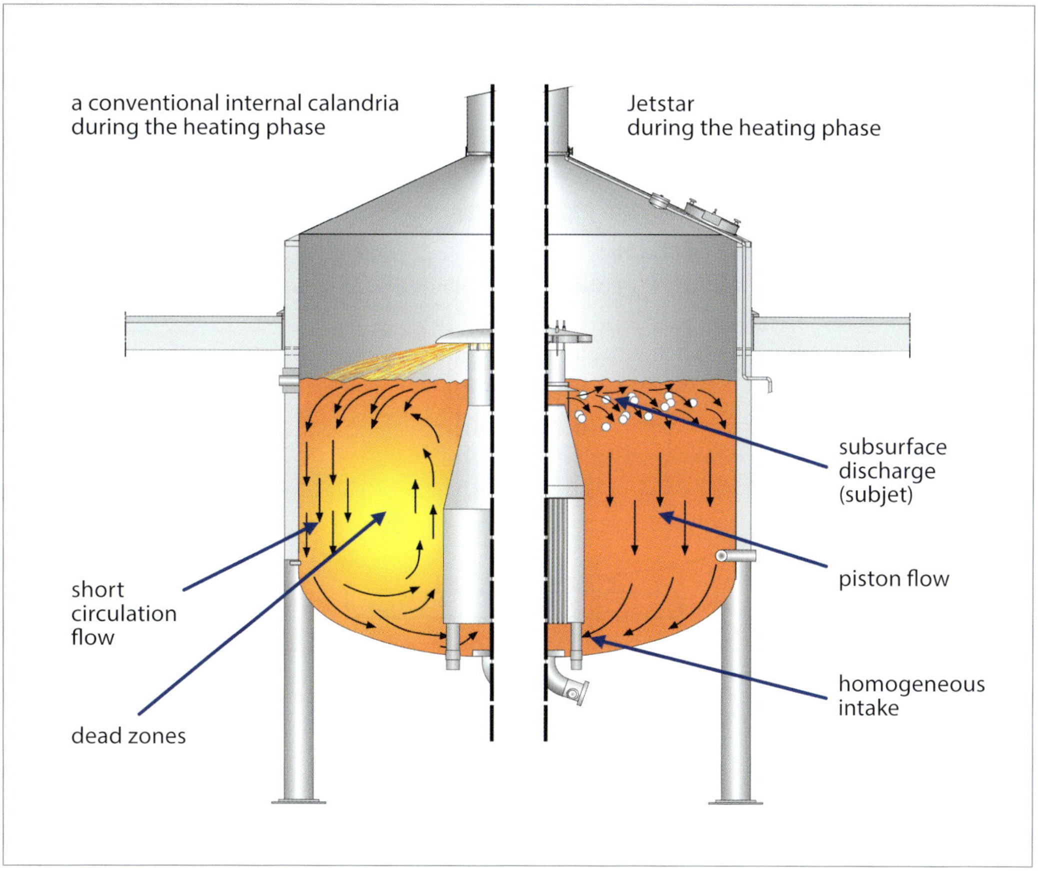

2.4.11 THE STROMBOLI INTERNAL CALANDRIA

The Stromboli internal wort boiling system (fig. 21) is an improved and optimized version of the Ecotherm wort boiling system. In this system, the wort is forced through a vertical pipe in the internal calandria by means of a frequency-controlled pump. A vacuum builds above the bundle of tubes, inducing suction at the point on the diagram marked "jet pump". As a result, a portion of the wort is conveyed upwards into the tubes of the internal calandria with the wort pumped up the central tube. The wort is dispersed evenly back over the surface upon passing through the calandria.

The forced flow prevents pulsation of the wort while it is being heated to boiling. In addition, it is also possible to lower the thermal stress on the wort by reducing the temperature of the heat transfer medium without adjusting the rate of recirculation. This serves to preserve the residual coagulable nitrogen in the wort. The forced flow and the increased number of run-off points at the base of the wort kettle guarantee a high degree of homogeneity during the heating phase and throughout the entire duration of the boil. Despite the low amount of total evaporation, ample volatilization of the undesirable aroma compounds in the wort takes place. In addition, the Stromboli is very efficient in its use of energy [27, 28].

Figure 21: Schematic diagram of a Stromboli wort boiling system (source: KRONES AG Werk Steinecker, Freising, Germany)

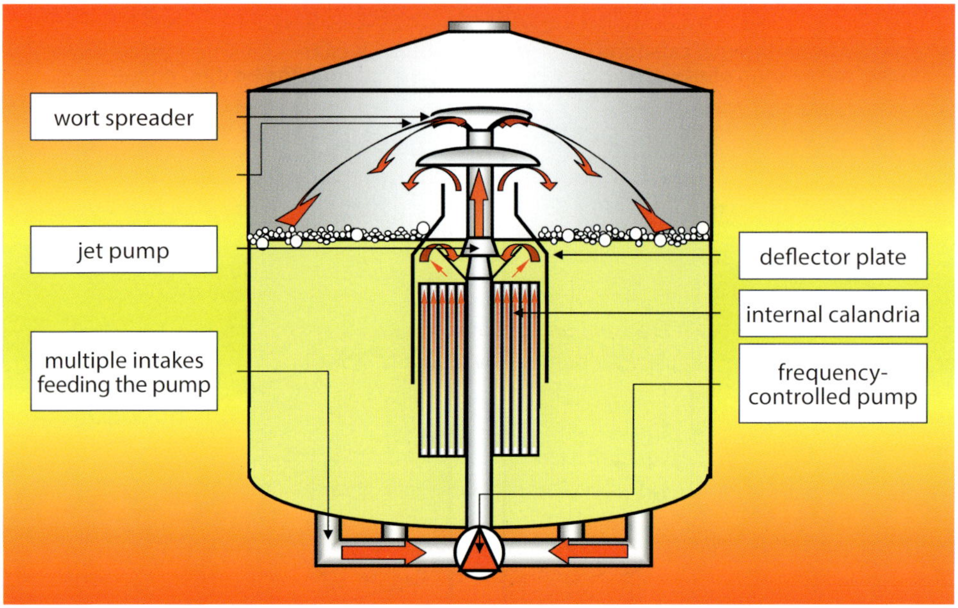

2.4.12 BREAK MATERIAL

2.4.12.1 Hot break material

Hot break separation after the wort has been boiled serves to clarify the wort and is beneficial for the quality of the finished beer. However, a certain amount of hot break is not only technically unavoidable in the wort but also desirable from a production standpoint [29, 30, 31, 32, 33, 34, 35]. Long chain fatty acids (> C16) and zinc, both of which can be found in the hot break, promote yeast vitality and accelerate fermentation. Particles of hot break facilitate the release of carbon dioxide during fermentation and thus increase fermentative activity (liberation of CO_2 from the yeast cells, an increased number of yeast cells in suspension, more rapid extract utilization). In this way, the hot break content influences both fermentation and the character of the beer.

Figure 22: Content of break material

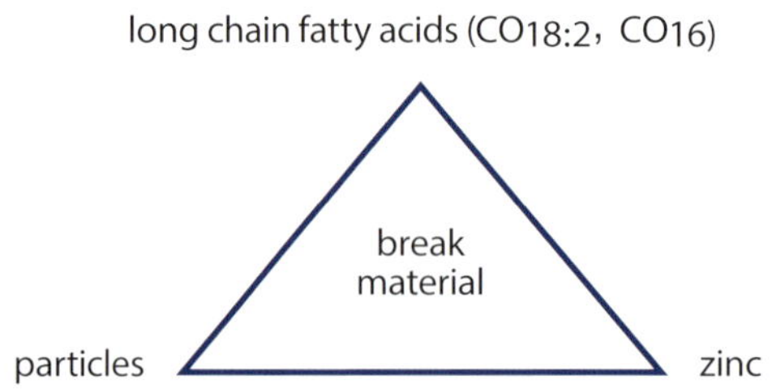

2.4.12.2 Cold break material

Cold break separation is a process step in wort production that not only increases costs but also poses a microbiological risk. In the literature, the necessity of removing cold break material remains a point of contention. Experimental trials in breweries examining methods such as flotation, cold wort filtration and cold wort centrifugation have shown that cold break separation is not necessary if the brewer ensures that the following conditions are fulfilled [36, 37]:

- hot break separation is successful
- beer quality is consistent
- yeast management is state-of-the-art
- the flavor profile of the beer is desired by the brewery

Although the flavor of the beer can be impacted by raising or lowering the concentration of cold break material in the wort, other quality parameters such as flavor stability, chemical-physical stability, reducing capacity (lag time) and foam stability are not negatively affected but, in fact, have a tendency to improve.

The changes in the beer are specific to each individual brewery in terms of flavor, aroma, bitterness and mouthfeel. The flavor of beer is generally more characterful with the retention of some cold break material. Aged beer produced with cold break material tends to be preferred in tastings. The reductive potential increases when the cold break material is left in the wort. The same is true for colloidal stability. Foam stability remains unaffected [36, 37].

2.4.13 VAPOR CONDENSATE

In order to avoid odorous emissions and also to save energy, the vapors arising from boiling the wort are usually condensed in a vapor condenser or following thermal or mechanical vapor compression in external or internal wort boiling systems. The amount of resultant vapor condensate corresponds to that of the total evaporation during wort boiling. In order to reduce the amount of wastewater in the brewery and the associated costs, options are available for recycling both treated and untreated vapor condensate.

Vapor condensate is composed of over 99 % water. It is free of salts and contains only organic compounds from the hops and malt. So far, 161 substances have been identified in vapor condensate. These belong to the following groups of chemical compounds: aldehydes, alkanes, alcohols, esters, furans, ketones and terpenes. The concentrations of the individual substances range from 5 to 338 µg/l [38, 39]. If the vapor condensate is yellow in color and it contains extract, it can be assumed that wort has been carried over into the vapor condensate through aerosol formation.

The constituents of the condensed vapor obtained from boiling wort can differ significantly from brewery to brewery. The reasons for this are as follows:

- The chemical oxygen demand (COD) of the vapor condensate is significantly higher if the highly concentrated coloring agent called *Farbebier* or natural lactic acid known as *Sauergut* has been added to the boiling wort.
- Hop additions do not significantly affect the COD of the vapor condensate.
- The average COD of the condensate used to produce pale beers is generally less than 300 mg/l without the addition of *Farbebier* or *Sauergut* to the wort during boiling.
- Other factors affecting the constituents of the condensate are the type of wort (beer style) and the wort color.

- Depending on the design of the wort boiling system, the COD reaches a maximum at the start of boiling or approximately 20 min thereafter.
- The pH of the vapor condensate is lower than that of the wort.

Thus far, reverse osmosis and activated carbon filtration have been studied as a means for treating wort vapor condensate. In brewing trials, wort vapor condensate, treated by means of reverse osmosis, was used as mash liquor and as sparge liquor. The quality of the fresh beer was comparable to that of beer from conventional production. A difference was only observed with aged beer samples. A decline in quality was noticeable in aged beer samples produced using purified wort vapor condensate. For this reason, large quantities of wort vapor condensate should not be allowed to enter the product stream [36].

Through active carbon filtration, the COD of the wort vapor condensate can be reduced by as much as 40 %. Quality at a level permissible for drinking water, brewing liquor or boiler feed water was not achieved. Consequently, active carbon filtration is not suitable for treating wort vapor condensate.

A number of possibilities have been investigated for the use of untreated wort vapor condensate (table 5).

Table 5: Options for the reuse of raw (untreated) vapor condensate

process	reusable proportion [%]
spraying out the false bottom of the lauter tun	33
final sparge during lautering	28
chilling water for evaporative condenser	100
use in alkaline cleaning solutions – bottle cleaning – CIP and single-use cleaning applications	 100 100

There are nevertheless a number of applications for wort vapor condensate due to its low hardness. The only problem it poses is its heavy COD load.

Only 33 % of the volume of wort vapor condensate is needed for rinsing out the lauter tun. Due to the presence of lauter rakes and channel formation in the grain bed, the vapor condensate percolates through the grain bed sooner than would be theoretically anticipated. In experimental trials conducted at a brewery, 28 % was the maximum amount of vapor condensate that could be utilized before it appeared in the run-off exiting the grain bed. All of the vapor condensate can be used in the preparation of alkaline cleaning agents if the caustic is refreshed often enough. Vapor condensate has a lower COD than the caustic solution, so the cleaning effect is not impacted.

Normally, the vapor condensate is cooled to 25–30 °C and disinfected with 6 ppm of chlorine dioxide. If no disinfectant is added to the condensate, vapor condensers and any intermediate storage vessels become susceptible to the development of slime and sludge deposits. In addition, adding chlorine dioxide prevents the development of offensive odors. The use of H_2O_2 has proven to be unsuitable for this purpose, since there is no depot effect with H_2O_2 and it is more expensive [41].

The subject areas of hops and wort acidification related to wort boiling are covered in the their respective chapters.

3 SUMMARY

Of the equipment described here, internal and external calandrias represent the two fundamental designs that form the basis of most modern wort boiling systems. Currently, many existing brewhouses are being modified and optimized through installation of supplemental equipment. Above all, this is attributable to the significant pressure to reduce costs and stiff competition within the brewing industry. New purchases are weighed very carefully. Technological or energy-related improvements must justify the price paid for them. In most cases, retrofitting turns out to be the least expensive means to modernize a brewhouse.

The influence of the wort boiling system on the flavor of the beer is difficult to evaluate as other factors contribute significantly to the character of a beer, such as the grain bill, how the beer is filled and particularly the yeast. When evaluating beer flavor, one must consider that boiling wort in a precise and targeted manner is a necessary prerequisite, but not a guarantee, that one will brew good beer. If, for example, the fermentation is less than ideal, one's endeavors in the brewhouse may come to nothing. Similarly, a well-functioning wort boiling system may also be unjustifiably considered inferior.
By properly adjusting the relevant parameters, modern wort boiling systems can produce beers with flavor profiles comparable to those brewed on conventional systems. Fine-tuning the wort boiling parameters (duration of the boil, evaporation, thermal stress) should definitely be undertaken with a new system. Since the process parameters can be altered quite easily with the wort boiling systems described above, this should not pose a problem.

Because all the wort boiling systems presented above meet existing technological requirements, factors such as energy management, cost-effectiveness and local conditions play a decisive role in their selection. In order to better understand a specific system's level of energy savings, one should take time to determine and record the values for total evaporation achieved at various boiling times.

4 OVERVIEW

Which wort boiling system is worthy of consideration?

- The type of system is largely dependent on a brewery's energy concept.
- Which existing equipment can continue to be employed? Is retrofitting possible?
- What kind of architectural and structural conditions exist on-site? How much space is available?
- What are the costs of purchasing and operating the system?
- A new system delivers a somewhat altered wort composition. Due to the high degree of flexibility with regard to adjusting the parameters of the boil, a modern wort boiling system is able to meet the desired objectives of the boiling process and the wort composition.

All of the wort boiling systems discussed above are capable of producing high quality beer. How does poor homogeneity of the wort affect the performance of the wort kettle? What benefits are apparent from handling the wort in a uniform manner?

- Poorly mixed wort creates dead zones, resulting in non-uniform temperatures in the wort kettle and thus inconsistent thermal treatment of the wort. This means that a certain portion of the wort is not boiled and undesirable aroma compounds are therefore not volatilized and eliminated. If this is the case, the beer exhibits a raw, husky character and a harsh, lingering bitterness.

- On the other hand, if the wort is well mixed, then the duration of the boil can be curtailed. Experimental trials in breweries have shown, for example, that additional recirculation using a wort spreader after only a short boil (in an internal calandria for approximately 40 min) at a low amount of total evaporation (approx. 3 %) makes it possible to achieve the reference values for the analytical wort parameters.

How does a wort boiling system affect total evaporation and therefore energy management?
or
What level of total evaporation is most practical with respect to energy recovery systems?

- With vapor compression:
  - Electricity costs are a factor to consider when implementing mechanical vapor compression. The total evaporation is of lesser importance from the standpoint of energy consumption. The investment costs for a mechanical vapor compressor are high. The initial outlay for thermal vapor compression is low but requires a boiler capable of delivering steam at 8–10 bar of gauge pressure, in order to prevent liquid from being carried over into the steam. Besides the condensate from the vapor condenser, the condensate from the entrained liquid in the steam also accumulates, and though it yields thermal energy, it also negatively impacts the quality of the wastewater (refer to 2.4.13). The temperature of the heat transfer medium (i.e. the vapor) should not fall below 120 °C in either case.
- Is there an energy reservoir available for the brewhouse as well as for other brewery equipment?
  - A vapor condenser is required. Is the energy reservoir able to be accessed by other brewery equipment (lauter wort pre-heater, bottling line)?
  - The total evaporation depends on how effective energy management is outside of the brewhouse. If total evaporation is too low (e.g. < 5 %), the hot water must be reheated. This is, in part, also contingent on the design of the energy reservoir.
  - Is there an energy reservoir available for the brewhouse?
  - A total evaporation of 4–5 % is optimal for the vapor condenser, especially when a lauter wort pre-heater is installed [11].
  - Heating the mash is only practical using the energy reservoir if the mash is not subjected to shear stress, as this would negatively impact the lautering process (a higher pressure difference, filterability suffers). The relevant heat transfer surfaces must be sufficiently large and efficient as well. The best choice in this case would be to install additional specially designed heating plates on the inner wall of the vessel. The variable speed agitator must be run at a suitable speed to facilitate satisfactory heat transfer without exerting shear stress on the particles in the mash.
- Is there no energy reservoir available?
  - Energy usage will decline as the total evaporation during wort boiling decreases, because the brewhouse consumes more thermal energy than any other part of the brewery.

What level of total evaporation is most practical from a technological point of view?

- Practical trials and pilot experiments, in which wort was subjected to only 2–3 % total evaporation, have shown that this level of evaporation is sufficient to produce beer with acceptable sensory characteristics. However, such a low level of evaporation is associated with certain disadvantages. The yield during lautering drops (the maximum volume of last runnings which can be used for mashing in the following batch is 2 % [11]) and in most cases too little hot water is produced, requiring subsequent heating of the water.

- The optimal range for total evaporation using any type of wort boiling system should be explored and tested. By doing so, one can determine on-site whether the analytical benchmark values, such as free DMS, thiobarbituric acid index, the various nitrogen fractions and wort aroma compounds (e.g. hexanal as an indicator for evaporation), all fall within their respective limits of acceptance and how the boiling process has influenced the flavor of the beer.
- Recovering the energy used to boil the wort through vapor compression allows for a higher total evaporation with respect to the energy requirements of the boiling process. However, with mechanical vapor compression, the rest of the hot water needed for daily operations must be heated directly at the brewery.

When can a wort pre-chiller be employed?

- A wort pre-chiller can be installed at any time, because it hardly has an impact on the management of hot water at the brewery.
- Pre-chilling the wort brings additional production benefits such as less thermal stress on the wort (TBI), less cleavage of DMS-P/accumulation of DMS and a positive impact on the hop aroma.

At what pressure or temperature should wort be boiled?
The higher the temperature is, the more rapidly the chemical reactions occur.

- Early on, boiling wort at high temperatures was problematic from a technical perspective. Strong fouling inside the heat exchangers negatively impacted the flavor of the beer and routine cleaning had to be performed more frequently. This issue has since been resolved by lowering the temperature for wort boiling to a maximum of 135 °C. Installing an additional heat exchanger at stage 3 allows cleaning to be performed at any time now.
- Replacing a system designed to boil at a constant temperature slightly above atmospheric pressure with the dynamic version of the same system increases the evaporative efficiency.
- The equipment utilized for boiling at atmospheric pressure is simpler in design.
- Evaporation at atmospheric pressure is not necessary and can occur in a vacuum [29]. An evaporator capable of withstanding negative pressure is required. The design should be relatively uncomplicated and be realized at a reasonably low cost. In this case, the wort kettle would serve only as a storage vessel, allowing for the continued use of older wort kettles.

What are the advantages of bringing about evaporation of the volatile compounds in the wort downstream from the whirlpool?

- Free DMS and other undesirable aroma compounds can be driven out of the wort. The degree of flexibility in modifying the individual parameters of a wort handling system allows for the factors affecting the boiling wort or the hot wort held at a constant temperature to be adjusted. For instance, undesirable aroma compounds can be volatilized and eliminated from the wort immediately prior to pitching the yeast; however, one disadvantage of this is that any aroma from late hopping in the brewhouse is driven out along with the other compounds.
- The malt quality becomes less of an issue (e.g. DMS-P).
- Once the wort has been chilled, no further thermal reactions (Maillard, cleavage of DMS-P) can take place. Concerning the products of these reactions, the wort composition is fixed, meaning that none of these reactions affect the aroma profile of the wort once evaporation and chilling are complete and the wort is ready for pitching.

What are the advantages of installing a lauter wort pre-heater?

- It saves time, since the wort is pre-heated in a heat exchanger on its way from the wort collection vessel to the wort kettle. There are also technological advantages, since a long heating phase in the wort kettle greatly reduces the amount of residual coagulable nitrogen. If the brewhouse is equipped with a vapor condenser with an energy reservoir, the hot water from the energy reservoir can and should be utilized.
- The energy recovered from the vapor produced during the wort boiling process can be reused close by elsewhere in the brewhouse (energy reservoir).
- The detrimental effects of wort pulsating inside the tubes of the heat exchanger diminish. The temperature of the wort inside the internal calandria and the temperature of the surrounding wort, which is still being heated and has not yet started to boil, are closer to one another if the lauter wort has been pre-heated. The wort inside the internal calandria is less likely to experience scorching and thus is subject to less thermal stress. Furthermore, the wort handling process is more uniform overall.

Problem	Cause	Measures
concentration of free DMS is too high in the pitching wort	• concentration of DMS-P in malt > 5 mg/kg • low evaporative efficiency • additional DMS formation in the whirlpool • microbial contamination	• measure the amount of DMS-P in the malt and reject batches with excessively high DMS-P concentrations if necessary • increase the rate of evaporation/amount of total evaporation • reduce the duration of the whirlpool rest; pre-chill the wort; perform subsequent evaporation or strip the wort
poor foam characteristics, poor flavor stability, dark wort color	• thermal stress	• boil the wort in a gentle manner by reducing the boiling time (monitor the concentration of free DMS!) • lower the rate of evaporation (monitor the concentration of free DMS!) • ensure that the boiling process is uniform • reference the information in the chapter on beer foam
pulsating internal calandria	• the temperature difference between the wort in the internal calandria and the wort surrounding it in the kettle is too large	• force the wort to flow (e. g. Stromboli) by means of a pump • preheat lauter wort with an additional heat exchanger
poor break formation (hot break)	• mechanical stress on the wort • wort pH is too high • wort lacking in nitrogenous or tannin compounds	• e. g. allow the agitator to run while the wort is held at a high temperature without evaporation (SchoKo) • a wort pH of 5.2 is optimal for break formation • utilize information about malt quality and select the appropriate mash program to achieve desired wort composition; hop additions introduce tannins, aiding in break formation • acidify the wort 10 minutes prior to the end of the boil to achieve sufficient break formation

5 REFERENCES

[1] Yamashita, H., Kühbeck, F., Hohrein, A., Herrmann, M., Back, W., and Krottenthaler, M.: Fractionated Boiling Technology: Wort Boiling of different Lauter Fractions. Brewing Science 59 (2006), no. 7/8, pp. 130–147.

[2] Krottenthaler, M.: Modern wort production. Brauwelt International 26 (2008), no. 3, p. 190.

[3] All of section 2.2.1 on evaporative efficiency is referenced from: Mezger, R.: Betrachtung moderner Würzekochsysteme bezüglich ihres Einflusses auf technologisch und physiologisch bedeutende Würzeinhaltsstoffe. Dissertation Technical University of Munich, 2006.

[4] Krottenthaler, M.: Entwicklung innovativer Technologien zur Optimierung der Würze- und Bierqualität. Habilitation, TU München 2007.

[5] Rübsam, H., Becker, T., Krottenthaler, M.: Optimizing brewhouse technology with reference to dimethyl sulfide. Brauwelt International 28 (2010), no. 3, pp. 138–145.

[6] Coors, G., Krottenthaler, M., Back, W.: Wort Pre-Cooling and its Influence on Casting. Brauwelt International 21 (2003), no. 1, pp. 40–41.

[7] Jacob, F. (ed.): Collection of Brewing Analysis Methods of the Mitteleuropäische Brautechnische Analysenkommission (MEBAK). Wort, Beer and Beer-based Beverages. Selbstverlag der MEBAK, Freising 2013.

[8] DIN 8777: 2017. Sudhausanlagen in Brauereien.

[9] Back, W., Forster, C., Krottenthaler, M., Lehmann, J. , Sacher, B., Thum, B.: New Research Findings on improving Taste Stability. Brauwelt International 17 (1999), no. 5, pp. 394–405.

[10] Anness, B. J., Bamforth, C. W.: Dimethyl Sulfide – a Review. Journal of the Institute of Brewing, 88 (1982), July/August, pp. 244–252.

[11] Narziß, L., Back, W., Gastl, M., Zarnkow, M.: Abriss der Bierbrauerei, 8. ed., Wiley-VCH, Weinheim 2017.

[12] Mezger, R., Krottenthaler, M., Back, W.: Modern Wort Boiling Systems – an Overview. Brauwelt International 21 (2003), no. 1, pp. 34-39.

[13] Kunze, W.: Technology Brewing and Malting, 6. ed., VLB Berlin, Verlagsabteilung, Berlin 2019.

[14] Hough, J. S., Briggs, D. E., Young, T. W.: Malting and Brewing Science 2nd ed. Chapman and Hall, 1982, Vol. II, p. 506.

[15] Denk, V.: In: European Brewery Convention, Manual of Good Practice, Wortboiling and Clarification. Fachverlag Hans Carl, Nürnberg 2002, pp. 97ff.

[16] Narziß L., Miedaner H., Schneider, F. P.: Weiterführende Untersuchungen zur Technologie der Würzekochung unter besonderer Berücksichtigung energiesparender Maßnahmen (Teil 3). Monatsschrift für Brauwissenschaft, 44 (1991), no. 9, p. 304.

[17] Hackensellner, T.: Würzebereitung mit dynamischer Niederdruckkochung: Energie- und Anlagentechnik – Teil 1. Brauindustrie, 86 (2001), no. 3, pp. 14–16.

[18] Bühler, T., Michel, R., Kantelberg, B., Baumgärtner, Y.: Die dynamische Niederdruckkochung – systematisch qualitätsoptimiert. Brauwelt, 142 (2003), no. 38, pp. 1173–1178.

[19] 6. Energietechnisches Seminar (2002). Thema: Würzekochung. Veranstalter: IGS Ingenieurbüro Dr. Georg F. Schu. Schirmherrschaft: Bayrischer Brauerbund e. V.

[20] Binkert, J., Haertl, D.: Neues Würzekochsystem mittels Expansionsverdampfung. Brauwelt, 140 (2001), no. 37 pp. 1494–1503.

[21] Mezger, R., Krottenthaler, M., Back, W.: Dividing Wort Boiling into two Phases. Using the "SchoKo" Boiling Process as an example. Brauwelt International 22 (2004), no. 1, pp. 29–32.

[22] Meura: Meura – Traditionally Pioneers since 1845 (CD-ROM).Tournai/Belgium 2001.

[23] Fohr, M.: Höhepunkte der Sudhaustechnik auf der Brau 2000. Brauwelt 139 (2000), no. 48, pp. 2090–2092.

[24] Krottenthaler, M., Hartmann, K., Back, W.: Use of a Flash Evaporator for Wort Treatment. Brauwelt International 19 (2001), no. 6, pp. 457–459.

[25] Weinzierl, M., Stippler, K., Wasmuht, K., Miedaner, H., Englmann, J.: Ein neues Würzekochsystem. Brauwelt, no. 5 (1999). pp. 185–189.

[26] Wasmuht, K., Weinzierl, M., Vasel, B.: Neue Ergebnisse zum Kochsystem Merlin®. Brauwelt, 139 (2000), no. 26, pp. 1057–1059.

[27] Stippler, K., Wasmuht, K., Gattermeyer, P.: Generationswechsel beim Innenkocher, Teil I und II. Brauwelt 140 (2001), no. 31/32 and 35/36 pp. 1265–1267 and pp. 1386–1397.

[28] Wasmuht, K., Weinzierl, M., Gattermeier, P., and Baumgärtner, Y.: Stromboli – ein "Vulkan" auf der Überholspur. Brauwelt 143 (2004) , no. 30/31, pp. 925–927.

[29] Mezger, R., Krottenthaler, M., Back, W.: Vacuum Boiling – A new Alternative for gentle Wort Processing in the Brewhouse. Brauwelt International 24 (2006), no. 1, pp. 22–25.

[30] Kühbeck, F., Back, W. Krottenthaler, M.: Release of Long-Chain Fatty Acids and Zinc from Hot Trub to Wort. BrewingScience 59 (2006), pp. 67–77.

[31] Kühbeck, F., Müller, M., Back, W., Kurz, T., Krottenthaler, M.: Effect of Hot Trub and Particle Addition on Fermentation Performance of *Saccharomyces cerevisiae*. Enzyme and Microbial Technology 41 (2007), no. 6/7, pp. 711–720.

[32] Kühbeck, F., Schütz, M., Thiele, F., Krottenthaler, M., Back, W.: Influence of Lauter Turbidity and Hot Trub on Wort Composition, Fermentation, and Beer Quality. Journal of the American Society of Brewing Chemists 64 (2006), no. 1, pp. 16–28.

[33] Kühbeck, F., Back, W., Krottenthaler, M.: Influence of Lauter Turbidity on Wort Composition, Fermentation Performance and Beer Quality – A Review. Journal of the Institute of Brewing 112 (2006), no. 3, pp. 215–221.

[34] Kühbeck, F., Back, W., Krottenthaler, M.: Influence of Lauter Turbidity on Wort Composition, Fermentation Performance and Beer Quality in Large-Scale Trials. Journal of the Institute of Brewing 112 (2006), no. 3, pp. 222–231.

[35] Kühbeck, F., Back, W., Krottenthaler, M., Kurz, T.: Particle Size Distribution in Wort during Large Scale Brewhouse Operations. Aiche Journal 53 (2007), no. 5, pp. 1373–1388.

[36] Dickel, T., Krottenthaler, M., Back, W.: Studies into the influence of the cold break insertion on beer quality. Untersuchungen zum Einfluss des Kühltrubeintrages auf die Bierqualität. Monatsschrift für Brauwissenschaft 53 (2000), no. 5/6, pp. 95–100.

[37] Dickel, T., Krottenthaler, M., Back, W.: Investigations Into the Influence of Residual Cold Break on Beer Quality. Brauwelt International 20 (2002), no. 1, pp. 23–25.

[38] Fohr, M., Meyer-Pittroff, R., Krottenthaler, M., Back, W.: Stand der Technik bei der Verwertung von Brüdenkondensat. Brauwelt, 137 (1998), no. 48, pp. 2387–2393.

[39] Fohr, M., Meyer-Pittroff, R., Krottenthaler, M., Back, W.: State-of-the-art in utilisation of vapour condensate. Brauwelt International, 17 (1999), no. 5, pp. 360–365.

[40] Back, W., Krottenthaler, M., Vetterlein, K.: Zurück zum Ursprung. Aufbereitung von Brüdenkondensat und Weiterverarbeitung im Sudhaus. Brauindustrie 83 (1998), no. 2, pp. 81–86.

[41] Fohr, M., Meyer-Pittroff, R., Krottenthaler, M., Back, W.: Verwertung von Brüdenkondensat ohne Aufbereitung. Der Weihenstephaner, 66 (1998), no. 1, pp. 25–28.

ACIDIFICATION WITH NATURAL LACTIC ACID

1 INTRODUCTION

The acidification of malt, mash and wort through microbiological means has been utilized for many years to adjust the pH and thus lower values to the range preferred for the brewing process, a practice associated with benefits for both production and quality [1, 2, 3].

Earlier, the production of acidulated malt (lactic acid concentration: 2–4 %) was considerably more widespread than it is today. The addition of approximately 3 kg of this malt per 100 kg of grist (~ 3–6 % of the grain bill) lowers the pH of the mash by 0.1. The production of acidulated malt is effected through spraying or soaking the batch of malt (green or kilned malt) with *Lactobacillus delbrueckii* inoculated steeping liquor (48 °C/24 h) and subsequently drying it gently at approximately 65 °C.

This somewhat complicated and laborious method has largely been replaced by the continuous production of natural lactic acid using unhopped wort as a substrate. This is referred to in German as *Sauergut*, a term now familiar to many brewers in the English-speaking world. The process for natural lactic acid production stemming from methods applied in the mashing process at distilleries has been described by Joergensen [4], although, in this case, protecting the mash against microbiological spoilage was the primary objective. In 1896, Leichmann [5] named the "cultured lactic acid bacteria" he isolated from such substrates, *Bacillus delbrueckii*. Henneberg was the first to describe bacteria found in *Sauergut* at a brewery [6, 7]; however, the physiological and biochemical traits of these microorganisms were very similar to the cultures present in distillery mashes. It was later determined that lactic acid bacteria have their origins in germinating grain, green malt and malt. They grow well in unhopped wort and thrive at an optimal temperature of 47–48 °C, producing (D)-lactic acid.

According to the by-laws governing the brewing industry in Germany, in the production of pale beers, mash or wort may be acidified using lactic acid of microbiological origin produced by lactic acid bacteria of the type that naturally exists on malt. These bacteria must also be cultured according to processes specifically approved for this purpose [8].

However, the most recent taxonomic classification of these lactic acid bacteria has shown that they are not *Lactobacillus delbrueckii* but rather *Lactobacillus amylolyticus* [9]. In most breweries, lactic acid production is achieved naturally using this species. The combination of attributes exhibited by strains TL3 and TL5[1] has proven to be most practical for lactic acid production in a brewery.

Individual reports concerning lactic acid production using other species of microbes can be found in the literature (e.g. *L. casei*, *L. amylovorus*, *L. helveticus*, *Pediococcus acidilactici*, *P. pentosaceus*, *P. dextrinicus*), and yet they are quite often associated with processing difficulties and/or qualitative deficiencies. The heterofermentative species *Lactobacillus fermentum* is sometimes mentioned in conjunction with the production of *Sauergut* but it is not well suited for the process as only 50 % of the total fermentable sugar present in the wort is converted to lactic acid. Moreover, *L. fermentum* generates alcohol, acetic acid and CO_2. Although these compounds evaporate. Although they evaporate during the wort boiling process, they still represent extract losses.

[1] **supplier: Chair of Brewing and Beverage Technology, TU Munich-Weihenstephan**

2 TECHNIQUES USED IN THE PRODUCTION OF *SAUERGUT*

Very conveniently, the production of lactic acid through microbiological means can be carried out underneath or in close proximity to the brewhouse. The required substrate, that is, unhopped first wort, is always readily available there. In addition, the higher ambient temperatures present below or near the brewhouse are advantageous for incubating the bacteria at approximately 48 °C. Insulating the *Sauergut* reactor vessels is unnecessary, and the substrate requires no supplementary heating.

2.1 LACTIC ACID CULTURES

The bacterial strains most often employed today in the production of *Sauergut* are strains of *Lactobacillus amylolyticus* isolated from barley or malt, which exhibit a very advantageous set of traits. These bacteria are Gram-positive, catalase-negative, oxidase-negative, homofermentative, thermophilic microaerophiles with the capacity to ferment not only maltose but also dextrins and starch. These lactic acid bacteria are classified in a separate group from *Lactobacillus amylovorus*, a related species, due to genotypic differences and have been thoroughly characterized [9]. *Lactobacillus amylolyticus* cells occur as elongated rods, which can reach lengths of 20–50 µm [3][2], and are found individually or in short chains. Since they are very sensitive to hops, lactic acid production must be carried out using the first wort or lauter wort prior to boiling. Due to this sensitivity, which can be as low as 2–4 EBC bittering units, one must be aware that any accidental transfer of hop bitter substances, e.g. through rinsing the hot break material with the grain bed in the lauter tun to increase efficiency, can negatively impact the bacterial culture and thus the production of lactic acid. This trait, however, does have one great advantage, namely that *Lactobacillus amylolyticus* cannot spoil hopped wort and beer. Table 1 summarizes the most important attributes of *Lactobacillus amylolyticus* strains for the production of natural lactic acid.

Table 1: Characteristics of *Lactobacillus amylolyticus* relevant to beer production

1. dominant growth in beer wort (rapid establishment)
2. high tolerance to acidity
3. homofermentative (2 moles of lactic acid produced from one mole of glucose)
4. able to grow at high temperatures (up to 52 °C)
5. fermentation of dextrins and starch
6. produces high proportions of L(+)-lactic acid
7. no beer spoilage (bacteria are sensitive to hops, no growth at temperatures < 30 °C)
8. no formation of amines (e.g. histamine) and other toxins
9. no formation of diacetyl and other undesirable substances
10. cultures are easy to cultivate and maintain

[2]images of *Lactobacillus amylolyticus*: Back, W.: *Colour Atlas and Handbook of Beverage Biology*, Fachverlag Hans Carl, Nuremberg 2005, p. 43.

2.2 THE ADVANTAGES OF USING NATURAL LACTIC ACID IN THE FORM OF *SAUERGUT*

The use of *Sauergut* to adjust the pH of the mash and/or the wort is accompanied by a host of other benefits as well. By lowering the pH of the mash, important cytolytic and proteolytic enzymes are activated while the activity of other less desirable enzymes, such as lipases, lipoxygenases, polyphenol oxidases, peroxidases as well as β-glucan solubilase, is undermined (cf. Mashing, Flavor Stability).

Acidification of the wort induces a more extensive precipitation of proteins and ultimately results in improved chemical-physical stability. The flavor stability, head retention and the microbiological stability of the finished beer are also positively influenced by the addition of *Sauergut*. Further production, sensorial and microbiological benefits are provided in table 2.

Besides numerous improvements to the quality of the beer, one must be aware of the fact that the isomerization of hop bitter substances occurs to a lesser extent at a lower pH. In order to avoid reduced hop utilization and poor bitterness yields, *Sauergut* should first be added at the end of the wort boiling process. This also has a positive effect on the cleavage of DMS-P and eventually leads to lower concentrations of free DMS in the chilled wort.

Table 2: Advantages of biological acidification through the addition of *Sauergut*

1. **Technological improvements**
   - enzyme availability (all important enzymes are activated through the souring of the mash, with the exception of α-amylase)
   - nutrient availability (the availability of zinc is also increased)
   - protein separation (better flocculation and sedimentation)
   - redox potential (lowered susceptibility to O_2 exposure)
   - fermentation progress (the pH drops more rapidly, better trub separation, higher degree of attenuation in the fermentation cellar)
   - filtration (better clarification, improved efficiency through lower amount of solids to be filtered out)
2. **Sensory improvements**
   - flavor (rounder, fuller and softer flavor profile)
   - hop bitterness (pleasant, does not linger)
   - drinkability (fresh, sparkling character)
   - foam (stable, fine bubbles)
   - color (paler and fresher)
   - chemical-physical stability (reduced tendency to form protein haze)
3. **Physiological and health benefits**
   - metabolism (stimulates metabolic activity)
   - digestion (lactic acid has a positive impact)
   - immune mechanisms (among other things, better protection against illness and disease, the growth of pathogenic microbes is suppressed)
4. **Reduced susceptibility of the beer to microbiological infection**
   - lower pH values:
     a) *Pectinatus* and *Megasphaera* usually do not grow at pH values below 4.4 (4.5)
     b) Potential beer-spoiling microbes usually do not grow at pH values below 4.5
     c) As the pH value sinks, so does the rate of growth of obligate beer spoilers
   - higher final degrees of attenuation: this lowers the amount of easily fermented sugars available for utilization by beer-spoiling microbes
   - greater selection pressure by the yeast: the growth advantages present for the yeast allows it to suppress the growth of competitors such as beer spoilers (important for primary fermentation and yeast propagation)

In table 3, the effects of acidification through the addition of *Sauergut* to standard-strength lager beer is compared with those that did not undergo acidification. While there was little difference in the qualitative assessment of fresh beer samples of both types, the results for beer subjected to forced aging were not similar. The samples of aged beer produced using acidification with natural lactic acid were deemed to be much better. The sensory analysis of the aged beers (cf. Sensory Analysis) and the evaluation of the beer's acceptance were both positive, and aging compounds were present in lower concentrations.

Table 3: The influence of biological mash acidification on beer quality

Mash-in temperature: 62 °C	**without acidification**	**with the addition of *Sauergut***
pH	4.61	4.55
color [EBC]	5.8	5.6
tannins [mg/l]	14.9	28.4
polyphenols [mg/l]	177	178
polymerization index	4.3	3.9
foam (ROSS and CLARK)	117	118
warm days (forced aging test)	2	4
aging compounds (fresh samples) [µg/l]	106	108
aging compounds (artificially aged samples) [µg/l]	266	173
lag time [min]	37	43
Sensory analysis:		
DLG tasting of fresh samples	4.02	4.24
tasting of artificially aged samples	2.4	1.8
acceptance of artificially aged samples [%]	70	90

The advantages of acidification with *Sauergut* versus acidification with food grade lactic acid derived from chemical processes are summarized in table 4.

Table 4: Comparison of the addition of *Sauergut* vs. food grade technical acid for adjustment of pH

	Acidification with *Sauergut* (*L. amylolyticus*)	**Acidification with food grade acid** (lactic acid/phosphoric acid)
Microbiological advantages		
suppresses growth of *Pectinatus, Megasphaera* and potential beer-spoiling microbes, growth of obligate beer spoilers is restricted by low pH values	+++	+++
better fermentation, higher degree of attenuation	+	–
greater selection pressure exerted by the yeast	++	–

	Acidification with *Sauergut* (*L. amylolyticus*)	Acidification with food grade acid (lactic acid/phosphoric acid)
Production advantages		
cost savings	++	–
availability of enzymes	++	–
enzyme activation	++	+
availability of nutrients	+++	+
protein flocculation/elimination	+	+
redox potential	++	+
progression of fermentation	++	+
filtration	+	(+)
Sensory improvements		
flavor: rounded, full, soft	+	–
sensory analysis of aged samples	++	+
hop bitterness	+	(+)
carbonation/drinkability	+	+
foam	+	(+)
color	+	+
chemical-physical stability	++	++

2.3 CULTIVATION AND PROPAGATION OF LACTIC ACID BACTERIA

The propagation of cultures used in *Sauergut* production is best carried out in diluted first wort (10–12 % w/w) at a constant temperature of 48 °C. The cultures should not remain at high incubation temperatures upon reaching the logarithmic growth phase and at a pH less than 3.5 (max. of 3 days); otherwise the cells begin dying rapidly.

Therefore, to sustain the culture over the weekend, it is recommended that fresh, unhopped wort (50–90 %) be added and the temperature be reduced < 30 °C. For longer periods (e.g. a suspension of the regular brewing schedule, culture maintenance), a freshly cultivated culture can be stored at 5–10 °C for a maximum of two months.

Contamination of *Sauergut* with foreign microbes is rare due to the high incubation temperatures and the very low pH values (pH 2.9–3.3). Only certain yeast strains capable of respiration (e.g. *Candida kefyr, Candida magnoliae*) are able to grow at temperatures of up to 47 °C, causing fermentation problems (evolution of gas) and off-flavors (e.g. ethyl acetate).

The growth of these yeasts can be suppressed quite well by raising the temperature to 50 °C and injecting gaseous CO_2 into the medium or by flooding the surface with the gas to form a protective layer. A CO_2 atmosphere also fosters the growth of the lactic acid bacteria and helps prevent the oxidation of *Sauergut*.

Rare cases of contamination with butyric acid bacteria (e.g. *Clostridium butyricum* or *Clostridium acetobutylicum*) have been reported. This is, however, only possible at higher pH values, if for example, the lactic acid bacteria did not grow properly due to a low cell count, a weakened culture or if temperatures remain over 52 °C (the maximum temperature for viability) for an extended period.

2.4 PRODUCTION OF *SAUERGUT* IN THE BREWERY

Several different procedures can be employed in the production of *Sauergut*. Fermentors (the number is dependent on the brewing schedule), dimensioned at two to four times the volume of the required *Sauergut* addition, are combined with a bulk storage tank of sufficient capacity to accommodate the volume of *Sauergut* required daily. Another alternative would be to have three tanks, each with a capacity corresponding to the daily volume of *Sauergut* required, alternating as a lactic acid bacteria propagation tank, an acid production tank and a bulk storage tank.

Table 5: Examples of procedures employed in *Sauergut* production

Volume of natural lactic acid required daily: 32 hl, 8 batches

Variation A:
bulk storage tank 37 hl (32 hl + approx. 15 % headspace), acidification continues in bulk storage tank

4 fermenters, each with a volume of 9 hl (8 hl + approx. 15 % headspace), removal of lactic acid (4 hl) and addition of fresh wort (4 hl of first wort, diluted correspondingly) to each fermenter at 12-hour intervals

Variation B:
3 tanks, each with a volume of 37 hl (32 hl + approx. 15 % head space)

<u>Day 1:</u>
tank 1: supply tank from which the lactic acid is removed
tank 2: propagation tank (addition of fresh wort corresponding to the brewing schedule)
tank 3: acidification tank (thorough acidification of the entire volume of wort)

<u>Day 2:</u>
tank 1: becomes the next propagation tank
tank 2: becomes the acidification tank
Tank 3: becomes the supply tank

Figure 1: Equipment for the production of *Sauergut*, consisting of two fermentors and one storage tank

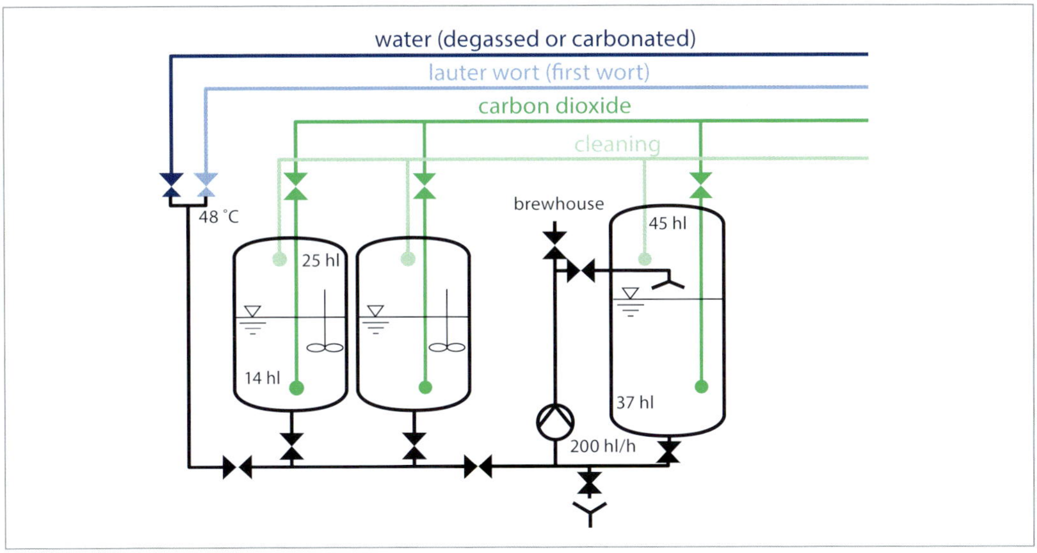

In the simplest cases, two vessels or even one is sufficient; however, this substantially diminishes the flexibility of the system. When removing *Sauergut* from the propagation vessel and refilling with first wort, one must ensure that the pH increases to a value above 4.0. At lower pH levels, the lactic acid bacteria cease to grow and no longer produce lactic acid. The *Sauergut* is simply diluted based on its lactic acid content. This means that correspondingly large volumes of fresh first wort must be added. If 50 % of the *Sauergut* is removed, an equivalent volume of unhopped wort is added to replenish it (1 : 1) according to the brewing schedule, as long as the interval between batches is 4 to 6 hours. If the interval is shorter, then either a second fermenter or a second vessel is necessary to allow complete acidification. Otherwise, the acidity must be regularly monitored and the addition of *Sauergut* adjusted as required.

A *Sauergut* reactor is usually installed in its own room on the hot side of the production in the brewery. The vessels are most often insulated stainless steel equipped with a heating element and regulated with a thermostat. An agitator and the necessary equipment and fittings for purging with CO_2 should be included in the design.

The water and wort lines should be equipped with a thermostat-controlled mixing device (temperature of the mixture: 48 ± 1 °C). Normally, the first wort is diluted to an extract concentration of 10–12 % (w/w) which is appropriate for the lactic acid bacteria. As the extract concentration nears 16 % (w/w) and above growth of the bacteria begins to decline rapidly. At least one of the fermentors must be equipped with a cooling system, in order for the lactic acid bacteria to be properly maintained over a weekend and during breaks in brewhouse operations. The bulk storage vessel does not need to be heated, since the acidity increases even as the temperature drops. The storage tank should be filled from the top and mixing the contents should be avoided. This ensures a thorough acidification of the fermenting *Sauergut*. By the time the medium in the bottom of the vessel is completely acidified, the whole batch is available for use.

Sauergut generally has a strong acidity of around 1.5 % and a low pH of approximately 2.9; both offer protection against microbiological infection. This acidity also is accompanied by an inherent self-cleaning effect. Thus, keeping the vessels and the lines clean requires little to no effort.

All of the vessels should be furnished with a fitting for connecting a CO_2 gas line. The gas can be injected into the vessel through a so-called sparger, that is, through a porous metal or ceramic device mounted inside the vessel, or in-line using a carbonation stone possessing fine pores while the *Sauergut* is being transferred or by purging the vessel with CO_2 prior to filling. The water may also be carbonated prior to its use in diluting the first wort.

Continuous production with immobilized lactic acid bacteria in a bioreactor has also found practical application in the industry [10]. This method of operation allows for a high level of efficiency and uncomplicated process monitoring. Operations are conducted using a fixed bed process in a reactor containing DEAE cellulose for immobilization of the bacteria (fig. 2). The system can run for several months without requiring regeneration. Even after one to two weeks of downtime, the immobilized bacteria can easily be reactivated.

Figure 2: Process diagram for the continuous production of *Sauergut*

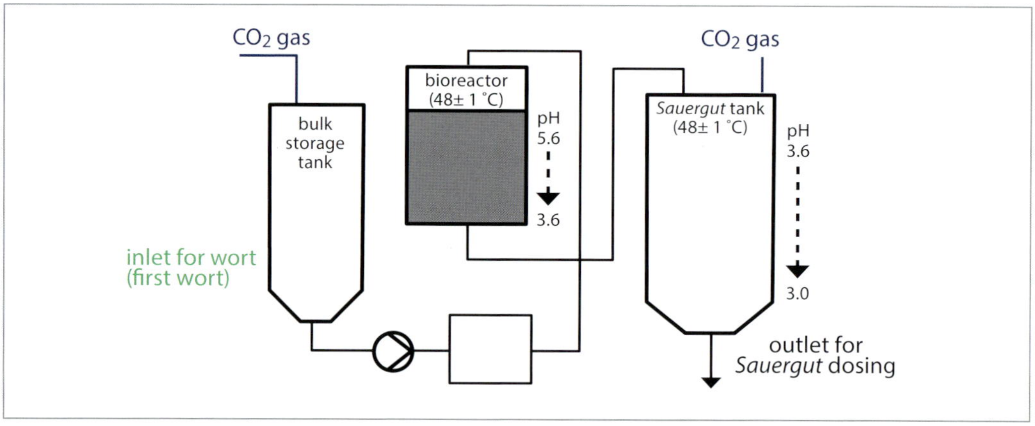

2.5 CALCULATING THE REQUIRED VOLUME OF LACTIC ACID

Generally speaking, 580 g of lactic acid per metric ton of malt grist is necessary to lower the pH by 0.1. This corresponds to approximately 10 g of lactic acid per hl of mash. In order to bring about an equivalent drop in the pH of wort, half the amount of lactic acid is required, that is, 5 g of lactic acid per hl of wort [11].

Table 6: Calculating the volume of *Sauergut* required for mash acidification

Sample calculation: (please note all tons (t) indicated are metric tons)

mash pH of 5.9 → 5.6 for a grain bill of 5 tons

amount of lactic acid required: 580 g/t grist x 5 t grist x 3 (per 0.1 pH unit) = 8,700 g

lactic acid concentration: 1.1 % mass to volume; amounts to 11 g/l

addition of *Sauergut*/batch: $\frac{8{,}700\text{ g}}{11\text{ g/l}} = 790\text{ l}$

for 8 batches/day; 8 x 790 l = 63.2 hl

Table 7: Calculating the volume of *Sauergut* required for wort acidification

Example: (please note all tons (t) indicated are metric tons)

wort pH of 5.5 → 5.2 for a grain bill of 5 tons

amount of lactic acid required: 290 g/t grist x 5 t grist x 3 (per 0.1 pH unit) = 4,350 g

lactic acid concentration: 0.8 % mass to volume; amounts to 8 g/l

addition of *Sauergut*/batch: $\frac{4{,}350\ \text{g}}{8\ \text{g/l}} = 544\ \text{l}$

for 6 batches/day; 6 x 544 l = 32.6 hl

3 SUMMARY

Biological acidification is normally achieved through the use of the thermophilic bacterial species *Lactobacillus amylolyticus*. This microbe brings numerous benefits to the brewing process, positively impacting beer quality. The bacteria can be managed without difficulty due to their strong growth at 48 °C and their rapid conversion of maltose, dextrins and starch to lactic acid.

Sauergut, the German term for diluted first wort fermented by these thermophilic lactic acid bacteria, can be employed to lower the pH of the mash and wort while also providing numerous other advantages, such as the activation of desirable enzymes and the inhibition of undesirable ones, as well as better flocculation, fermentation and filtration. These benefits extend to the quality of the finished beer with enhanced liveliness and drinkability and a more pleasant bitterness resulting in a harmonious overall flavor profile. Furthermore, the flavor stability, head retention and the microbiological stability of the finished beer are improved. Comparisons of beer samples produced with acidification with *Sauergut* versus lactic acid produced by chemical means clearly show the inherent advantages of *Sauergut*.

In cultivating lactic acid bacteria, long periods of downtime at high temperatures and a low pH must be avoided. Foreign microbes are normally only a problem at low temperatures or due to incorrect handling of the microbes.

Sauergut production is relatively uncomplicated and only requires a few vessels. If possible, the removal of the *Sauergut* and the addition of diluted first wort (approx. 12 % w/w) should be synchronized with the brewing schedule and occur in a ratio of 1 : 1. It is important that the addition of the first wort results in an overall pH above 4.0, since the growth of the bacteria can quickly come to a halt at a lower pH.

The cultivation of the lactic acid bacteria should take place under a CO_2 atmosphere, since they are, aside from being thermophilic also microaerophilic and thrive in the absence of oxygen. Additionally, the *Sauergut* is protected against oxidation, and inhibits the growth of yeast strains belonging to the genus *Candida*, which are detrimental to the lactic acid production process.

Numerous methods have been developed for producing *Sauergut* in a batch process as well as using immobilized lactic acid bacteria in a bioreactor.

4 OVERVIEW

Phenomenon/Problem	Possible cause	Corrective measures
decrease in acidity measured throughout the production week	not enough wort (first wort) was added to the *Sauergut*; the increase in pH (preferably > 4.0) not high enough	increase the ratio of wort (first runnings) to *Sauergut* (should be a minimum of 1:1)
gas generation in the *Sauergut*	infection and fermentation with *Candida* yeast strains	gradually raise the temperature over several days to 50 °C; aerate with CO_2 gas
cultures are dying off	downtimes too long (> 4 days) at high temperatures and low pH values without the addition of wort	cool on weekends or during breaks in brewing (5–10 °C) and add wort
lactic acid bacteria exhibit poor growth or the fermentation is not progressing	bitter compounds from hops are present in the wort used to produce *Sauergut*; concentration of the wort (first runnings) is too high (> 16 %)	no early utilization of hot break material in the lauter tun; dilute the first wort (not possible with mash filters)
lactic acid fermentation is not progressing	temperature is too high (> 51 °C); initial pH is too low (< 4.0); cell count of the lactic acid bacteria is too low or the culture is weak	observe correct procedures for producing *Sauergut*
dead rods or lactobacilli observed in samples of harvested yeast, unfiltered beer or liquid enrichment cultures viewed under the microscope	dead cells from the lactic acid culture are carried over into brewing process	does not pose any disadvantages, in the case of questionable results, continue to incubate for an additional day, if there is no increase, then the cells are dead
poor saccharification	low α-amylase activity due to excessive mash acidification (pH values < 5.4)	either carry out a moderate mash acidification or none at all
lautering problems	poorly modified malts, high mash-in temperatures with non-homogenous malts (β-glucan solubilase activity)	low mash-in temperature (45 °C) and acidification of the mash

Phenomenon/Problem	Possible cause	Corrective measures
low rates of isomerization	pH value too low during wort boiling	add *Sauergut* at the end of the boil; either carry out a moderate mash acidification or none at all
poor fermentation due to zinc deficiency	zinc deficit in the raw materials	acidify the mash to a pH of 5.4–5.2
slow fermentation, slow degradation of diacetyl	FAN concentration is too low	mash acidification and rest at 48 °C
insufficient drop in pH during fermentation	excessive mash acidification and rest at 50 °C, phosphatase activity is too strong causing excessive formation of buffering compounds	utilize a *hoch-kurz* mash (HKM) program; omit mash acidification if there is a rest at 50 °C
final degree of attenuation is too high	α-amylase and β-amylase activity is too strong	reduce the rest at 62 °C; acidify the mash in order to slow the activity of α-amylase
poor flavor stability	ingress of oxygen during wort preparation, avoid contact with oxygen especially during mixing and lautering, excessive proteolysis, strong lipoxygenase and peroxidase activity	either carry out a moderate mash acidification or none at all; utilize a *hoch-kurz* mash (HKM) program; review the selection of raw materials
poor foam characteristics	excessive proteolysis	no mash acidification, only wort acidification; mash in at a higher temperature (> 60 °C) or at 55 °C with a short rest
susceptibility to infection	the pH of the beer is too high (e.g. buffering in the wort is too strong as a result of mash acidification)	acidify the wort (pH 5.0–5.2)

5 REFERENCES

[1] Back, W.: Biologische Säuerung. Monatsschrift für Brauwissenschaft 41 (1988), no. 4, pp. 152–155.

[2] Narziß, L., Back, W., and Leibhard, M.: Optimierung biologischer Verfahren zur Herstellung von alkoholfreiem Bier mittels geeigneter Kulturen. Brauwelt 129 (1989), no. 45, pp. 2206–2214.

[3] Back, W.: Colour Atlas and Handbook of Beverage Biology. Fachverlag Hans Carl, Nuremberg 2005, pp. 42–45.

[4] Jørgensen, A.: Die Mikroorganismen der Gärungsindustrie. 5th ed., Berlin 1909.

[5] Leichmann, G.: Über die im Brennereiprozeß bei der Bereitung von Kunsthefe auftretende spontane Milchsäuregärung (Bacillus Delbrücki).
In: Zentralbl. Bakteriol. Parasitenk. Abt. II, 20, 1896. pp. 281–285.

[6] Henneberg, W.: Zur Kenntnis der Milchsäurebakterien der Brennereimaische, der Milch und des Bieres. (*Bacillus Delbrücki* (Leichmann), *Bacillus delbrückii* var. alpha n. sp., *Pediococcus lactis acidi* (Lindner) – *Bacillus lactis acidi* (Leichmann) – *Saccharobacillus pastorianus* (v. Laer) – *Saccharobacillus pastorianus* var. *alpha* n. sp., *Saccharobacillus pastorianus* var. *berolinensis* n. sp., *Bacillus lindneri* n. sp.). Wochenschrift für Brauerei 18 (1901), pp. 381–384.

[7] Henneberg, W.: Bakteriologische Untersuchungen an säuernden und gärenden Hefemaischen. (Ein Beitrag zur Kenntnis des Verhaltens des *Bacillus delbrückii* bei verschiedenen Temperaturen). Zeitschrift für Spirituosenindustrie 28 (1905), pp. 26–29.

[8] BierSTG i. d. F. vom 01.01.1993.

[9] Bohak I., Back, W., Richter, L., Ehrmann, M., Ludwig, W., Schleifer K. H.: *Lactobacillus amylolyticus* sp. nov., Isolated from Beer Malt and Beer Wort. System. Appl. Microbiol. 21 (1998), pp. 360–364

[10] Pittner, H., Back, W.: Continuous Production of Acidified Wort for Alcohol Free Beer using immobilized Lactic Bacteria. Technical Quarterly 32 (1995), no. 3, pp. 163–168.

[11] Narziß, L., Back, W.: Die Technologie der Würzebereitung. 8th ed., Wiley-VCH, 2009.

YEAST TECHNOLOGY AND FERMENTATION

1 INTRODUCTION

The quality of the yeast and the fermentation process both have a significant impact on the quality of the finished beer. Yeast in poor physiological condition can lead to the following problems:

- delays in fermentation
- elevated diacetyl values
- incomplete reduction of diacetyl and carbonyl compounds
- filtration problems and turbidity
- a slower and insufficient drop in pH during fermentation
- higher pH values in beer
- greater risk of microbiological contamination
- poor foam stability
- fluctuations in beer flavor
- initial flavor impression – sulfury or yeasty
- unpleasant, bitter aftertaste
- inadequate flavor stability

In less extreme cases, the physiological condition of the yeast results in a slower rate of fermentation and has an influence on aroma compounds. Fluctuations in the time required for fermentation may, for example, lead to capacity issues. Despite the fact that the physiological condition of yeast is known to impact beer quality, it is an aspect of the brewing process that is often neglected. Unfortunately, yeast management has often been limited to simply preparing the volume of yeast required.

In light of the fundamental role that yeast plays in beer production, it is essential that the necessary steps be taken to assess the condition of the yeast using specific criteria, optimize yeast management practices and correctly handle cropped yeast.

2 YEAST MANAGEMENT AND FERMENTATION

2.1 YEAST VIABILITY AND VITALITY

Yeast vitality is the most common means for evaluating the physiological condition of a brewing yeast population and is generally understood to be the ratio of living to dead cells. However, this term is also applied in the evaluation of the physiological condition of living cells. In an effort to avoid potential misunderstandings, it is helpful to distinguish between the two. Viability denotes the proportion of living cells, whereas vitality describes the physiological condition of the living yeast cells, including their activity and capacity for fermentation.

2.1.1 YEAST VIABILITY

Determining the ratio of living to dead cells in pitching yeast is an important test, since dead cells reduce the cell count and diminish fermentation performance. This brings about a change in the aroma profile [1, 2]. The yeast cells autolyze, releasing substances such as proteinases and fatty acids into the medium. Increased proteinase activity results in poor foam stability (cf. Beer Foam). In addition, flavor stability is also negatively impacted as a result of a lower reducing capacity on the part of the yeast cells as well as the release of medium chain fatty acids into the medium [3]. The pH value of the finished beer may also be higher, making the beer more susceptible to microbiological contamination.

The methods typically employed to determine yeast viability either involve monitoring cell growth or the use of dyes [4, 5].

The reference methods for determining cell viability involve establishing whether the cells are alive (e.g. cultures cultivated on streak plates or microscope slides). However, these are not suitable as routine procedures in a brewery laboratory since longer incubation periods are needed to obtain the results [6].

When staining cell samples, it is important to ensure that the cell count is not too high (approx. 50 cells/visible field) and that the contact time is neither too long nor too short. Approximately 200 to 500 cells should be counted in order to guarantee a statistically relevant result. A Thoma counting chamber can be used for this purpose, as it provides better orientation. The cells are counted using a microscope with brightfield illumination.

When using fluorescent dyes in conjunction with a Thoma counting chamber, the microscope slide must also be exposed to UV light with low transmission, so that the measurement quadrants and the non-fluorescing cells are also visible.

In the brewing industry, methylene blue staining is the most common method for determining the ratio of living to dead cells, mainly because it is simple to perform and yields practical results. However, problems arise when cells, which have been killed through exposure to heat, do not absorb the stain either. For this reason, the method is sometimes purported to be less reliable. Several variations on the methylene blue method are described in the current literature. One tried-and-tested method is the use of a methy-lene blue solution acidified with citric acid (preparation: mix a solution of 0.02 g methylene blue + 100 ml distilled H_2O with a solution of 4 g trisodium citrate + 100 ml distilled H_2O) [6]. Comparable results can also be achieved by using methylene violet (stain: methylene violet 3RAX; prepare the solution as per instructions for methylene blue above).

Furthermore, a range of fluorescent dyes is also available for determining the ratio of living to dead cells. Each dye possesses very different individual characteristics (such as binding to nucleic acids or the ability to diffuse across membranes). Some have proven to be reliable indicators for determining the ratio of living to dead cells. They are also able to distinguish cells killed by the application of heat. Their disadvantage lies in the fact that some of the dyes are very expensive or even toxic. Moreover, a fluorescent microscope with the proper range of filters is required. Dyes that are considered reliable include acridine orange [7], berberine [8] and oxonol (dye: $DiBAC_4(3)$), among others [9].

2.1.2 YEAST VITALITY

Analytical testing and evaluation of the complex physiological condition of yeast is very difficult. For this reason, conclusions regarding yeast vitality are drawn, for example, from methods for quantifying particular compounds found in yeast (such as glycogen, a storage carbohydrate). During the lag phase of cell growth, glycogen plays a key role in the synthesis of fatty acids and sterols. Various authors have obtained considerably different results regarding the correlation between absolute glycogen content and the course of fermentation [10, 11, 12, 13, 14]. One can therefore conclude that simply measuring the compounds present in yeast cells does not provide a clear assessment of their physiological condition.

Other methods are based on the measurement of yeast metabolic activity (e.g. reducing capacity or ability to produce CO_2). However, it has been shown that an individual yeast compound or a specific metabolic process is far from representative of the complex interplay of biochemical reactions inherent to yeast physiology. By contrast, the ICP method (a measure of the intracellular pH) yields very reliable results by including numerous central metabolic pathways. The intracellular pH is primarily dependent on the activity of ATPase in the plasma membrane, which plays a decisive role in the transportation of nutrients and trace elements across the cell membrane. Moreover, the activity of key enzymes is influenced by the intracellular pH. Thus, there is a correlation between intracellular pH and the physiological condition of the yeast [15, 16, 17]. The measurement range lies between pH 5.2 and pH 6.7. Here, a pH of more than 6.0 reflects good yeast vitality, whereas pH values below 6.0 (particularly below 5.6) are an indication of low vitality (zinc deficiency, target 0.15 mg/l). Although the method provides valuable insight, it is extremely time-consuming, labor-intensive and requires a fluorescence spectrophotometer. Since the original method was introduced, a modified method has been developed that requires less time and is easier to perform [18]. As a fluorescence spectrophotometer is not considered a standard piece of brewery lab equipment, this method has yet to be incorporated into routine analyses.

A simple, yet equally reliable method is to measure the formation of CO_2. In this method, for example, the increase in internal pressure in a bottle filled with wort and yeast is measured with a bottle manometer (possible scenario: transfer 100 ml wort and 10 g centrifuged yeast to a 0.33 l bottle, place on a magnetic stirrer, attach the manometer to the bottle and monitor the pressure increase over the course of two hours). Maintaining the temperature by placing the bottle in a water bath, for example, is highly recommended but not absolutely essential. Given the conditions described above, a good result would be approximately 3 bar of pressure after two hours; values below 2 bar are considered inadequate [19]. There is no direct correlation between viability and vitality. The number of dead cells does not change very much if yeast with high levels of viability and vitality is stored at very low temperatures over an extended period of time. However, a measurable decline in vitality (ICP) is observed after just 2–3 days (fig. 1). This relationship is different for active populations, for example, that are propagated at very high temperatures. A large proportion of dead cells (approx. 10–20 %) is often documented in such cases. The vitality of the remaining cells is very high despite the high number of dead cells. This level of vitality is also reflected in their excellent performance during fermentation.

Figure 1: Changes in viability (MB = methylene blue) and vitality of a yeast population during storage at 0 °C and 14 °C

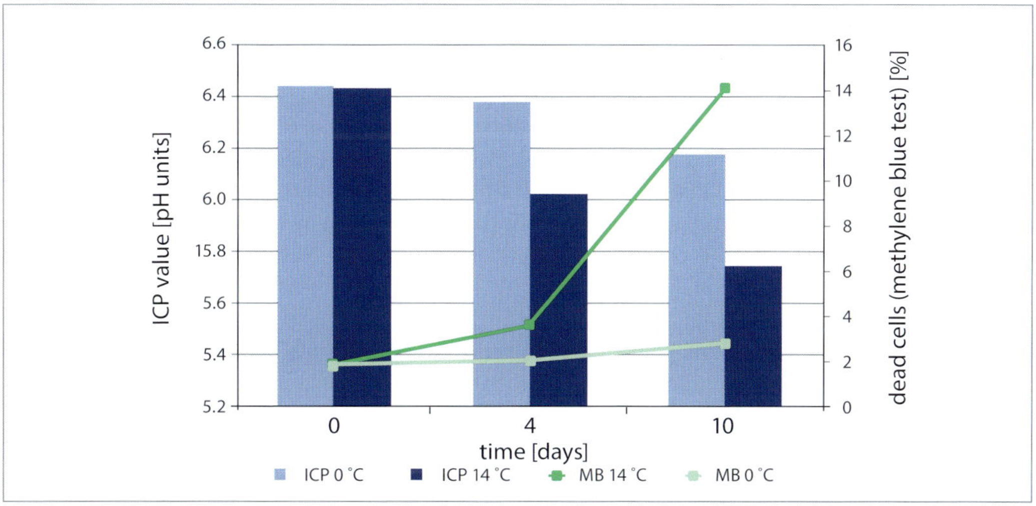

It follows that viability and vitality are both relevant parameters for fermentation and therefore should be monitored on a regular basis. In practice, measuring the viability (methylene blue test) is sufficient, provided that the yeast is cropped early enough and is stored at low temperatures for only a short period (see section 2.5). In any case, measuring the viability of the pitching yeast should never be omitted, as it provides timely insight into fermentation and subsequent processes through to the finished beer. If a batch of yeast exhibits poor viability or low vitality, it must be withdrawn from the production process.

2.2 YEAST CULTIVATION AND ASSIMILATION TECHNIQUES

2.2.1 GENERAL INFORMATION REGARDING YEAST ASSIMILATION

Yeast assimilation, a special technique for propagating yeast, differs from other techniques of yeast production in that the growth conditions are ideal in every possible way. Cells in all physiological growth phases are present during conventional propagation processes [20] and classical yeast cultivation techniques [21]. By contrast, with the assimilation technique, the resulting cells are solely in the logarithmic phase at the time the yeast is pitched (fig. 2). The three areas shown in figure 3 (p. 131) must be coordinated with one another to achieve this.

Figure 2: Changes in cell count during yeast cultivation as a function of temperature

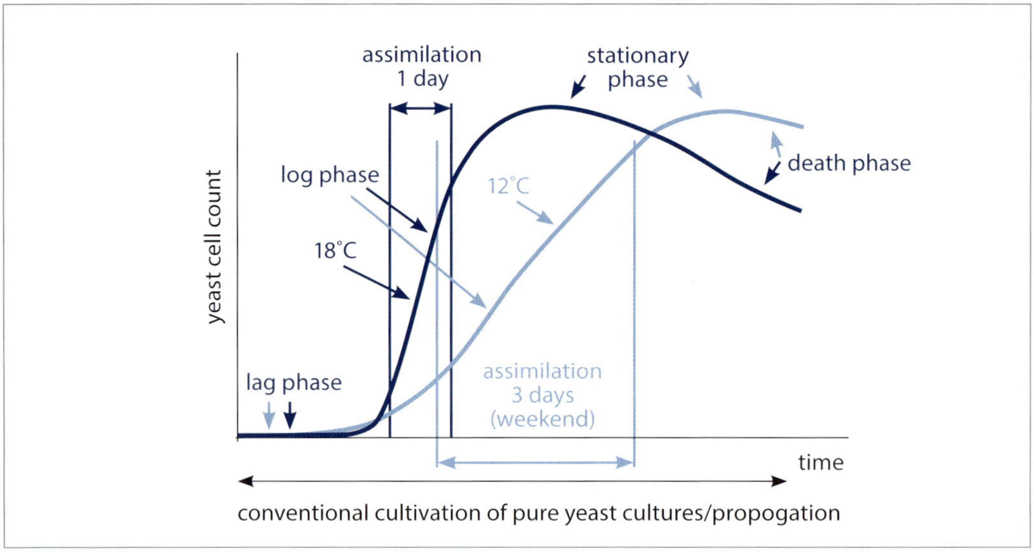

As these parameters can vary significantly from brewery to brewery, it is necessary to make adjustments to ensure that propagation occurs in an optimal manner, yielding a suspension of cells with a high vitality. A properly functioning assimilation system is characterized by a high rate of growth in conjunction with low extract consumption. The extract concentration should drop by 3–3.5 % w/w with the cell count increasing from 10 million cells/ml to 100 million cells/ml. It is important to ensure that foam formation is kept to a minimum, as this interferes with the propagation process and limits the function of propagation equipment.

However, foam formation can never be completely eliminated because intensive mixing and aeration of the medium are required. Only under these conditions can a sufficient supply of nutrients and oxygen be guaranteed for the yeast. Thus, a compromise must be found between ideal nutrition for the yeast culture and the lowest possible amount of foam formation. The Crabtree effect is an essential factor when determining the intensity of aeration required. Various metabolic pathways are inhibited when the nutrient medium contains high concentrations of sugar. This effect can already be observed at very low concentrations of sugar (> 0.1 g sugar/l) [22].

Figure 3: Three important areas that must be coordinated for optimal propagation to occur

Vessel geometry (assimilator)	**Technology** (process integration)	**Growth requirements of the yeast** (assimilator)
• tank • mixing conditions • pump capacity • aeration system • pipe diameter • inlet angle • flow rate at inlet	• brewing schedule • volume of cast-out wort • pitching method • fermentation temperature • fermentation method • yeast crop	• wort characteristics • temperature • mixing conditions • aeration • pressure • time

This means that aerobic growth and reproduction does not take place to the full extent possible, despite having well-aerated wort. Based on this information, in theory, the volume of oxygen required for yeast propagation decreases significantly. Thus, very intense aeration is not absolutely essential [23].

Foam formation can be kept to a minimum through the installation of a recirculation loop with a sufficiently large pipe diameter. This allows adjustment of the flow rate (≈ 1 m/s), and carefully coordinated aeration and/or recirculation sequences can be employed as well. Depending on pipe design, ideal flow rates can also be consistently achieved with the help of a variable frequency pump. This prevents any incidence of shearing forces brought about by incorrectly designed piping. Shearing or axial distension forces may cause cells to become deformed or elongated. A variable frequency pump can also be used to control the volume of the yeast added to the system or any solutions required for cleaning. Shearing forces may also arise from changes in pipe diameter or any equipment with sharp edges on its inner surfaces. This is often encountered with aeration nozzles and static mixers.

When aerating the medium during recirculation, the aerated yeast suspension should flow into the propagation system at a so-called "mixing angle" of approximately 23° relative to the radius (fig. 4). This keeps the yeast in suspension and allows the contents of the tank to be supplied with oxygen. By contrast, a tangential feed (angle of the whirlpool inlet) does not thoroughly mix the contents of the tank. If this is the case, then only the yeast cells located on the periphery of the tank are supplied with sterile air, preventing the rest of the cells in suspension from receiving adequate amounts of oxygen [24].

Figure 4: Vessel inlet angles

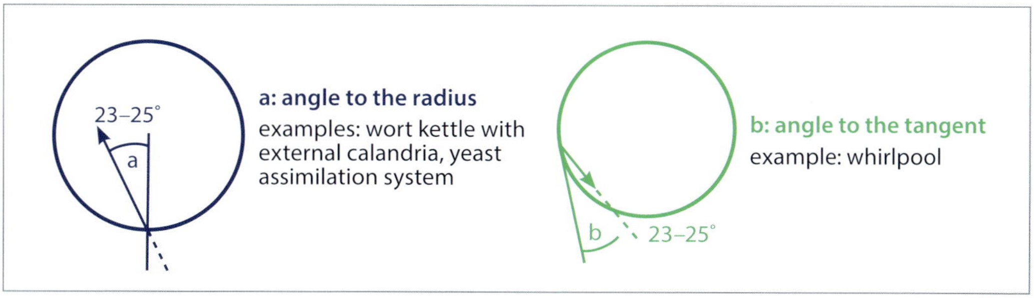

2.2.2 PROCEDURES FOR YEAST ASSIMILATION

As with other yeast propagation techniques, assimilation begins with an inoculum from a pure yeast culture that has been grown up in the brewery laboratory. Subsequently, cast-out wort from the brewery (without further sterilization) is added to the yeast, bringing the yeast cell count to 10–15 million cells/ml. Depending on the propagation temperature (approx. 12–18 °C for bottom-fermenting yeast or 20–25 °C for top-fermenting yeast), the suspension can reach a concentration of about 100 million cells/ml after 24 hours. Higher cell counts (up to 200 million cells/ml) are not desired, as once such high cell counts have been reached, the yeast cells are no longer in the log phase of growth and are beginning to transition into the stationary phase. This leads to a decline in the propagation rate and a lower yield factor [23]. The yeast suspension, with a cell count of approximately 100 million cells/ml, can now be used for pitching. A small volume is left in the assimilation tank and is topped up with wort again. This "repeated fed-batch" method, as it is known, keeps the yeast suspension in the log phase indefinitely. With this method, a steady supply of fresh assimilation yeast for pitching can be produced within short intervals [25].

The yeast assimilation process can be regulated by means of temperature, apparent extract concentration and cell count. In wort brewed to an original gravity of 12.0 % w/w, if the extract values (E_a, apparent extract) decrease to below 7.0 % w/w (at a cell count significantly lower than 100 million cells/ml), the air being venting from the system is noticeably saturated with CO_2. This is an indication that the amount of air supplied per cell is insufficient (fig. 5a). However, this may also result from improper mixing. In such cases, a portion of the population begins to ferment the wort, causing the vitality of the population as a whole to decline. Foam starts to accumulate as a result of fermentation. Thus, additional aeration is problematic due to the risk of excessive foaming. Figure 5b shows how propagation progresses in a well-designed yeast assimilation system.

Figure 5a: Yeast assimilation with suboptimal settings

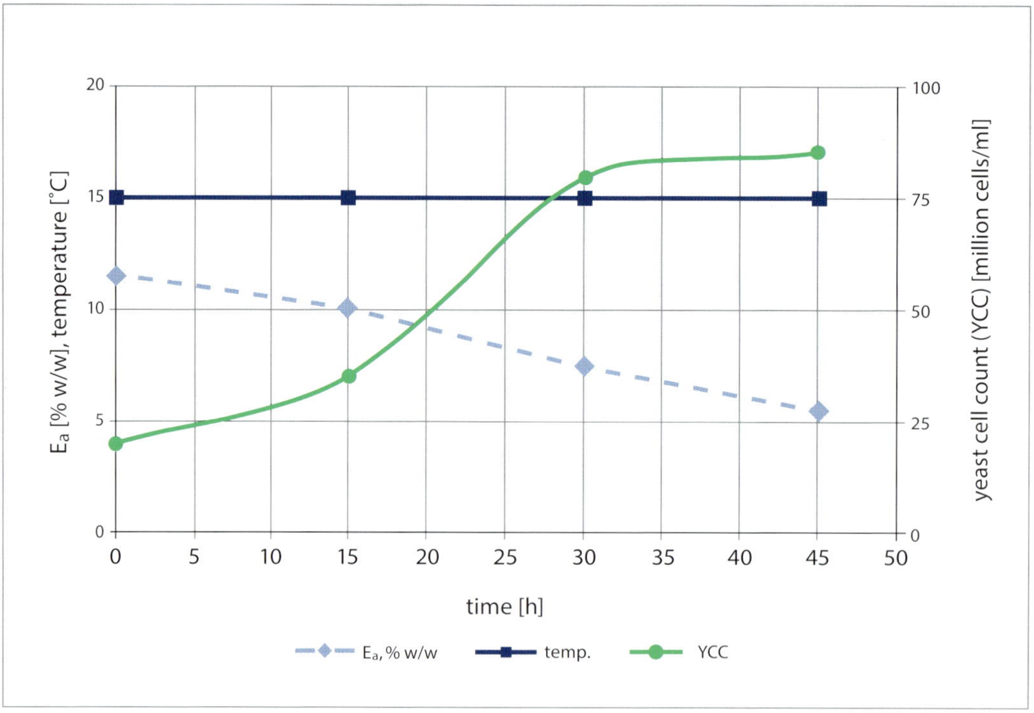

Figure 5b: Yeast assimilation with the proper settings

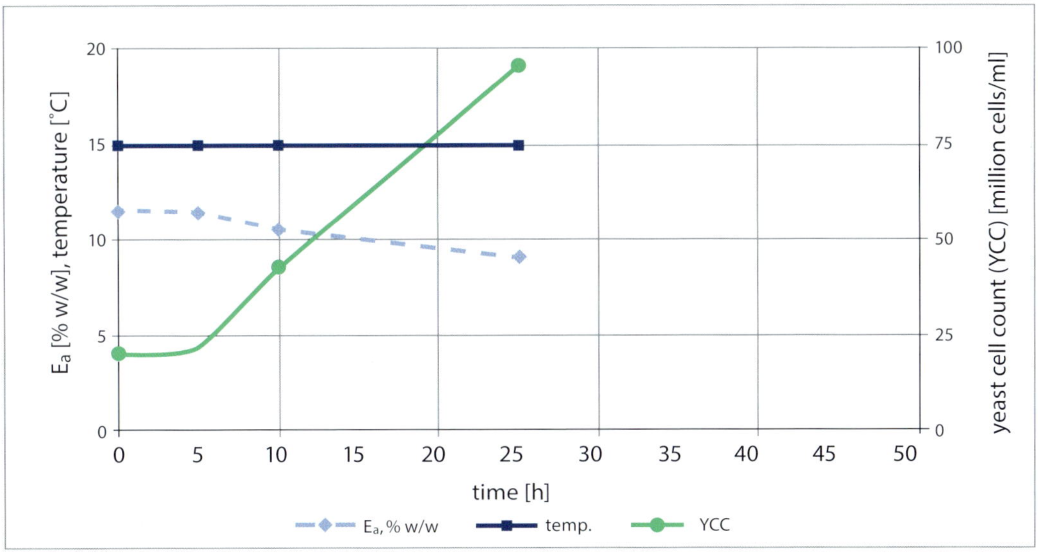

Short interruptions in production (breaks in the brewing schedule, weekends, etc.) can be compensated by extending the log phase of growth to ensure that an adequate volume of vital cells is available at the required time. This can be achieved with a lower initial cell count or by reducing the temperature. If longer pauses are necessary (breaks in the brewing schedule), the yeast culture must be stabilized no later than the end of the log phase. This involves ceasing aeration of the medium and simultaneously cooling to 0–3 °C. Under these conditions, a period of up to two weeks can be bridged without suffering any significant loss in activity.

2.2.3 DESIGNING A YEAST ASSIMILATION SYSTEM

Prior to planning and designing the yeast cellar, the brewery must develop a clear concept for yeast management. Principally speaking, two options exist: either the exclusive use of assimilation yeast (especially for top-fermenting beers) or a combination of assimilation yeast and cropped yeast. In the latter case, a mixture at a ratio (with regard to cell count or yeast concentration) of about 20–30 % assimilation yeast and 70–80 % cropped yeast has proven to be favorable. High levels of viability and vitality can also be seen in the cropped yeast thanks to continuous replenishment of the population through introduction of young yeast cells. Section 2.5 covers the most important aspects of handling cropped yeast.

Aside from economic aspects (e.g. lower investment costs), utilizing a combination of assimilation and cropped yeast is also beneficial for technological reasons, such as accelerating the initial fermentation phase, producing beer of consistently high quality (sensory attributes, foam) and obtaining higher concentrations of SO_2 (flavor stability).

Once a specific method has been devised, the operational framework and the yeast requirements need to be coordinated. The assimilation tanks should then be constructed accordingly (fermentation process, headspace, the volume of yeast retained in the tank for new inoculation, etc.) [26].

Table 1: Plans for design and implementation of an assimilation system are based on a brewery that produces only lager beer.

Operating conditions:

batch volume: 250 hl number of batches/tank: 3

fermentor volume: 750 hl batch interval: 4 h

composition of the yeast mixture: 30 % assimilation yeast + 70 % cropped yeast

cell count in the assimilation system: 100 million cells/ml after 24 h

pitching yeast cell count: 12 million cells/ml

proportion of yeast remaining in the assimilation tank: 10 %

Calculating the dimensions of the assimilation tank:

required volume of assimilation yeast: 12 million cells/ml * 0.30 = 3.6 million cells/ml

assimilation volume: $\frac{750\ \text{hl} * 3.6\ \text{million/ml}}{100\ \text{million/ml}} = 27\ \text{hl}$

headspace: 30 %

Total volume (+ volume in tank bottom): 27 hl ÷ 0,9 = 30 hl

required tank volume:

headspace of 40 %: 30 hl ÷ 0.6 = 50 hl

headspace of 50 %: 30 hl ÷ 0.5 = 60 hl

number of tanks required: 2

An assimilation cycle is required every 12 hours (3 batches/tank x 4 h brewing interval; the entire volume of yeast is always pitched with the first batch of wort). Since an assimilation cycle takes 24 h, two assimilation tanks are necessary to supply each tank with 30 % assimilation yeast.

It is important that the assimilation system is amply dimensioned to offer the flexibility needed in unexpected situations (e.g. a change in the brewing schedule, lower cast-out wort volume, etc.). Once the volume of yeast and the tank sizes have been determined, auxiliary equipment for the assimilation system can be selected, such as the type of aeration (internal aeration, aeration during recirculation, aeration device), diameter of the pipes (for aeration and recirculation), pump design, etc. Within this context, it is also necessary to decide on the degree of automation (e.g. regulation of the air supply) and the connections to existing systems (lines for wort removal, yeast addition or CIP).

Figure 6 shows the elements of an assimilation system along with the peripheral (required/optional) devices.

Figure 6: Schematic of a yeast assimilation system

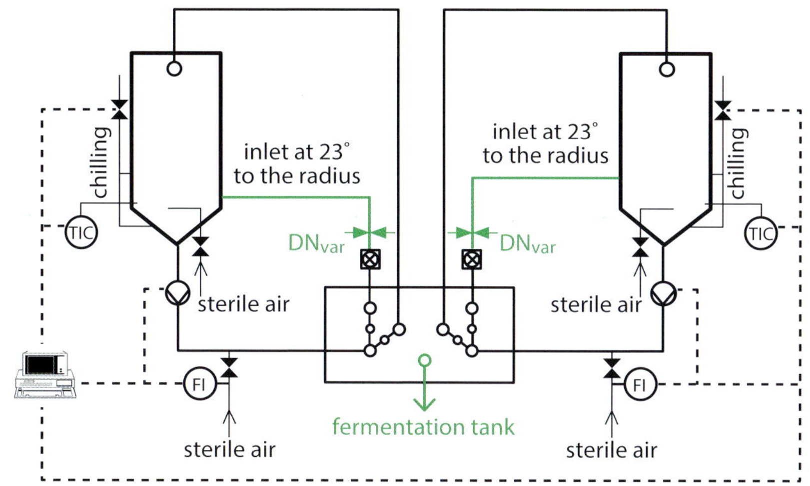

TIC = temperature indicator controller
DN_{var} = pipe diameter, variable
FI = flow indicator

IDM (inductive flowmeters) or level transmitters (which measure differential pressure) are suitable for regulating volume and can also be used to transfer wort to the system or dose the correct amount of yeast. By utilizing a variable frequency pump, the rate of recirculation can be easily adjusted to suit the respective conditions (e.g. flow conditions, flow rates, etc.). Excessive foaming or an inadequate supply of dissolved oxygen can be avoided by adjusting the level of aeration to the growth rate of the yeast population. In addition, the extract concentration E_a should not be allowed to fall below 7.0 during the assimilation process.

An assimilation system that has been specially adjusted to brewery operations will function reliably and provide flawless pitching yeast. Utilizing this yeast brings with it numerous advantages for the downstream processes as well as for the beer quality. The consistently high level of yeast vitality helps reduce the length of fermentation, and the pH drops more rapidly, lowering the overall pH downstream in the brewing process (fig. 7).

Figure 7: The change in pH during primary fermentation conducted with different cell counts and variations in the physiological condition of the yeast population

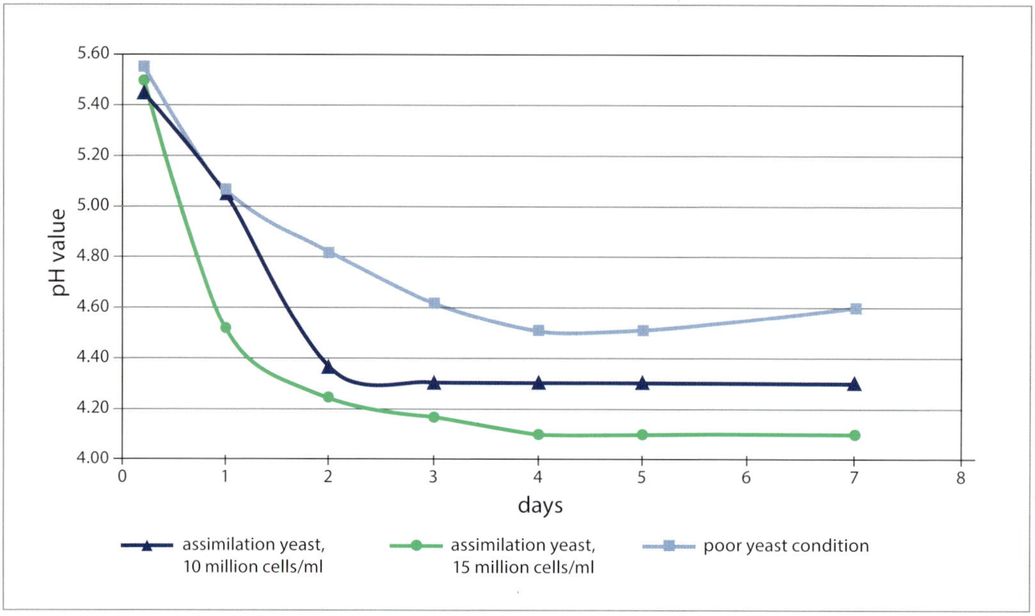

The concentrations of diacetyl and other undesirable fermentation by-products are lower when using assimilation yeast. There is also less of a risk of contamination with beer spoilers due to the high level of yeast vitality. The beers are characterized by a consistent and particularly high level of quality (sensory attributes, stability).

The combination of assimilation yeast and cropped yeast (mostly comprised of cells used for the first time), for example, at a ratio of 20:80 (based on cell count), reduces investment costs without negatively impacting quality. However, this does presuppose that the cropped yeast is handled properly (see section 2.5).

2.3 YEAST PITCHING TECHNOLOGY AND SO_2 FORMATION

Breweries carry out fermentation in vastly different ways. Fermentation is impacted by brewing equipment and techniques as well as the individual goals pursued by each brewery. Within this context, possible goals may constitute achieving a particular aroma profile or shorter fermentation times.

The myriad possibilities available will not be discussed in detail here. Instead, several important aspects will be reviewed.

Measures taken to influence the fermentation process should strive to fulfill the following objectives:

- consistent fermentation times
- more rapid extract conversion and pH drop
- desired (higher) degree of attenuation
- consistent beer quality
- high level of flavor stability
- adherence to reference values (extract, pH, diacetyl, and others)

The most influential factors are wort composition, temperature, cell count at pitching and wort aeration. Since wort composition is determined by the raw materials and brewhouse operations, the other parameters – individually and in combination – influence fermentation. Certain substances are important for the yeast and positively impact fermentation. As such, these should be present in the wort in sufficient concentrations (table 2). The value for free amino nitrogen (FAN) is commonly expressed as a percentage of the total nitrogen (21–22 %, or 23–24 % when using adjuncts).

Table 2: Important substances for yeast

Substance	Concentration
calcium	10–20 mg/l
iron	generally positive; extremely high concentrations are toxic
FAN	20–24 % of the total nitrogen content
magnesium	> 40 mg/l
zinc	0.10–0.15 mg/l

Figure 8a: The influence of fermentation temperature and yeast vitality on the concentration of acetaldehyde in the finished beer

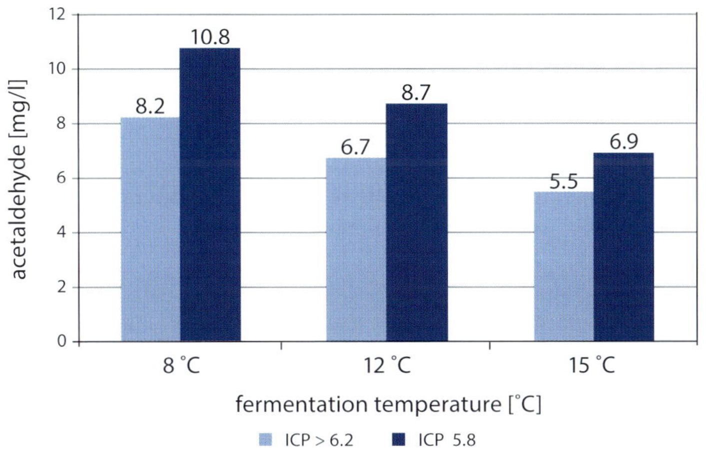

Figure 8b: The influence of extract concentration and yeast vitality on the concentration of acetaldehyde in the finished beer

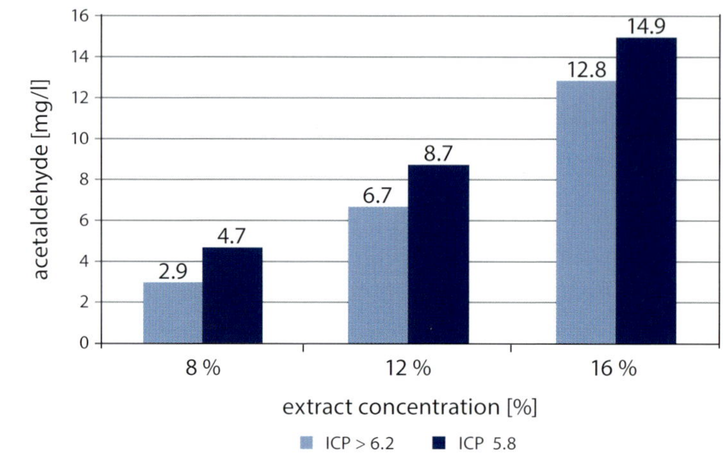

The individual yeast strain has a significant influence on the fermentation process and on the aroma profile. Yeast strains can vary dramatically in their requirements (e.g. the demand for nutrients and oxygen) as well as the aroma compounds they form. The following information pertains to the yeast strains most commonly used in breweries.

Potential variation with regard to the pitching parameters is limited, if only one batch is brewed and the yeast is pitched with the wort in one vessel (horizontal or vertical). The temperature directly determines the rate of the initial stages of fermentation (and those in the entire fermentation process) as well as the formation of fermentation by-products. The relationship between higher fermentation temperatures and shorter fermentation periods is attributable to the fact that the temperature optimum for yeast is 25 to 30 °C. However, at higher temperatures, more fermentation by-products are produced which are not necessarily desirable, especially in bottom-fermented beers.

In figures 8a and 8b, the key factors influencing the formation of fermentation by-products are shown using acetaldehyde as an example. This also includes the influence of fluctuations in yeast vitality (measured as the ICP value). However, according to the standard criteria applied in practice, this does not necessarily mean that the yeast is unsuitable, since only 3–7 % of the cells were found to be dead (methylene blue test) in these particular trials.

The volume of yeast pitched also has an effect on the fermentation rate and the amount of fermentation by-products formed. When a larger volume of yeast is pitched, fewer new yeast cells arise because there is more competition for nutrients and less oxygen available per cell. Pitching a smaller volume of yeast promotes yeast reproduction.

The formation of esters is also linked to yeast reproduction. In the literature, reports indicate a correlation between the formation of esters and acetyl-S CoA availability. Acetyl-S CoA is essential for the synthesis of fatty acids and is therefore relevant for yeast proliferation. If reproduction is inhibited, then acetyl-S CoA is available for the production of esters [27, 28]. Unfortunately, this explanation does not adequately describe the complex relationships found in day-to-day practice [29].

If aeration is performed in a conventional manner, with sterile air, aerating to the point of saturation is recommended (8–10 mg O2/l, depending on the temperature of the wort) for good yeast propagation and a rapid rate of fermentation. Alternatively, much greater oxygen solubility can be achieved (> 20 mg O2/l) by aeration with pure oxygen. However, this leads to excessive yeast growth, causing changes in the aroma profile and lowering fermentation efficiency. If oxygen concentrations are too high, they can have a toxic effect. A large proportion of the extract is utilized to produce new biomass and is therefore not available for alcoholic fermentation. Aside from this, flavor stability may also be affected.

Additionally, wort aeration plays a significant role in the formation of sulfur dioxide (SO_2) by yeast. SO_2 is a substance with a strong antioxidant effect, which positively impacts the flavor stability of the beer (cf. Flavor Stability). One should be aware that in some countries, concentrations of SO_2 above a certain level must be declared on the label (e.g. concentrations above 10 mg/l in Germany). As the formation of SO_2 depends heavily on the individual yeast strain (fig. 9), the conditions and operations specific to each brewery must be reviewed.

Figure 9: Production of SO_2 by two different yeast strains during primary fermentation of a batch of wort without subsequent aeration; the cell count provided represents the total number of cells and not only those in suspension (after 2 days: methionine entirely consumed; after 3 days: air entirely consumed; after 4 days: nutrients and substances required for growth entirely consumed)

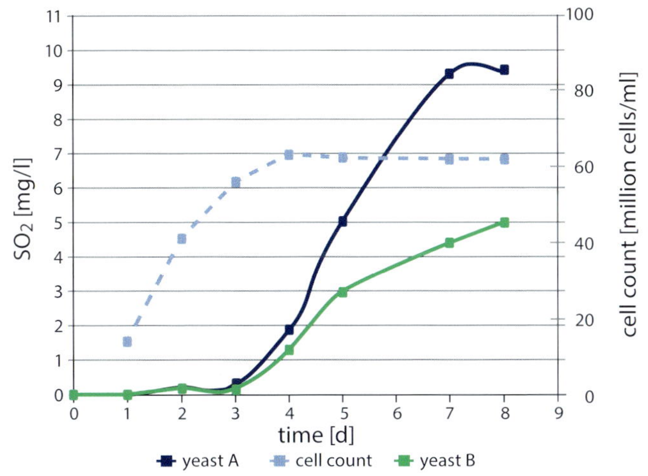

While bottom-fermenting yeast (*Sacch. pastorianus* var. *carlbergensis*) can produce more than 20 mg/l of SO_2 under certain conditions, top-fermenting yeast strains (*Saccharomyces cerevisiae*) only generate very low levels of SO_2 (< 2 mg/l).

However, even 5-7 mg/l of SO_2 can substantially improve flavor stability. The parameters influencing SO_2 formation must be taken into account in order to achieve these values and maintain them over the course of all fermentations. Principally, the formation of SO_2 increases dramatically when cultivation conditions become unfavorable. Conditions such as minimal aeration, high original gravity and low levels of yeast vitality will lead to higher SO_2 values. The volume of yeast pitched is also a significant factor, because a high level of extract per cell as well as less oxygen per cell promotes the production of SO_2. The interplay between these parameters is expressed in figure 10.

Figure 10: Factors influencing SO_2 formation

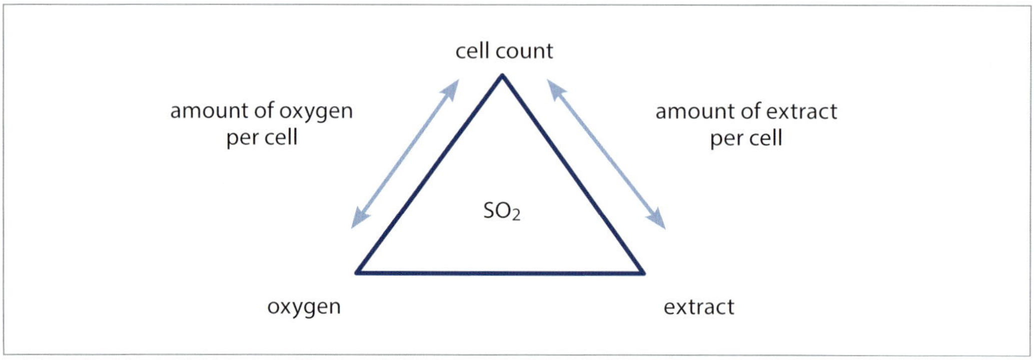

In practice, a fermentation tank is usually filled with a combination of several batches of wort over time, which creates a more complex situation from the very beginning. Pitching and aeration are able to be carried out at various times and in different combinations.

Ideally, the yeast should be pitched with the first batch of wort. This ensures that all of the yeast cells are exposed to the same conditions with regard to nutrient availability and oxygen over the course of the pitching process. Yeast added at a later time is at a great disadvantage in the competition for nutrients compared to the yeast already in the wort which has become quite active. This disadvantage causes yeast cells added at a later point to suffer from reduced vitality and to exhibit signs of premature aging [30, 31].

The wort should be intensively aerated in order to ensure efficient fermentation. Normally, when pitching the yeast with the first batch of wort and then adding subsequent batches to the tank, normally all batches should be aerated to the same degree. This allows for a large amount of cell proliferation all the way up until the last batch of wort. After exhausting the supply of oxygen, the yeast begins to ferment the wort. If the intervals in wort production are too long, the yeast is constantly switching between aerobic and anaerobic metabolism, leading to delays in fermentation and unchecked formation of fermentation by-products.

Given the fact that the oxygen requirements and rate of oxygen uptake depend strongly on the strain of yeast, the ratios must be tailored to suit each particular situation.

Aside from a specific strain's ability to generate SO_2, the point at which a new aerated batch of wort is added also has a significant impact on SO_2 formation. Very high levels of SO_2 can only be achieved if the first batch of wort is aerated. By contrast, if subsequent batches are aerated and are added 2 to 6 hours later, the final SO_2 concentration will be especially low (fig. 11).

Figure 11: Relationship between SO_2 formation and the time at which a second aerated batch is added to the fermentation tank (total capacity: two batches)

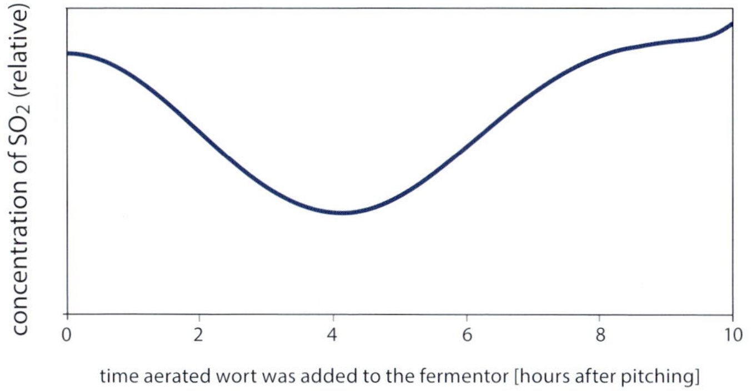

An alternative to aerating only the first batch of wort is to aerate all subsequent batches to a lesser degree, which will also lead to increased concentrations of SO_2. When determining the volume of pitching yeast and the intensity and time at which the wort is aerated, a compromise must be found between SO_2 formation and the fermentation rate, because all of the factors that positively influence fermentation tend to suppress the formation of SO_2.

The following strategy can be adopted to achieve consistently high levels of SO_2:

- pitch the entire volume of yeast with the first batch:
  spread the entire amount of aeration across the individual batches of wort
  - option 1: 1st batch: 50 %, 2nd batch: 50 %, 3rd and 4th batches: no aeration (for a brewing schedule producing one batch every 2 hours)
  - option 2: aerate all batches equally and moderately (a higher overall concentration of SO_2 will be achieved)
- divide the total volume of yeast among all of the batches (only do so if pitching tanks are employed): pitch each batch separately with yeast and transfer to a cylindroconical fermentation tank three hours later without any additional aeration.
  Since all of the methionine is consumed within 3 hours, SO_2 will continue to be formed without interruption.

Figures 12a and 12b depict how the amount of yeast pitched, aeration and yeast vitality influence the formation of SO_2.

Figure 12a: The influence of the amount of yeast pitched and the yeast vitality on SO_2 formation by the yeast strain Hebru

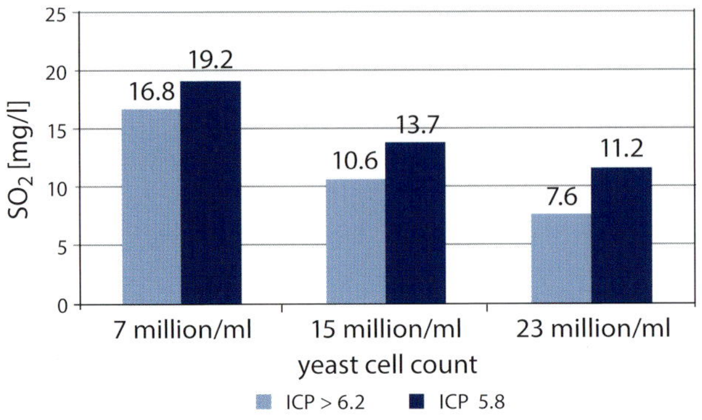

Figure 12b: The influence of aeration and the yeast vitality on SO_2 formation by the yeast strain Hebru

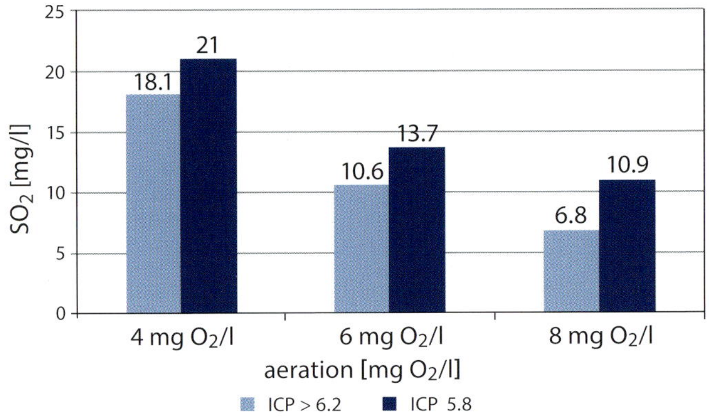

Unfortunately, yeast populations propagated under aerobic conditions form very little SO_2 due to their excellent physiological condition. This also applies to assimilation yeast that has been cultivated under optimal conditions. In order to achieve consistently high levels of SO_2 when using assimilation yeast (for the purpose of attaining good flavor stability), a blend of harvested yeast and assimilation yeast is recommended. The assimilation yeast portion guarantees an efficient fermentation, whereas the harvested yeast ensures the generation of adequate SO_2 concentrations. The SO_2 content as well as the length of fermentation can be modified by varying aeration levels in combination with different ratios of assimilation and harvested yeast.

Vital assimilation yeast tolerates low levels of aeration without affecting the length of fermentation. If desired, higher levels of SO_2 can also be achieved through the exclusive use of highly active assimilation yeast in conjunction with reduced aeration or lower pitching yeast rates. Despite the less favorable growth conditions at the beginning, fermentation progresses normally (must be coordinated with the length of time required for fermentation).

2.4 FERMENTATION

Fermentation conditions are primarily determined by yeast pitching techniques, which in turn, heavily influence the rate of fermentation and the development of aroma compounds in beer. The wort composition (including the original gravity) and the yeast strain should be viewed as set factors, whereas the pitching yeast (cell count, viability and vitality) and the amount of aeration are factors to be determined prior to pitching. The remaining parameters are temperature and pressure.

Although the yeast is pitched at a defined temperature initially, it may be adjusted over the course of the fermentation process. As noted previously, the optimum temperature for most yeast strains is between 25 and 30 °C. Therefore, higher temperatures will speed up the fermentation process (fig. 13).

Figure 13: Fermentation of 12 % wort with the yeast strain Hebru in different physiological conditions: the time required for fermentation until $E_a = 4$ % at different temperatures

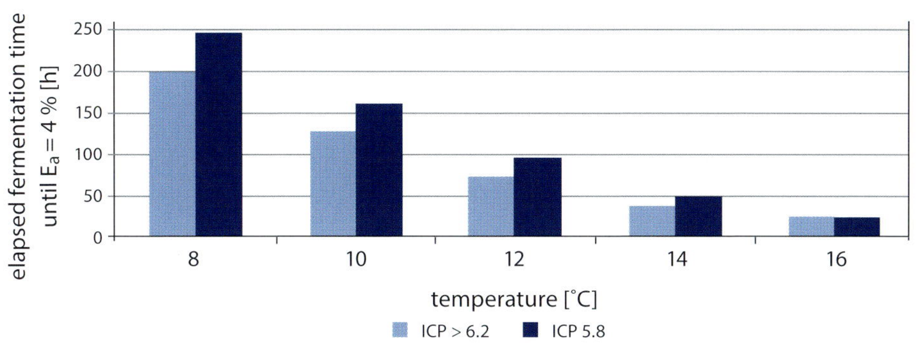

Despite the fact that the optimum temperature for yeast lies above 20 °C, there are several reasons for avoiding high fermentation temperatures:

- greater foam formation
- more headspace in tanks is required
- higher pressures
- the fermentation process is difficult to control

Particularly high fermentation temperatures summarily lead to an increase in higher alcohols and esters (particularly 2-phenylethanol), although this does not necessarily apply to individual compounds in these groups. An increase in certain fermentation by-products can either be reduced or prevented through the application of pressure during fermentation.

Regarding vicinal diketones, a higher fermentation temperature leads to an earlier and higher maximum; however, the degradation process also progresses more rapidly.

Higher temperatures have a particularly negative effect on the yeast located at the bottom of the tank in the cone. Because these yeast cells no longer have access to nutrients, the cells age and rapidly undergo autolysis at high temperatures. As temperatures in the yeast sediment layer are far higher than those in the fermenting green beer, even cooling the cone cannot stop the physiological condition of the cells from deteriorating. This is because aggregates of yeast cells form a well-insulated layer of sediment, making it nearly impossible to adequately cool the yeast in the center of the cone. For this reason, it is advisable to crop the yeast early on in the process.

Lastly, pressure also influences fermentation, although there is a distinction made among three kinds of pressure. One is the osmotic pressure exerted by the extract concentration of the wort (original gravity). The second is the hydrostatic pressure created by the geometry of the vessel (column of liquid), while the third is the partial pressure of the CO_2 applied to the fermenting medium. Principally, all three types of pressure affect fermentation and the production of metabolites.

The osmotic pressure affects yeast propagation as well as the production of metabolites. As the extract content increases, the concentration of metabolites such as higher alcohols, esters or SO_2 increases linearly, or even disproportionately, in some cases.

The hydrostatic pressure is primarily determined by the level of liquid in the fermentation tank. Various tank shapes and heights result in different aroma profiles, although the factors of convection and CO_2 accumulation also play a role. By comparison, the hydrostatic pressure alone does not have a noticeable effect on the yeast cells. For example, it has been shown that when high pressure is applied to food, a pressure in excess of 800 bar is necessary to reliably kill yeast and other microorganisms.

The parameter actually controlling pressure is the CO_2 content of the fermenting medium. The concentration of carbon dioxide influences the course of fermentation and the production of metabolites [32, 33].

However, application of overpressure should be avoided at the beginning of fermentation, as it hinders the proliferation of yeast cells, offsetting the time gained from fermenting at higher temperatures. Initially, fermentation should be allowed to commence in a pressure-free environment or under slight overpressure. At an apparent attenuation of approximately 25 %, a small amount of pressure (0.3 bar) can be applied. Once the level of attenuation has reached about 50 %, the pressure can be adjusted based on the following criteria: temperature ÷ 10 – tank height ÷ 20 (e.g. given a temperature of 18 °C and a tank height of 14 m → the overpressure should be 1.1 bar). This pressure is maintained until the final attenuation has been reached, at which point it is lowered to 0.6–0.7 bar, in order to fine-tune the CO_2 content to the target concentration. A high fermentation temperature is maintained until the diacetyl concentration reaches 0.12 to 0.15 mg/l. After that, it can be cooled to a suitable temperature for lagering [34, 35]. The corresponding fermentation diagram is shown in figures 14a and 14b.

Figure 14a: A graph of fermentation conducted under pressure at 13 °C. The arrows indicate the points at which either the yeast was harvested or the yeast sediment was purged from the vessel.

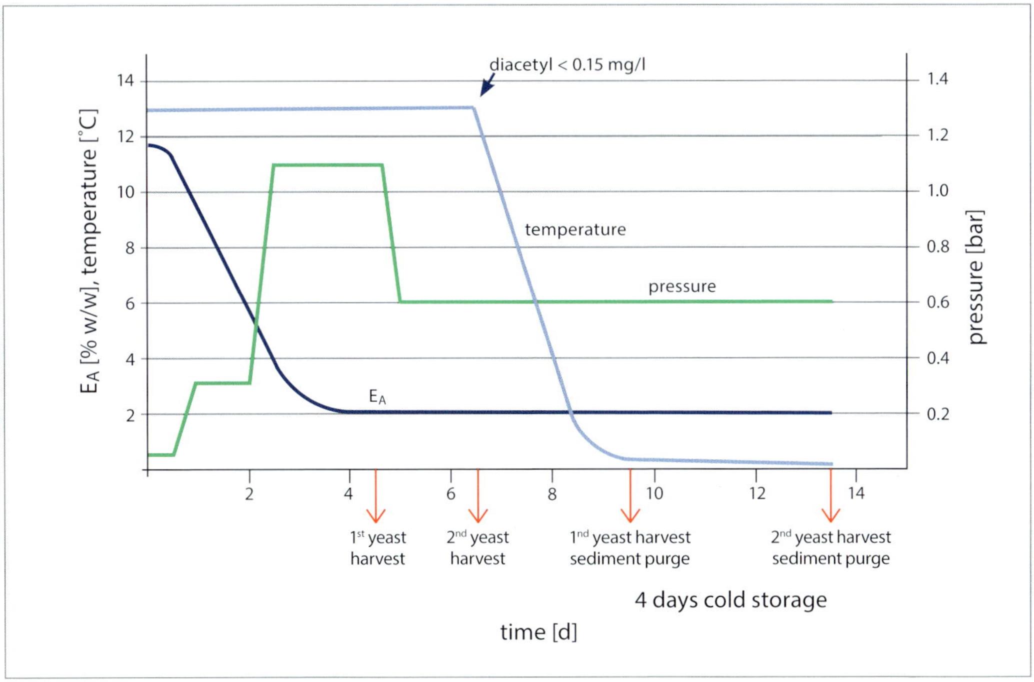

Figure 14b: A graph of fermentation with assimilation yeast (12 Mio/ml) or a mixture of assimilation yeast and cropped yeast (e.g. 30 % : 70 %).

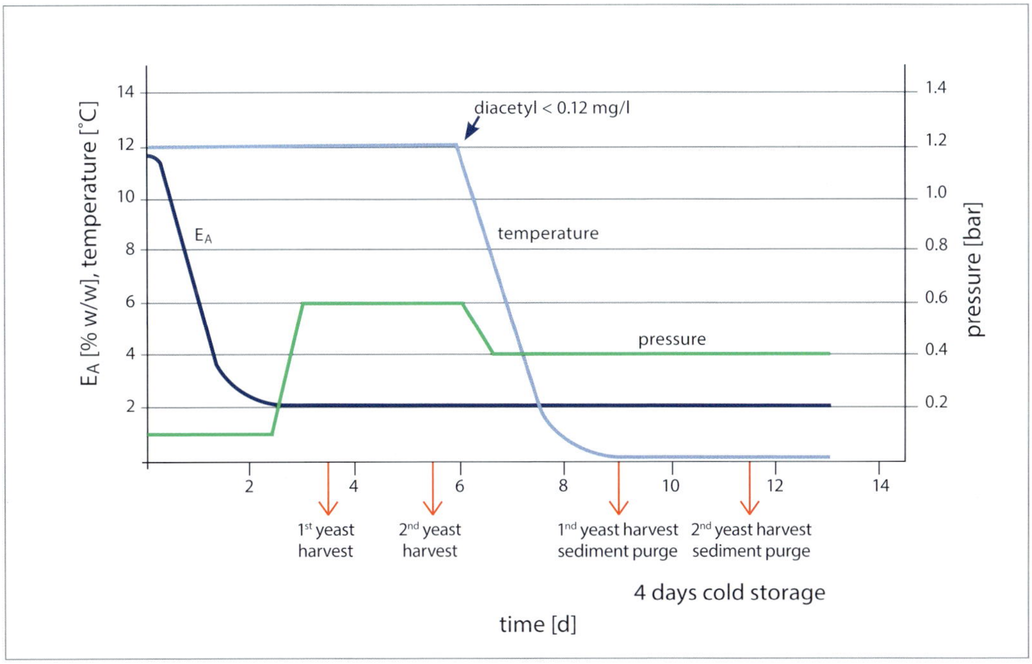

Early yeast cropping is very important when fermenting under pressure. In this case, the yeast is not only exposed to higher temperatures but also to additional pressure. Cropping the yeast twice and proper handling can compensate for the additional stress on the yeast. If pitching yeast possesses an extremely high level of vitality, there is no significant loss of activity in later generations. This was demonstrated in the research results shown in figure 15. After fermentation under pressure, the physiological condition (ICP value) of the yeast and that of later generations was measured at 6.2 (the first time the yeast is pitched = fermentation with the pure culture), which is considered to be very good.

2.5 PROPER HANDLING OF CROPPED YEAST

As described above, yeast that has dropped out of suspension by the end of primary fermentation no longer has access to nutrients and is subjected to high concentrations of CO_2 and ethanol, as well as greater pressure at higher temperatures. The latter two sources of stress are typically encountered in cylindroconical tanks. This yeast must be removed from the fermenting medium at the correct time to avoid negatively affecting the fermentation process or the cropped yeast. Cropping yeast multiple times, as described previously, is performed after the final degree of attenuation has been reached up to the time when the tank is cooled. Following this procedure guarantees cropped yeast of a consistently good quality.

Figure 15: ICP values measured in yeast collected from a fermentation tank (3000 hl) under pressure, after utilization in the first, second and third fermentation at 14 °C

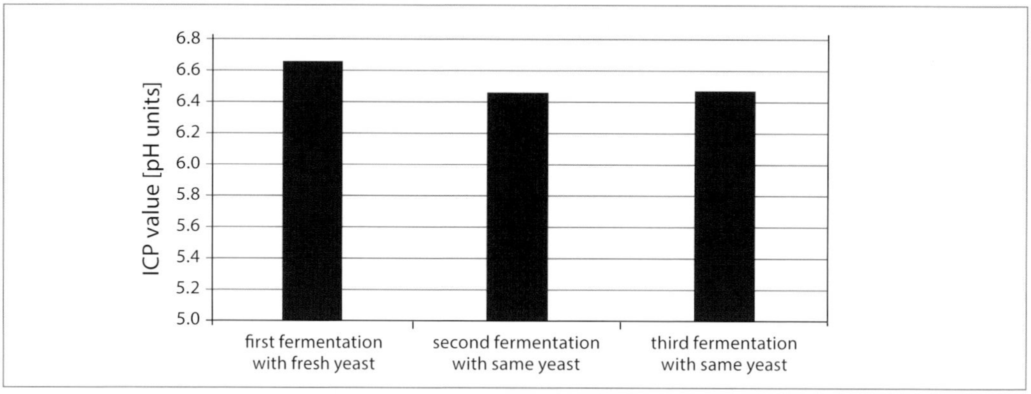

Additional aspects must be considered when cropping yeast. Gentle pressure release is essential so that carbon dioxide and alcohol along with certain undesirable fermentation by-products can be excreted by the yeast cells. The yeast may not, under any circumstance, be actively aerated at this time. Due to a lack of available sugar, the yeast would utilize its stored carbohydrates, thus weakening the cells. This, in turn, would lead to premature autolysis and result in an accumulation of medium chain fatty acids, proteinases and other undesirable substances in the beer. Moreover, any Gram-negative bacteria that may be present in the medium, particularly the ubiquitous acetic acid bacteria, are activated by aeration. This limits filterability (cf. Filterability – Issues with Turbidity) and can lead to fermentation problems. In any case, the importance of this issue should not be underestimated, as acetic acid bacteria (e. g. *Gluconobacter frateurii*) can grow at very low temperatures (as low as 5 °C), with material from autolyzed cells, ethanol and oxygen providing optimal nutrients and growth conditions. Naturally, yeast requires intensive aeration at pitching in order to induce metabolic activity.

Research has shown that when yeast is handled properly (fig. 16) it is possible to increase filter capacity/operating time by 40–50 %. Foam values (NIBEM) improve from 250 to 270, the FAN content is 10 %

lower and the decanoic acid content decreases by approximately 30 %, which contributes to a notable improvement in flavor stability.

If yeast is to be stored, it is immensely important to cool the suspension down to a temperature of 0–3 °C, otherwise the cells will consume their carbohydrate reserves. The loss of vitality at higher temperatures is shown in figure 1 (p. 129). When cropping yeast (fig. 16), there is evidence that the use of a yeast screen (CIP-compatible) can be beneficial. Aside from thinning out inactive yeast cells, it provides a type of passive aeration, enabling further liberation of CO_2 and other undesirable substances. An external yeast chiller should be employed to cool the cropped yeast, since tank cooling requires a substantial amount of time to bring the temperature down. Due to the consistency of the yeast, it is also impossible to achieve uniform cooling. Adding 20 to 50 % water (possibly already added prior to passing the yeast through the screen) also has a positive effect when storing yeast, particularly if it must be stored for more than 2 days. The added water gives the yeast a more fluid consistency, making it much easier to homogenize the suspension, to find the right temperature balance and to accurately determine the cell count. Furthermore, the concentration of negative metabolites is diluted, especially the ethanol content (up to 10 % by volume in dense yeast suspensions). Without question, prolonged exposure to such levels of ethanol has a negative impact. For this reason, storing yeast under water provides a certain advantage compared to storage under beer.

Figure 16: Handling and storage of cropped yeast (P_F = pressure during fermentation, P_0 = atmospheric pressure)

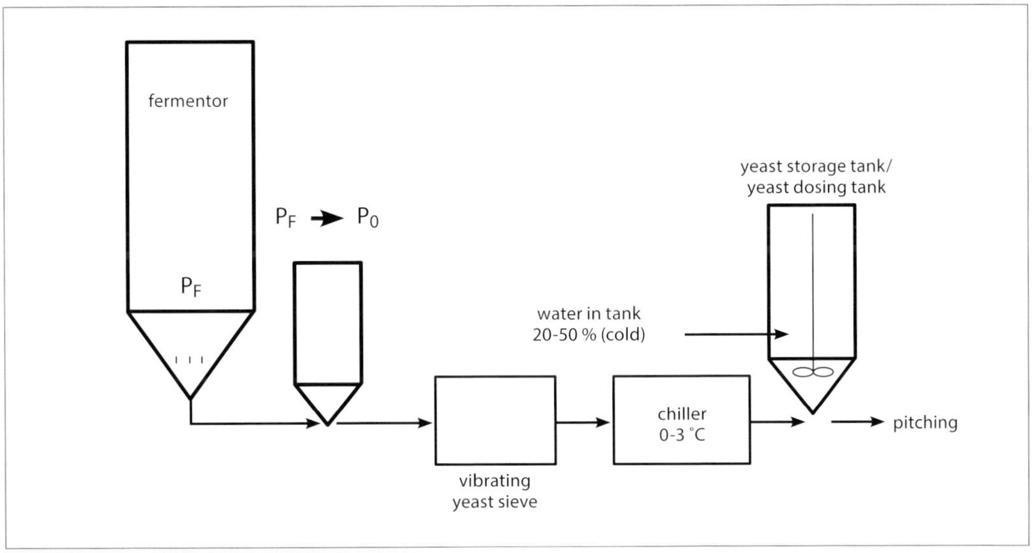

2.6 MATURATION AND LAGERING

Traditionally, the stages of fermentation are divided into primary fermentation, maturation and lagering. According to Narziss, the maturation process should be viewed as "a rounding out and refinement of the flavor, an improvement in the aroma as well as an enhancement of the beer's wholesome character" [35]. When using certain targeted fermentation processes, such as fermentation under pressure as described in section 2.4, the process of maturation is already integrated into primary fermentation. After reaching the final degree of attenuation, the temperature is maintained until the concentration of diacetyl drops below the flavor threshold value. The lagering phase, brought on by a subsequent reduction in temperature, primarily serves to clarify the product and thus improve the colloidal stability of the beer. To that end, a minimum lagering period of approximately one week, at –1 °C for example, should be carried out to ensure that the desired precipitation and sedimentation occurs. Sending the beer through a chiller during the transfer to the lager tank is beneficial. The beer can be rapidly chilled to the lagering temperature while also avoiding stratification.

It is likewise advantageous to use a green beer centrifuge to achieve a defined cell count (0.5–1 million cells/ml). This reduces the amount of autolyzed cell material and simplifies subsequent filtration. However, from time to time, a higher cell count can be worthwhile. For example, yeast cells adsorb very fine chill haze particles and protein/tannin complexes. When the cells sediment out, these substances are effectively removed from the beer. In certain years, problems with poor filterability emerge. In individual cases, the yeast cell count in the lager tank can be increased as a countermeasure when this occurs. In the production of traditional lager beer, the beer is transferred before all of the extract is fermented to allow additional physical and biochemical processes to take place. Aside from clarification and maturation of the flavor profile, the residual extract is fermented to achieve the desired pressure through generation of carbon dioxide. Thus, the temperature should not be too low, otherwise secondary fermentation will cease. In turn, this will prevent the desired degree of attenuation from being reached and the elimination of diacetyl will not be complete. The ideal temperature range throughout maturation and lagering is between +3 °C and –2 °C, though it is necessary to reduce the temperature gradually over the course of the process. Equally as relevant is the volume of yeast pitched and the physiological condition of the cells, as the transition is made from primary fermentation to maturation. If the cell count is too high, the residual extract is fermented too rapidly, i.e. before the actual maturation process has ended. Secondary fermentation will progress very slowly if there is not enough yeast or if the cells are in poor condition. This will then be carried over to an equally inadequate maturation (e.g. with high values for diacetyl and apparent extract) [35]. Insufficient degradation of diacetyl, even at high maturation temperatures, is usually due to a low cell count in suspension at the end of the fermentation process. The cells in suspension absorb diacetyl, which had earlier formed outside of the cell as 2-acetolactate, and further metabolize it (fig. 17). From a different perspective, a low cell count at pitching (e.g. 10 million cells/ml of yeast exhibiting a high vitality) facilitates a more rapid degradation of diacetyl than do yeast pitched at higher initial cell counts (such as 25 million cells/ml). More or less, the same applies to 2,3-pentanedione, although this compound has a much higher sensory threshold and therefore is less relevant for beer quality. Its formation is comparable to that of diacetyl from α-acetohydroxybutyrate.

A higher 2,3-pentanedione content combined with a high diacetyl content suggests inadequate maturation or very poor yeast quality. A high diacetyl content in conjunction with a low concentration of 2,3-pentandione usually indicates the existence of a *Pediococcus damnosus* (or less commonly, *Lactobacillus*) contamination.

In conventional brewing processes, the secondary fermentation/maturation phase can also be achieved through the addition of *kräusen*. The green beer is transferred with an extract contact of approximately 3–6 % above the final degree of attenuation and mixed with 6–10 % kräusen (apparent attenuation = 25–35 %, yeast cell count = 40×10^6). The young cells in the kräusen facilitate a more rapid maturation, yet the finished beer contains less FAN, C_6–C_{10} fatty acids and almost no yeast proteases. This results in enhanced foam characteristics and a higher flavor stability as well as better filterability.

Figure 17: The formation and degradation of diacetyl

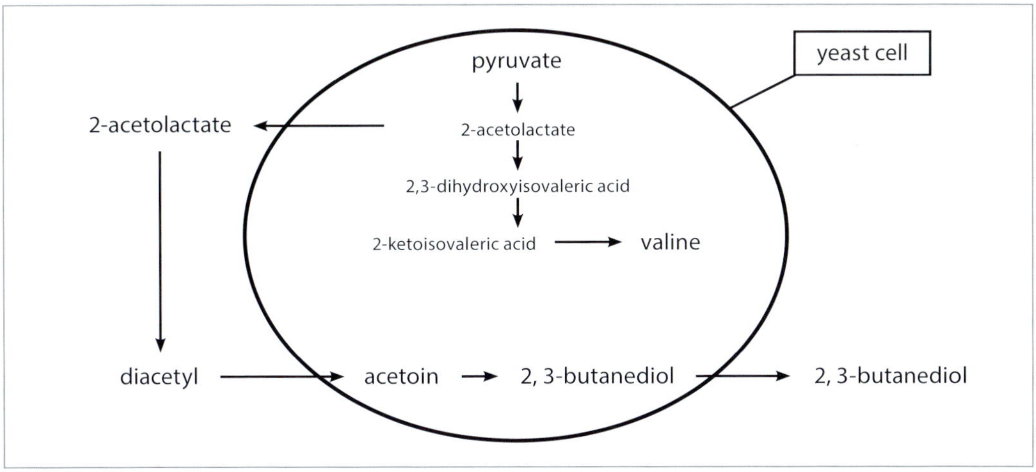

3 SUMMARY

Fermentation is a fundamental step in the brewing process. It not only determines the final character of the beer but also its beneficial qualities. The quality of the yeast plays a significant role during this phase. After all, less vital yeast can result in fermentation delays, unwanted turbidity, filtration problems, lingering bitterness as well as lower foam and flavor stability. As such, yeast propagation, cropping and storage are all of crucial importance.

In order to obtain good yeast from the outset, careful monitoring of yeast propagation is essential. When propagating yeast using the assimilation method, the growth conditions (e.g. temperature and wort composition), the specific aspects of an individual brewing operation (e.g. brewing schedule, volume of cast-out wort or fermentation process), as well as vessel geometry (tank with peripherals) should all be optimally attuned to one another. This will ensure that production yeast in the log phase of growth with the highest possible vitality is always available when required.

During fermentation, other factors aside from the physiological condition of the yeast, such as the volume of yeast pitched, the amount and timing of aeration, along with the temperature, also play a defining role in fermentation and in the formation of metabolic products. This means that the production of esters and higher alcohols, as well as SO_2 (important for the flavor stability of beer) can be influenced as needed.

After fermentation has ended, the yeast is recovered. In doing so, it is important to avoid any negative impact on the yeast. This is why it is essential to adhere to scheduled times for yeast cropping, proper yeast handling techniques, the duration of lagering and especially a suitable storage temperature for the yeast.

Yeast that is kept in good condition has a positive effect on every subsequent step of beer production. Therefore, it is also important to be able to evaluate the condition of the yeast. For this purpose, two independent parameters are determined – namely viability (proportion of dead yeast cells) and the vitality (physiological condition of the living yeast cells). Determining both parameters is a crucial prerequisite in targeted yeast management programs. In order to ensure a consistent and high level of quality, poorly performing yeast must be completely eliminated from the process.

4 OVERVIEW

Problem	Cause	Corrective/preventative measures
different proportions of dead cells in samples taken from the same batch of yeast and tested with the same staining method	the yeast was not mixed thoroughly; the yeast preparation was too dense (causing poor distribution of the stain); the number of cells counted is too low; the stain was left on the cells for an incorrect amount of time (too short or too long)	improve mixing conditions and shake sample well; prepare the microscope slide so that approx. 50 cells are the in the visible field; count out at least 200 cells; stain the cells with methylene blue for two to five minutes (this applies to many staining agents)
many cells appear light blue in color or are colorless after staining; a flash-pasteurized yeast suspension was stained with methylene blue	methylene blue is unsuitable for the evaluation of yeast cells killed by the application of heat	fluorescent dyes can be used as an alternative
only low yeast cell counts (< 100 million/ml) are achieved during assimilation or propagation despite high extract consumption	cells are not all in the log phase: the reasons are a lack of oxygen (less common) and/or poor mixing in the propagation vessel (the nutrients and oxygen are not uniformly distributed; a portion of the population is in the fermentation stage)	mix thoroughly and/or supply additional oxygen; increase the recirculation rate; test the pump and the agitator to determine if it is possible to thoroughly mix the medium
extremely long lag phase during yeast reproduction	initial yeast cell count is too low; the temperature is too low for yeast propagation; poor yeast quality from the outset	ensure an initial cell count of 10 to 20 million cells/ml; assimilation temperatures should range from 10 to 18 °C (weekends: 6–10 °C); use a fresh yeast culture
strong flocculation at the end of yeast reproduction (after ceasing recirculation or agitation); the yeast forms thick aggregates and quickly sediments out	final degree of attenuation is either reached or almost reached; yeast reproduction was allowed to progress too long or the yeast was not in the log phase; too much reproduction is unfavorable for the condition of the population	the drop in extract should not exceed 3–4 % with an increase in yeast cell count from 20 million/ml to 100 million /ml
deformed cells during yeast reproduction	shearing stress and cell distortion is created by unsuitable agitators, recirculation systems, aeration systems or equipment operation at incorrect settings	lower the speed of agitation or recirculation; examine the pump (run the pump at the operating point)

Problem	Cause	Corrective/preventative measures
high percentage of dead cells (> 5 %) at the end of yeast propagation	the propagation conditions were unfavorable or the propagation process was allowed to go too far (over 110 million cells/ml); the initial yeast culture was either in poor condition or not enough new cells were formed because the initial cell count was too high	optimize propagation conditions, e.g. at very high propagation temperatures, the shift from vital to dead cells occurs rapidly; reduce extract consumption to 3–4 %; cells that were dead in the culture prior to propagation will remain in the culture (improve the quality of the yeast used for propagation; reduce the initial cell count in order to carry fewer dead cells over into the propagation process)
very few yeast cells are produced during fermentation (the pitching volume is close to the volume of yeast harvested)	too much yeast was added and/or the yeast was in poor condition; although a large pitching volume results in more rapid fermentation, the condition of the yeast population also deteriorates more quickly, so that almost no reproduction occurs (the aroma profile is also different, since fermentation by-products are often dependent on yeast reproduction); yeast in poor condition intensifies these effects	improve the condition of the initial yeast culture used for propagation (introduce a fresh yeast culture); reduce the volume of yeast added; increase aeration
poor values for foam stability; harsh, bitter aftertaste	the yeast strain was unsuitable (deficient) or too much yeast was added, causing autolyzed cell material and proteinases to be released into the medium	improve the condition of the initial yeast culture used for propagation (introduce a fresh yeast culture); lower the volume of yeast added; increase aeration
stuck fermentation	not enough yeast was added (very few cells found in a sample collected from the fermenting wort); the level of aeration was insufficient (enough yeast present but almost no cell division after 24 hours); poor yeast condition; deficiencies in wort composition (zinc), microbial contamination, a combination of these factors	if problems are observed early enough (too little yeast pitched, poor yeast condition or insufficient aeration), they can be resolved by the addition of more yeast (must be of a high vitality) and intense aeration (recirculate if necessary)
variable fermentation performance, fluctuations in beer aroma and SO_2 content	lack of uniformity in the yeast addition, the amount of aeration and the condition of the yeast	improve yeast pitching by determining the cell count in the yeast tank and pitch the correct volume of yeast based on this information (the cell count is more accurate if the yeast suspension is diluted with H_2O in the yeast tank and homogenized); improve yeast condition by testing the viability and disposing of low quality batches of yeast

Problem	Cause	Corrective/preventative measures
high proportion of dead cells in the harvested yeast	poor condition of the yeast used initially; the yeast was harvested too late (early yeast cropping is particularly important with high fermentation temperatures and fermentation conducted under pressure)	use yeast in a better physiological condition or bring in a fresh yeast strain; ensure that the yeast is harvested at the proper time
a long period of time is required to reduce the temperature in the yeast tank to lagering temperature	the temperature is too high for harvested yeast and/or incomplete homogenization of the yeast suspension (in some cases, due to lack of cooling capacity)	homogenization can be improved by utilizing a better mixing system or agitator and by diluting the yeast with water; the yeast can be cooled more rapidly by adding chilled water; the optimal solution is to harvest yeast through a yeast chiller
fermentation is slow after a long period of storage, despite a low number of dead cells	when yeast is stored at low temperatures, even over several weeks, the proportion of dead cells does not increase substantially; however, the vitality begins to deteriorate after several days at very cold temperatures	avoid storing yeast for more than 4 days (in an emergency, more than a week) pre-propagate the yeast with aerated wort

5 REFERENCES

[1] Thiele, F.: Einfluss der Hefevitalität und der Gärparameter auf die Stoffwechselprodukte der Hefe und auf die Geschmacksstabilität. Dissertation, Technical University of Munich (Weihenstephan), (2006).

[2] Wackerbauer, K., Krämer, P., Toussaint, H. J.: Technologische Parameter der Esterbildung. Monatsschrift für Brauerei, 3 (1980), pp. 91–99.

[3] Eils, H.-G., Eidtmann, A., Back, W.: Maßnahmen zur Verbesserung der Hefetechnologie und Umsetzung in der Praxis. Brauwelt 140 (2000), no. 15, pp. 590–595.

[4] Heggart, H., Margaritis, A., Stewart, R. J., Pilkington, H., Sobczak, J., Russell, I.: Measurement of Brewing Yeast Viability and Vitality: A Review of Methods. Master Brewers Association of the Americas Technical Quarterly 37 (2000), pp. 409–430.

[5] Lentini, A.: A review of the various methods available for monitoring the physiological status of yeast. Ferment 6 (1993), pp. 321–327.

[6] Back, W.: Farbatlas und Handbuch der Getränkebiologie Teil II. Getränke-Fachverlag Hans Carl, Nürnberg 2000, pp. 193–194.

[7] Back, W.: Colour Atlas and Handbook of Beverage Biology. Fachverlag Hans Carl, Nuremberg 2005, pp. 6–7.

[8] Peladan, F., Leitz, R.: New method for the differential staining of dead and living cells of yeast and bacteria. Proceedings of the 23th European Brewery Convention Congress, Lisbon, 1991, pp. 481–487.

[9] Lloyed, D., Dinsdale, G.: From bright field to fluorescence and confocal microscopy, Brewing Yeast Fermentation Performance, K. A. Smart. Ed. Blackwell Science, Oxford, UK, 2000, pp. 3–9.

[10] Pickerell, A. T. W., Hwang, A., Axcell, B. C.: Impact of yeast handling procedures on beer flavour development during fermentation. Journal of American Society of Brewing Chemists 49 (1991), pp. 87–92.

[11] Quain, D. E.: Studies on the yeast physiology-impact on fermentation performance and product quality. Journal of Institute of Brewing 95 (1988), pp. 315–323.

[12] Cantrell, I. C., Anderson, R. G.: Yeast performance in production fermentations. Proceedings of the 19th European Brewery Convention Congress, London, 1983, pp. 481–488.

[13] O'Conner-Cox, E. S. C., Majara, M. M., Lodolo, E. J., Mochaba, F. M., Axcell, B.: The use of Yeast Glycogen and Trehalose Contents as Indicators for Process Optimisation. Ferment 7 (1996), pp. 321–328.

[14] Sall, J., Seipp, J. F., and Pringle, A. T.: Changes in Brewer's Yeast During Storage and the Effect of These Changes on Subsequent Fermentation Performance. Journal of American Society of Brewing Chemists. 46 (1988), pp. 23–25.

[15] Imai, T., Nakajima, I., Ohno, T.: Development of a New Method for Evaluation of Yeast Vitality by Measuring Intracellular pH. Journal of American Society of Brewing Chemists 52 (1994), no. 1, pp. 5–8.

[16] Back, W., Imai, T., Forster, C., Narziß, L.: Hefevitalität und Bierqualität, Monatsschrift für Brauwissenschaft 51 (1998), no. 11/12, pp. 189–195.

[17] Hori,T., Imai,T., Yasuda,Y.,Takeuchi, T., Ohkochi, M.: The Relationship Between Yeast Vitality and Yeast Handling. Master Brewers Association of the Americas Technical Quarterly 37 (2000), pp. 31–35.

[18] Thiele, F., Back, W.: Measurement of Yeast Vitality using a Modified Version of the Intracellular pH Measurement (ICP). BrewingScience 58 (2005), no. 1/2, pp. 2–5.

[19] Müller-Auffermann, K., Silva, W., Hutzler, M. Jacob, F.: Evaluation and determination of an alternative analysis method for rapid determination of yeast vitality. Brewing Science 67 (2014), no. 5/6, pp. 72–80.

[20] Narziß, L., Back, W., Gastl, M., Zarnkow, M.: Abriss der Bierbrauerei, 8th ed., Wiley-VCH, Weinheim 2017, p. 247.

[21] Weinfurtner, F.: Richtlinien für die Hefereinzucht, Biologische Brauerei-Betriebskontrolle, Desinfektionsmittelprüfung. Verlag Hans Carl GmbH, Nürnberg (1957).

[22] Thiele, F., Back, W.: Mikrobiologie des Bieres. Back, W.: Mikrobiologie der Lebensmittel. Getränke. B. Behr's Verlag GmbH & Co., Hamburg (2008).

[23] Nielsen, O.: Yeast Propagation – Oxygen and Beer Yeast, a Complicated Relationship. Proceedings of the 29th European Brewery Convention Congress, Dublin, 2003, pp. 166–174.

[24] Vetterlein, K., and Back, W.: Praxiserfahrung mit der Assimilationstechnik. Der Weihenstephaner 67 (1999), no. 3, pp. 134–137.

[25] Back, W., Bohak, I., Ackermann, T.: Optimierte Hefewirtschaft. Brauwelt 133 (1993), no. 35, pp. 1960–1963.

[26] Lehmann, J., and Back, W.: Praktische Umsetzung der Hefeassimilation. Brauindustrie 82 (1997), no. 4, pp. 225–228.

[27] Boulton, C., Quain, D.: Brewing Yeast and Fermentation. Blackwell Science Ltd., Oxford 2001, pp. 422–432.

[28] Narziß, L., Back, W., Gastl, M., Zarnkow, M.: Abriss der Bierbrauerei, 8th ed., Wiley-VCH, Weinheim 2017, p. 243.

[29] Boulton, C., and Quain, D.: Brewing Yeast and Fermentation. Blackwell Science Ltd., Oxford 2001, pp. 121–125, 399–402.

[30] Back, W., Bohak, I., Ackermann, T.: Optimised Yeast Management. Brauwelt International 12 (1994), no. 2, pp. 136–139.

[31] Back, W. (ed.): Mikrobiologie der Lebensmittel – Getränke. Behr's Verlag, Hamburg 2008. p. 147.

[32] Nielsen, H., Hoybe-Hansen, I., Ibaek, D., Kristensen, B. J., Synnesvedt, K.: Pressure Fermentation and Wort Carbonation. Master Brewers Association of the Americas Technical Quarterly 24 (1987), pp. 90–94.

[33] Boulton, C., and Quain, D.: Brewing Yeast and Fermentation. Blackwell Science Ltd., Oxford 2001, pp. 430–432.

[34] Miedaner, H.: Optimisation of Fermentation and Conditioning in the Production of Lager. Proceedings of the 14th European Brewing Convention Congress, Salzburg, 1978, pp. 110–134.

[35] Narziß, L., Back, W., Gastl, M., Zarnkow, M.: Abriss der Bierbrauerei, 8th ed., Wiley-VCH, Weinheim 2017, pp. 274–293.

FILTERABILITY – ISSUES WITH TURBIDITY

1 INTRODUCTION

With the exception of wheat beers and unfiltered specialty beers such as *Kellerbier*, beer drinkers expect brilliant, clear beers that retain this clarity even after an extended period of storage.

Various filtration methods are employed to achieve a brilliant appearance of the filtered beer, but they can also be used to remove any undesirable microorganisms from the beer. Brewers must strike a fine balance between the contrasting demands of filter performance and microbiological safety.

Starting with mass filters, filtration technology has developed over the years, culminating in the precoat filtration common today. Precoating with kieselguhr is an integral part of the current state-of-the-art filtration performed with candle filters, horizontal filters or plate and frame filters. Nonetheless, kieselguhr and the issues with its disposal have come under criticism. For some time now, this has been the impetus behind efforts to replace kieselguhr with organic filter aids such as cellulose, and to establish other methods for filtering beer without filter aids, e.g. membrane filtration.

The filterability of beer is a major challenge encountered not only in the development of new filtration techniques, but also in the daily routine of production. Filterability can fluctuate greatly within a batch; however, variation is generally more common batch to batch. Problems caused by poor filterability significantly impact the cost of filtration in the form of the additional personnel required and higher consumption of filter aids. In extreme cases, filtration problems could even compromise the timely delivery of beer to customers.

Therefore, it should come as no surprise to find that researchers have been dedicating their efforts to this topic for some time now. Numerous scientific articles [1, 2, 3, 4, 5] and dissertations [6, 7, 8, 9, 10, 11, 12] are available on the subject. In most of these publications, poor filterability is associated with the pressure climbing too rapidly at the filter inlet, thus curtailing filtration operations. The authors tend to focus to a lesser extent on the actual goal of filtration, namely the clarification of beer and the degree of success achieved. However, an objective assessment of the filtration of beer is only possible if both the increase in pressure at the filter inlet (based on the filtration volume, the area available for filtration and the duration of filtration) and the clarification effect are measured in EBC turbidity units. This has given rise to the following definition for the filterability of beer:

Filterability is a characteristic affected by the constituents in beer, which in turn, determines the quality of the filtration process. The pressure increase at the filter inlet and the turbidity of the beer at the filter outlet are both parameters for evaluating the quality of filtration.

2 ATTRIBUTES INFLUENCING BEER FILTERABILITY

The effects of individual groups of compounds and microorganisms, including a variety of common techniques are described in the research publications mentioned above. The impact of polysaccharides, proteins, polyphenols, melanoidins, minerals, yeast and bacteria are examined in each process step, from the mill to the filter. This information is summarized in the two figures below. The various compounds influencing filtration, grouped according to their origin, can be found in figure 1.

Figure 1: Substances affecting the filterability of beer

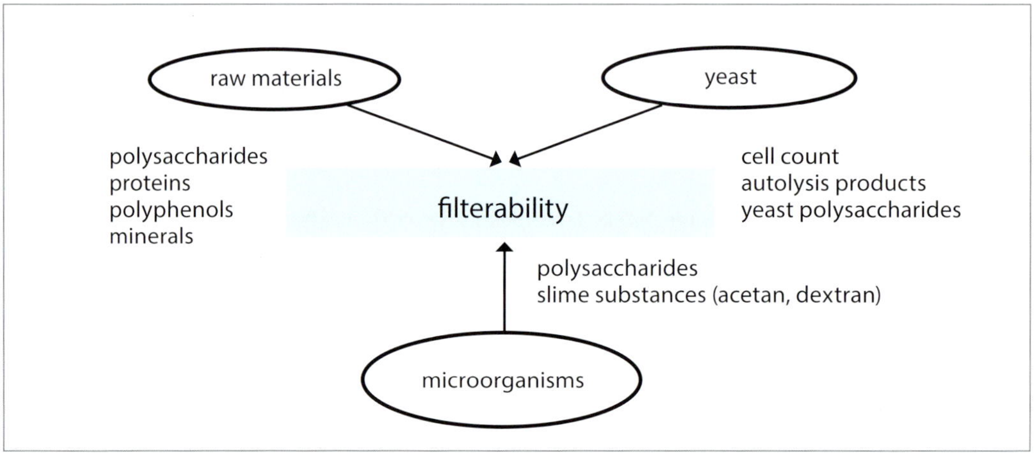

The ramifications of each step in the brewing process on the filterability of beer is described in figure 2, tracing the path of the beer as it makes its way through the brewery.

Figure 2: The impact of process technology on the filterability of beer (image: Deutscher Brauerbund)

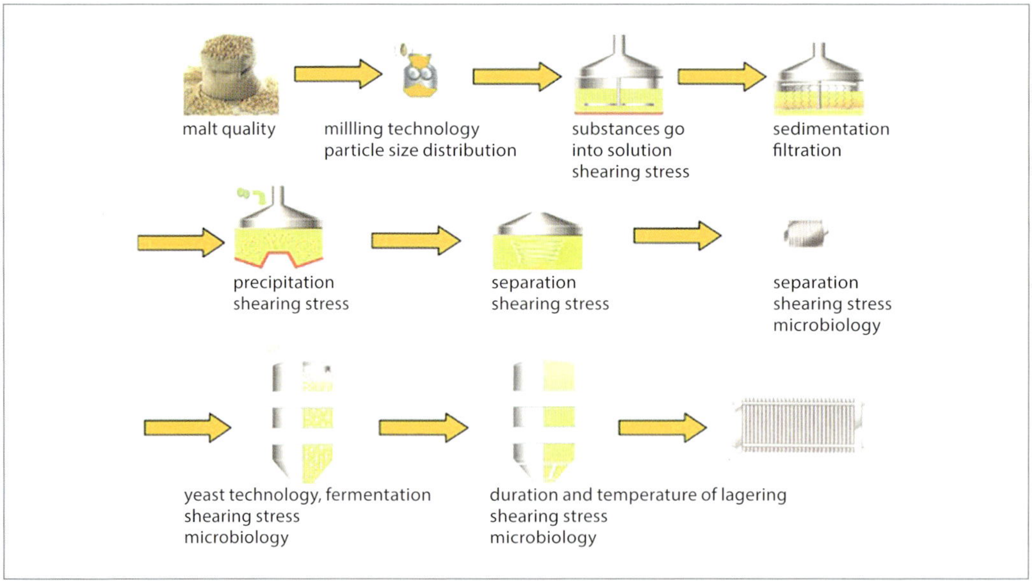

2.1 THE IMPACT OF MALT QUALITY ON BEER FILTERABILITY

The quality of the malt is an important characteristic in determining the filterability of beer. The degree and type of modification in the malt plays a decisive role in the quantity of high molecular weight polyphenols and proteins dissolved in the wort. In essence, these substances indicate whether there will be a risk of poor filterability. Nevertheless, malt quality alone is not entirely responsible for disruptions in filtration. These problems are due to a combination of factors including the quality of the raw materials and the choice of brewing techniques. A review of the qualitative attributes of barley malt and wheat malt (cf. Malt) provides suggestions for adapting brewing techniques to certain malt specifications and examines the corresponding effects on filterability.

2.2 THE IMPACT OF BREWHOUSE OPERATIONS ON FILTERABILITY

Each step in the brewhouse has an impact on the solubility or precipitation of polysaccharides and proteins and consequently, on the filterability of the beer. Depending on the substance in question, this can result in an increase in pressure or higher turbidity during filtration.

Technical deficits occurring during the processing steps affecting the gelatinization and solubility of the starch can lead to high photometric iodine values (due to the presence of high molecular weight α-glucans) and increased turbidity values. Particularly vulnerable steps in the process include the grist composition (mill settings), the pH of the mash, the mash temperatures and rests during saccharification, the temperature of the sparge water and the point at which the hot break material is added to the lauter tun during sparging [13]. With the exception of the grist composition, the photometric iodine test can be used as a quality control tool to monitor the individual steps in the process. Samples can be collected at the following points: the first wort, the run-off after sparging for the first time, the lauter wort in the kettle after sparging is complete (prior to the boil), the cast-out wort and the hot break material if it was added to the lauter tun during sparging.

A large number of the disruptions during filtration are the result of excessive spikes in pressure at the filter inlet due to the presence of β-glucans. Key characteristics of β-glucans include elevated viscosity of the wort and beer as well as a propensity for gel formation. Both characteristics are dependent on the concentration and the molecular weight of the β-glucans in addition to shearing and tensile forces exerted on the molecules during the production process. β-Glucan gel formation, as it is termed, is a complex process that is significantly more difficult to quantify analytically than the free β-glucan content. The β-glucan gel is actually not a gel in the narrow scientific sense. In order to be considered a true gel by definition, all of the β-glucan molecules would have to spontaneously form an interconnected network and changes would have to be measured in the rheological behavior of the gel as the wort or beer is subjected to shearing forces. β-Glucans exhibit this type of behavior only above concentrations of 4 %, that is, 40 g/l, or higher [14]. However, in a brewery, concentrations of more than 500 mg/l are rarely encountered even with poor cytolytic modification. This eliminates the possibility of spontaneous gel formation, as well as the opportunity to make direct measurements in a stepwise quality control program. The modified filter test according to Raible is an indirect method for measuring the β-glucan content of wort and beer.

Analysis results from a stepwise quality control program (cf. section 2.4) would make it possible to detect and monitor how the β-glucan content affects filterability throughout the entire brewing process while taking factors such as time, shearing forces and ethanol content into account. Elevated turbidity values caused by the presence of β-glucans are not normally expected.

In the brewhouse, protein content is primarily influenced by the mash temperatures, the duration of the rests and the intensity of the boil. No conclusive evidence has been drawn from protein analysis of the wort indicating a correlation between proteins and an increase in pressure at the filter or the turbidity of the beer, which would allow predictions to be made. In contrast to polysaccharides, the content of high molecular weight proteins changes significantly after boiling. This is brought about by the change in the pH during fermentation, as a result of characteristics of individual yeast strains, storage conditions and stabilization measures on the cold side.

2.3 THE IMPACT OF FERMENTATION AND STORAGE ON THE FILTERABILITY OF BEER

Yeast characteristics play an important role in the filterability of beer. Depending on the individual strain, yeast cells subjected to stressful conditions (high gravity, high fermentation temperatures) may excrete glycogen, a storage compound [15]. Glycogen has a similar structure to that of amylopectin, but is more highly branched in its molecular structure, making it very difficult to measure with the photometric iodine test. Glycogen cannot be filtered out of the beer and causes turbidity values of up to 1 EBC (measured at an angle of 90°, MEBAK II 2.15.21).

Yeast performs a key function in the clarification of beer. Major factors influencing clarification include the surface of the yeast cell, the point at which the green beer is run through a separator (if this is employed), the yeast cell count and the time spent in the lager cellar. To date, it has not been possible to measure the influence of the yeast on filterability. Furthermore, conflicting evidence has been reported in the literature, especially with regard to lagering, green beer separation and yeast cell count. Additional research is required in this area, particularly in relation to proteinaceous particles.

Products of yeast autolysis or microbial contamination (e.g. *Enterobacter* strains capable of forming slime capsules or acetic acid bacteria) can also negatively affect filterability. It has been demonstrated that the polysaccharides formed by these microorganisms such as acetan or dextran in the form of β-glucans can cause an extreme rise in pressure during filtration, even at very low concentrations [16].

Contamination, with *Gluconobacter frateurii* for example, at concentrations as low as 10^5 to 10^6 cells/ml (a few cells visible in the microscope field of view at 400x magnification) can bring about a steep decline in filterability. Figure 3 depicts slime capsules (acetan) formed by these microorganisms, which may reach proportions many times larger than the cell itself.

Figure 3: Bacterial flora typically found with yeast in unfiltered beer: encapsulated acetic acid bacteria; the coccobacilli frequently appear in the aggregated form as sarcinae

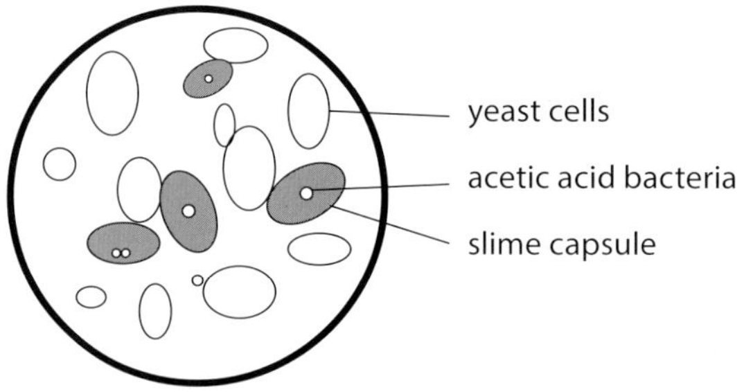

2.4 MONITORING THE FILTERABILITY OF WORT AND BEER AS PART OF A STEPWISE QUALITY CONTROL PROGRAM

As mentioned previously, filterability is not exclusively determined by the polysaccharides in solution originating from the malt dissolved in the wort. At various points in the production process, filterability can be negatively affected by strong shearing forces applied to wort and beer during processing as well as through the influence of yeast and bacteria on the cold side. However, the effects of this problem are first manifested at the filter inlet or filter outlet.

Performing a stepwise analysis [11] of wort and green beer samples for quality control allows the filterability to be monitored throughout the production process. This is the first possible opportunity to measure the effect of the three sources of polysaccharides (malt, yeast and bacterial contamination). The sample collection points for such a stepwise quality control regimen are listed in figure 4.

Figure 4: Quality control regimen for collecting samples at various steps in the brewing process for the purpose of solving filterability problems (image: Deutcher Brauerbund)

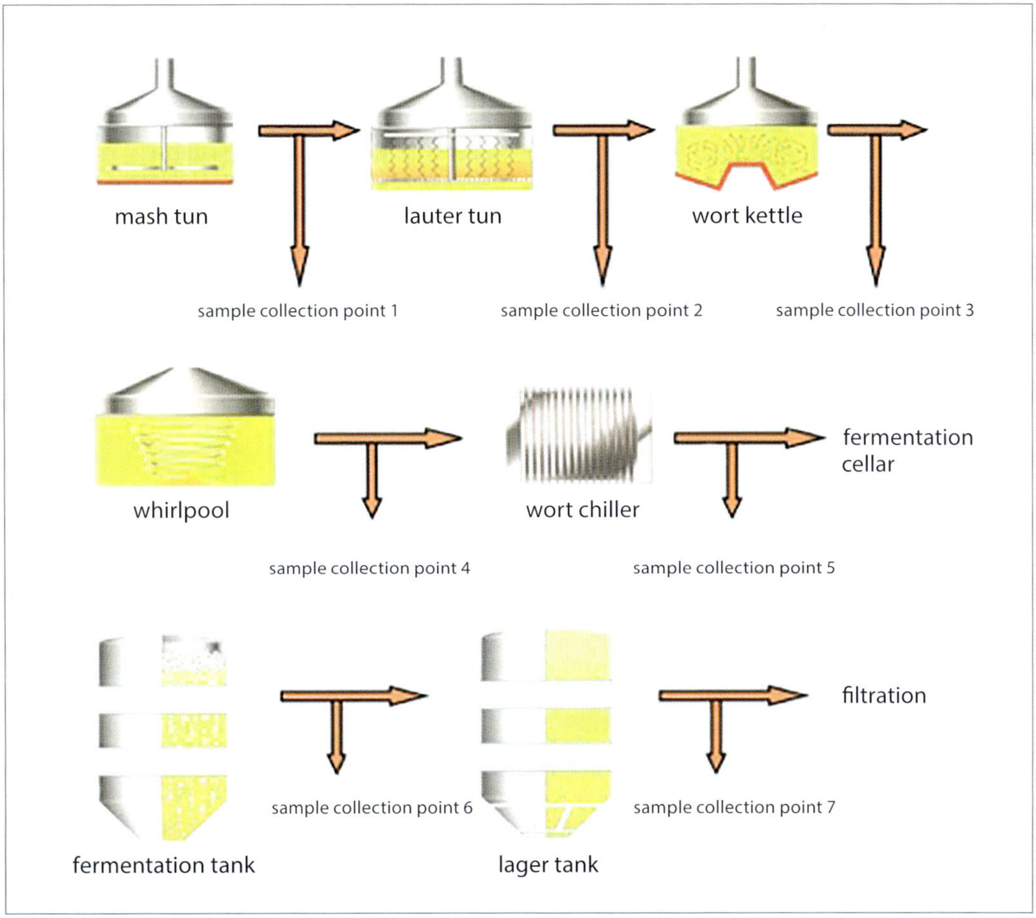

Table 1: Results from analyses conducted in five breweries as part of a stepwise quality control program

F_{spec} [hl/m²h]	Beer style	Cast-out wort	Downstream from whirlpool	Downstream from wort chiller	Green beer	Unfiltrate
brewery A	Export	4.9	4.9	4.8		5.6
brewery A	pale lager		5.9	5.7	6.1	5.9
brewery B	Pilsner	5.6		6.1		7.0
brewery C	pale lager	5.1		5.9	3.2	3.7
brewery D	pale lager	5.9		3.9		3.3
brewery E	Bock	3.1		3.0		3.2

Breweries A and B served as a reference for breweries producing beer with a high level of filterability (table 1). The wort samples from the export-style beer from brewery A exhibited somewhat lower values for filterability compared with those of the pale lager beers. However, there was only a slight difference in the specific filtrate volume of the beer. By contrast, brewery C experienced a severe filtration problem. It should be noted that poor filterability was first evident in the beer samples. According to values obtained from the preceding wort analysis, no filtration problems were expected. Wort aeration was found to be the source of the problem at this brewery. The wort was contaminated with Gram-negative bacteria and was also subjected to strong shearing forces during aeration. In brewery D, a reduction in the filterability of the specific filtrate volume was already detectable in the values for wort samples collected immediately downstream from the whirlpool. In this case, a cavitating hot wort pump was to blame. A *Bockbier* brewed by brewery E is a good example of filtration difficulties originating in the malt. Although other, lower gravity beers produced in this brewery were not problematic at this stage in the process, the higher gravity beers led to disruptions in filtration due to their higher wort concentrations and the quality of malt used.

The stepwise quality control regimen made it possible to identify the various factors behind poor filterability and showed that these issues are not always attributable to the quality of the malt.

2.5 THE INFLUENCE OF FILTRATION ON TURBIDITY IN BEER

Filtration is the last step of production in which beer quality can be positively influenced. The steps that follow (flash pasteurization, if applicable, and filling) must be designed so that the beer's level of quality is maintained. This means that all of the changes (e.g. adjusting the original gravity and color) carried out after filtration must be done carefully, so that the existing equilibria (e.g. the formation of calcium oxalate crystals) are not disturbed.

The clarification effect can be influenced through the application of a combination of various filters and filter aids. In most breweries, this is monitored through a dual angle light scattering measurement technique. Interpreting the results from this measurement is a complex task. In a simplified form, one can assume that particles smaller than 1 µm are detected at an angle of 90°, while the forward scattered light measurement is performed at angles of 25° or 11°, indicating a particle size larger than 1 µm. The turbidity measured at an angle of 90° is affected by the quality of the unfiltered beer and/or the combination of filter aids. These limit values are determined by the brewery and generally lie within a range of 0.5 to 1.0 EBC. Compounds in beer responsible for increases in turbidity can be detected with an enzymatic turbidity test (fig. 5) [18].

Particulates are primarily detected at an angle of 25° and this measurement is therefore used to identify breaches in kieselguhr filtration. For this reason, this turbidity value must be kept as low as possible. Nonetheless, throughout the filtration process, the transfer of miniscule amounts of kieselguhr to the beer is unavoidable. This problem is also encountered with PVPP filters. Here the effect is more pronounced, since the PVPP is mechanically reduced in size during regeneration. These small particles pass almost unhindered through the support layer. Therefore, it is recommended that a fine filter suited to this purpose be integrated downstream from the PVPP filter. Since most of the particles exiting the PVPP filter only have a diameter between 5 and 10 µm, the fine filter must be capable of removing particles of this size in an efficient manner. Filtration with sheet filters as well as candle filters with an absolute capacity to remove particles as small as 5 µm or two-stage candle filters has been shown to safely and satisfactorily remove these particles [19]. Filter aids can end up in the filled bottles if a filter does not have the appropriate capacity for the removal of these particulates. In bottled beer, these particles not only form sediment but can also trigger gushing in certain situations.

Gushing is a problem with multiple causes. Over the past few years, gushing has been grouped into two general categories, primary and secondary gushing. Primary gushing is caused by malt contamination (e.g. contamination by *Fusarium*), while secondary gushing is attributed to various technical and technological factors.

After comprehensive analysis of raw materials spanning many years, there has still been no proof of a clear correlation between the degree of *Fusarium* contamination in malt and gushing in the resultant beers. Current research is exploring various classes of compounds (proteins and hydrophobins) found in elevated concentrations in kernels contaminated with *Fusarium* [20, 21, 22, 23].

First, the CO_2 content and its binding capacity are essential criteria. However, the presence of particles and other nucleation sites, at which CO_2 can be liberated as a gas, are a prerequisite as well. Furthermore, certain physical or geometric conditions enhance the release of CO_2 through physical acceleration when a container is opened.

In reproducible trials as well as in ordinary commercially produced beers, it has been demonstrated that porous particles or those bearing sharp edges (diameter approx. 1–10 µm) such as calcium oxalate, calcium carbonate and dextrin crystals as well as filter aids (kieselguhr, perlite, activated charcoal) induce gushing if they are present in certain concentrations [24]. Even poor bottle quality (internal roughness) can have a negative effect. The carbon dioxide concentration, how the carbonic acid is bound and the diameter of the gas bubbles are crucial factors [25]. The basic composition of the beer (viscosity, surface tension, hydrophobicity, iron content, special proteins such as protein Z and LTP 1 [20]) plays a subordinate role compared to the physical factors.

Figure 5: Flow diagram for identifying sources of turbidity [17]

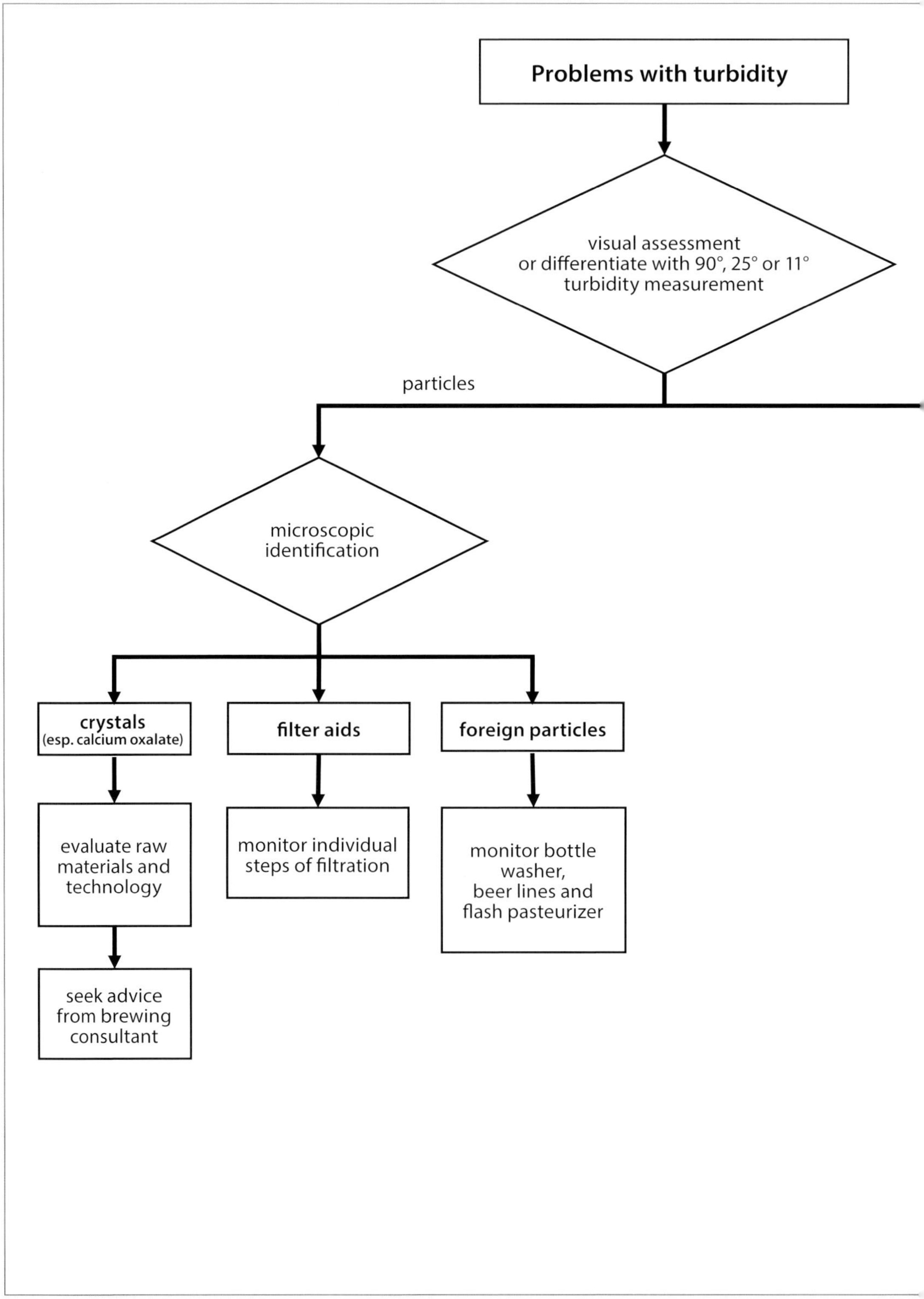

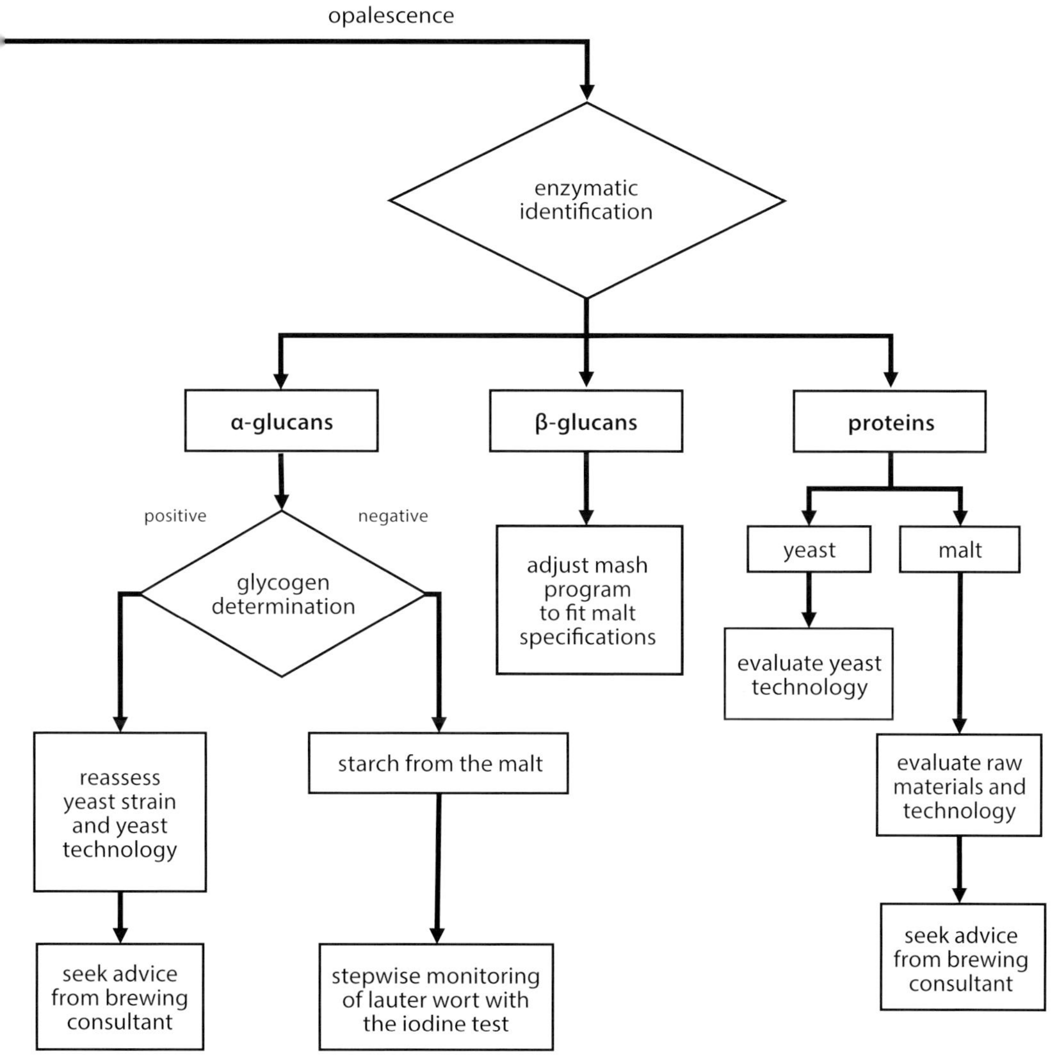
opalescence
enzymatic identification
α-glucans
β-glucans
proteins
positive
negative
glycogen determination
adjust mash program to fit malt specifications
yeast
malt
evaluate yeast technology
reassess yeast strain and yeast technology
starch from the malt
evaluate raw materials and technology
seek advice from brewing consultant
stepwise monitoring of lauter wort with the iodine test
seek advice from brewing consultant

The phenomenon of gushing can only be explained sufficiently given the assumption that multiple factors must be present simultaneously which together cause the threshold to be exceeded, consequently triggering gushing behavior. Thus, a critical potential must be assumed with regard to the tendency for a beer to exhibit gushing. Figure 6 shows the maximum and minimum potential risk for factors capable of triggering gushing.

Figure 6: Factors influencing the gushing potential of bottled beer (based on experience) [2, 3, 27]

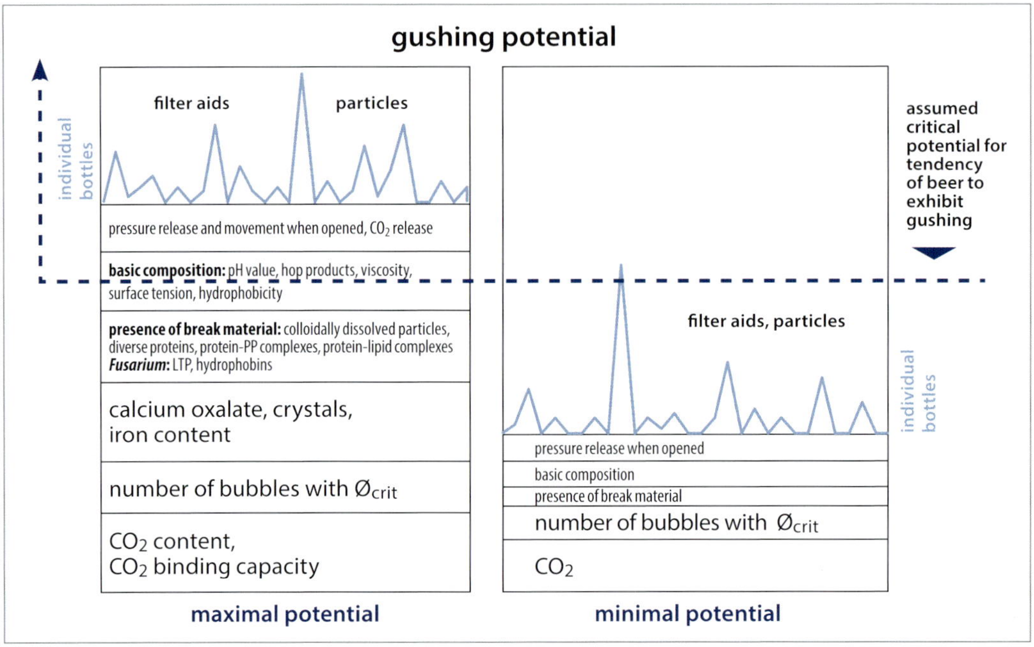

However, whether a beer will ultimately gush or not is dependent on the potential inherent to a particular beer and also whether it has exceeded the critical threshold for that potential. Thus, it would make sense for those involved along the process chain to strive to keep the various gushing factors to a minimum in order to reduce the potential for gushing. This phenomenon is frequently exemplified by bottles with completely different tendencies to gush despite the fact that they are in the same case of beer. The explanation for this would likely be as follows: although gushing factors are typically uniformly distributed throughout a given beverage, due to the pressure and flow conditions during filling, particles (oxalate crystals, filter aids) end up in different quantities in the individual bottles. This, in turn, results in some of the bottles exceeding the threshold for the minimum gushing potential while others do not [27].

3 SUMMARY

In this chapter, filterability was defined in terms of a pressure increase at the filter inlet and turbidity at the filter outlet. Adequate filterability of the beer is an important economic factor for breweries. A rapid rise in pressure at the filter inlet increases production costs, while turbidity problems negatively impact the beer quality. Pressure increases at the filter can be caused by malt quality, the mash program, shearing forces on the wort, yeast characteristics and microbial contaminants. Turbidity that is not removed through filtration is primarily attributable to the presence of α-glucans and proteins. Filter aids pose a further source of turbidity if they are not removed to a suitable level by means of filtration prior to packaging.

4 OVERVIEW

Problems with elevated turbidity levels

analyses: determination of turbidity by means of an enzymatic method (turbidity measured at a 90° angle), particle identification through membrane filtration and examination under a microscope (turbidity measured at a 25° angle), photometric iodine test

Problem	Cause	Measures/Prevention
elevated turbidity levels in filtered beer caused by α-glucans		perform stepwise quality control of wort samples with photometric iodine test
	α-amylase activity of the malt < 40 ASBC	check the malt specifications; increase the length of the α-amylase rest
	grist fractions are too coarse; whole kernels present during wet milling	check the gap distance between rollers in the mill
	incorrect rest temperatures during mashing, inhibition of α-amylase activity; mash pH is too low (< 5.4)	check mash temperatures and pH values
	overshooting the target temperature of the rest, inhibition of α-amylase	review mashing program
	mash-out temperature > 78 °C	mash out at a lower temperature
	temperature of sparge water > 78 °C	check water blending system for proper function
	yeast glycogen (high gravity brewing, high fermentation temperatures, dependent on individual yeast strain)	
elevated turbidity levels in filtered beer caused by proteins	malt modification	check malt specifications
	mash pH is too high > 5.8	consider mash acidification
	poor yeast quality	review yeast handling procedures
	dead cells in the pitching yeast	
	poor yeast sedimentation during the lagering process, too few yeast cells in the beer during lagering and therefore no separation and removal of substances causing turbidity	
	insufficient protein stabilization	check for proper dosage of stabilizing agent

Problem	Cause	Measures/Prevention
elevated turbidity levels in filtered beer caused by filter aids	poor precoat layer	adjust precoat and body feed at the filter for proper filtration of the beer
	damaged filter sheets	replace damaged filter sheets
	pressure surges	use a pulsation damper [21]
	inadequate post-filtration polishing	choose a filter with a higher capacity for clarification
elevated turbidity values in filtered beer caused by calcium oxalate crystals	temperatures during lagering were too high	temperatures of 0–1 °C should be reached in the lager tank
	lager period was too short	the crystals need time to grow to a size which can be removed by filtration (approx. 10 days)
	calcium uptake during and after filtration	select a kieselguhr product with a lower calcium content; adjust the gravity and color prior to filtration

Problems with pressure increases at the filter

analyses: filter test, β-glucan gel, viscosity, photometric iodine test

Problem	Cause	Measures/Prevention
poor filter test results for unfiltered beer		perform stepwise quality control of wort samples
possible mistakes made during wort production	β-glucan, malt quality	check specifications
	mash pH too high, inhibition of β-glucanases at mash-in temperatures < 50 °C	check mash acidification procedures
	incorrect rest temperatures, inhibition of β-glucanases	check the thermometer sensor for proper function
	overshooting the rest temperature, inhibition of β-glucanases	adjust mash program to compensate for the quality of the malt
	formation of β-glucan gel, shearing forces on the wort	check the blades on the agitator, pumps
poor filter test results for unfiltered beer; however, the filterability values for the wort are acceptable	formation of β-glucan gel, shearing forces on the yeast	check the yeast pump, green beer centrifuge
	yeast autolysis	
	microbial contamination in unfiltered beer (*Gluconobacter, Acetobacter, Enterobacter*)	perform stepwise microbiological quality control
	spikes in yeast cell count	avoid pressure surges; install buffer tank upstream from the filter; check procedure for purging yeast from the tank
	poor sedimentation of the yeast during lagering, too few yeast cells present during lagering and therefore no elimination of substances causing turbidity	
	carryover of silica gel	

5 REFERENCES

[1] Schimpf, F. W., Rinke, W., and Ehrke, H. F.: Untersuchungen über die Filtrierbarkeit des Bieres. Monatsschrift für Brauerei 22 (1969), no. 12, pp. 353–361.

[2] Hug, H., and Pfenninger, H.: Filtrationsschwierigkeiten – Ursachen und Möglichkeiten zur Behebung. Schweizer Brauerei-Rundschau 85 (1974), no. 7, pp. 133–152.

[3] Narziß, L.: Über die Filtrierbarkeit des Bieres. Brauwelt 115 (1975), no. 45, pp. 1491–1496.

[4] Schur, F.: Bierfiltration und Einflüsse auf die Bierfiltrierbarkeit. Proceedings of the 22nd European Brewery Convention Congress, Zürich 1989, pp. 371–383.

[5] Annemüller, G.: Über die Filtrierbarkeit des Bieres – Beurteilung und Einfluß der Inhaltstoffe. Monatsschrift für Brauwissenschaft 44 (1991), no. 2, pp. 64–71.

[6] Lindemann, B.: Bewertung der Kieselgurfiltration von Bier. Dissertation D 83, Fachbereich 13, no. 317, Berlin 1992.

[7] Wagner, N.: β-Glucan in Bier und Bedeutung dieser Stoffgruppe für die Bierfiltration. Dissertation D 83, Fachbereich 13, no. 288, Berlin 1990.

[8] Lindemann, A.: Untersuchungen der Struktureigenschafts-Beziehungen von β-Glucan bei der Bierherstellung. Dissertation D 83, Fachbereich 15, no. 055, Berlin 1995.

[9] Eiselt, G.: Untersuchungen über hochmolekulares β-Glucan. Dissertation D 83, Fachbereich 15, no. 050, Berlin 1995.

[10] Eßlinger, H. M.: Einflußfaktoren auf die Filtrierbarkeit des Bieres. Dissertation, Technical University of Munich (Weihenstephan),1985.

[11] Kreisz, S.: Der Einfluss von Polysacchariden aus Malz, Hefe und Bakterien auf die Filtrierbarkeit von Würze und Bier. Dissertation, Technical University of Munich (Weihenstephan), 2002.

[12] Kupetz, M.: Comperative Identification of Filtration-Inhibitory Substances in Membrane and Diatomaceous Earth Filtration of Beer. Dissertation, Technical University of Munich (Weihenstephan), 2017.

[13] Narziß L., Back, W.: Die Technologie der Würzebereitung. 8th ed. Wiley-VCH, Weinheim 2009.

[14] Clasen C., and W.-M., Kulicke: Zum Gelbildungsprozeß von (1-3)(1-4)-β-D-Glucanen, Monatsschrift für Brauwissenschaft 56 (2003), no. 9/10, pp. 161–171.

[15] Malcorps, Ph., Haselaavs, P., Dupire, S., and van den Eynde, E.: Glycogen released by the yeast as a cause of unfiltrable haze in the beer. Proceedings of the 27th European Brewery Convention Congress, Cannes 1999, pp. 831–838.

[16] Kreisz, S., Wagner, F., and Back, W.: The influence of polysaccharides of yeast and bacteria on the filterability of beer. Proceedings of the 28th European Brewery Convention Congress, Budapest 2001, pp. 226–235.

[17] Hartmann, K.: Bedeutung rohstoffbedingter Inhaltsstoffe und produktionstechnologischer Einflüsse auf die Trübungsproblematik im Bier. Dissertation, Technical University of Munich (Weihenstephan), 2006.

[18] Hartmann, K., Kreisz, S., Zarnkow, M., and Back, W.: Detection of Haze Particles. Proceedings of the 30th European Brewery Convention Congress, Prague 2005.

[19] Hartmann, K., Kreisz, S., Zarnkow, M., and Back, W.: Identifizierung von Filterhilfsmitteln nach den einzelnen Filtrationsschritten. Der Weihenstephaner 72 (2004), no. 4, pp. 141–143.

[20] Winkelmann, L.: Das Gushing-Puzzle eine Erfolgsgeschichte. Brauwelt 144 (2004), no. 25, pp. 749–751.

[21] Müller, M.P.: Beeinflußbarkeit des primären Gushings im technologischen Ablauf. Dissertation, Technical University of Munich (Weihenstephan), 2014.

[22] Stübner, M.L., G.; Vogel, R. F.; Niessen, L., Heterologous expression of the hydrophobin FcHyd5p from *Fusarium culmorum* in *Pichia pastoris* and evaluation of its surface activity and contribution to gushing of carbonated beverages. International Journal of Food Microbiology 141 (2010), pp. 110–115.

[23] Deckers, S.M., Gebruers, K., Baggerman, G., Lorgouilloux, Y., Delcour, J.A., Michaelis, C., Derdelinckx, G., Martens, J., Neven, H.: CO_2 -hydrophobin structures acting as nanobombs in beer. Part 1. A critical review of hypotheses and mechanisms. Brewing Science 63 (2010), pp. 54–61.

[24] Zarnkow, M., and Back, W.: Neue Erkenntnisse über gushingauslösende Substanzen. Brauwelt 141 (2001), no. 9/10, pp. 363–370.

[25] Fischer, S.: Bestimmung gushingverursachender Partikeln in Versuchslösungen und in Bier mittels Photonenkorrelationsspektroskopie. Schlussbericht für das Forschungsvorhaben R 314 (2000). pp. 9–11.

[26] Englmann, J.: Aktuelles aus der Betriebsberatung. 38. Technologisches Seminar, 34/1–34/4.

[27] Gastl, M., Zarnkow, M., and Back, W.: Gushing – ein multikausales Problem! Brauwelt 148 (2008), no. 32, pp. 896–897.

BEER FOAM

1 INTRODUCTION

Beer foam counts as one of the most important criteria for evaluating beer quality. Therefore, one of the primary tasks of a brewer is to produce beer with an excellent head of foam. Beer foam is not only important in the sensory realm; it also fulfills a technical role. As an immobilized cushion of gas on the surface of the beer, the foam layer prevents CO_2 from freely diffusing out of the beer, thus maintaining the carbonation level and drinkability of the beer over the period in which the beer is consumed.

2 ASPECTS OF BEER FOAM

2.1 THE PHYSICS OF FOAM

In order to understand how various compounds in beer influence the formation and stability of foam, it is helpful to review the four fundamental processes occurring in beer foam, which in essence are responsible for its physical attributes [1]:

- foam formation
- drainage
- coalescence
- disproportionation

Foam is a dispersed system characterized by a dispersant (beer) through which a disperse phase (CO_2) is distributed. As the beer is poured into a glass, the solubility equilibrium of the CO_2 falls below the threshold causing bound CO_2 to be released from solution. A small amount of ambient air also mixes with the CO_2 to create a layer of foam and a film of beer separates the foam lamellae. In the drainage process, the effects of gravity cause the liquid portion of the foam to flow along the foam lamellae and back into the beer. Drainage is primarily influenced by the factors of temperature and viscosity. As the liquid flows through the foam back into the beer, the thickness of the film between the lamellae shrinks, thus favoring the processes of coalescence and disproportionation. Coalescence refers to the destruction of the film of liquid between the foam lamellae. As the drainage process progresses, coalescence is induced. Certain compounds present in beer (hydrophobic amino acids, detergents) can also promote the effects of coalescence [1, 2]. Coalescence results in a coarsening of the foam and/or the collapse of the bubbles in the foam. Disproportionation involves the diffusion of gases between foam bubbles of different sizes. Diffusion of the gas is made possible by differences in the internal pressure of the bubbles. As a result, the smaller bubbles disappear, while the number of less desirable, larger bubbles increases. In particular, a non-homogeneous distribution of bubble sizes promotes the effects of disproportionation. The gas mixture (CO_2, N_2, CO_2/N_2) present in the beer affects gas diffusion, since the gases or gas mixtures possess different solubility characteristics in the liquid located between the foam lamellae, ultimately accelerating or delaying the diffusion process.

2.2 BIOCHEMICAL AND TECHNOLOGICAL ASPECTS OF BEER FOAM

2.2.1 COMPOUNDS IN BEER WITH A POSITIVE IMPACT ON FOAM

In addition to the physical effects of the formation and collapse of foam, a number of endogenous compounds in beer also have an impact on foam. Generally speaking, the interactions between hydrophobic polypeptides and hop bitter acids serve as the basis for foam stability in beer. Simpson and Hughes [3] explain this as the result of cross-linking brought about by hydrophobic and dipole forces. The polypeptides can be grouped according to their size into high molecular weight proteins (35–50 kDa) and low molecular weight proteins (5–15 kDa) [4]. The high molecular weight fractions include the protein Z (the so-called 40 kDa protein), which can be further subdivided into two protein isoforms, Z4 and Z7. The group of low molecular weight proteins essentially consists of lipid transfer proteins (LTP-1) and lipid binding proteins, which have yet to be investigated thoroughly [5].

Furthermore, a foam-stabilizing effect is attributed to the group of glycoproteins with surface-active properties [6]. Hop bitter acids consist mainly of α-acids and iso-α-acids, but other hop compounds can also promote foam stability. These include trans-isoadprehumulone [7] and reduced iso-α-acids (tetrahydro iso-α-acids) [8], the the latter of which may not be used under the laws regulating beer brewing in Germany.

Maillard products such as melanoidins also serve to enhance foam stability [9]. Much like α-acids or iso-α-acids, they can form polar complexes with proteins and/or counter the negative effects of lipids. Metal ions such as copper, zinc, calcium, magnesium, aluminum and iron are also purported to have a positive influence on foam stability. An insufficient supply of certain metal ions in the wort or poor yeast vitality results in a sluggish fermentation and may induce the excretion of foam-negative proteases into the beer. Where the *Reinheitsgebot* is not observed, alginate additives may be used to improve foam stability.

2.2.2 COMPOUNDS IN BEER WITH A NEGATIVE IMPACT ON FOAM

Lipids, in particular, have a negative effect on foam stability. Elevated concentrations of lipids may come into contact with the beer either as a consequence of flawed brewing techniques or in improperly cleaned beer glasses (lipstick, etc.) [10, 11]. Another negative effect stemming from shortcomings in the brewing process is the excretion of proteinase A into the beer by the yeast [12, 13, 14, 15, 16, 17]. Lipid transfer protein-1 has been identified as the substrate for proteinase A [18]. This protein is degraded by proteinase A and is therefore no longer capable of contributing to foam stabilization.

The effects of cleaning and disinfecting agents also have a negative influence on beer foam. These agents can come into contact with the beer if the equipment is not rinsed sufficiently as part of the CIP cleaning regimen [19]. Chemical residues can lower the surface tension of the beer, thus destroying the foam. Utilization of filter aids for protein stabilization (silica gels) in beer can also compromise head retention [20]. Problems with foam stability are occasionally reported for beers containing high concentrations of alcohol (strong beers) [21]. Ethanol accelerates the drainage process as well as the disruption of the liquid film between the foam lamellae and therefore increases coalescence and disproportionation.

2.3 INFLUENCE OF BREWING TECHNIQUES ON BEER FOAM

In the following section, several aspects of the beer production process are examined in more detail with regard to their influence on beer foam.

It has been shown that the crude protein content in malt should typically be in the range of 9.5–11 % for pale lager beers (cf. Malt). A soluble nitrogen content, also known as the Kolbach index, of 650 to 750 mg/100 g malt (dry matter) corresponds to a value for protein solubility of 38 to 42 % (based on the Congress mash). An excessively high Kolbach index is associated with lower foam quality. The content of protein Z does not decrease, but rather other foam-positive proteins such as hordeins or non-starchy polysaccharides like β-glucan or arabinoxylan, are removed [22]. For FAN concentrations, values in the range of 135 to 155 mg/100 g malt (dry matter, based on the Congress mash), which correspond to 20–22 % of the total nitrogen, are considered favorable (yeast nutrition).

In mashing, higher mash-in temperatures (> 60 °C) are required for head retention, since these temperatures are above the optima for proteolytic enzymes. Higher mash-in temperatures restrict the activity of lipoxygenases (LOX). LOX catalyze the incorporation of oxygen into polyunsaturated fatty acids, converting them to the corresponding unsaturated, conjugated hydroperoxy acids. This is the first step in the oxidative metabolism of fats, ultimately generating degradation products (trihydroxy fatty acids) that negatively affect foam stability [23]. An extended rest at 72 °C has also been shown to be beneficial, since more glycoproteins go into solution without being degraded [6]. Additionally, greater amounts of foam-positive proteins (40,000 Da) are released at 70–72 °C as the duration of the mash rest (30–60 min) is extended.

The boiling process directly impacts foam stability. Foam-positive proteins are better preserved through a gentle wort boiling process (short boiling times, moderate temperatures, low shearing forces, and low-temperature heating media). The value for coagulable nitrogen is an important metric in wort boiling, and it should lie in the range of 2 to 3 mg/100 ml for lager beers and 3.5 to 4 mg/100 ml for Southern German wheat beers. However, a sufficient degree of evaporation is necessary in order to eliminate free DMS (< 100 µg/l, measured in the pitching wort) from the wort. Acidification of the wort with natural lactic acid generally leads to better values for foam (enhanced fermentation, more non-isomerized alpha acids), although more coagulation and precipitation occurs closer to the isoelectric point.

Furthermore, the type and timing of the hop additions are important. Tannin-free hop extracts, such as CO_2 extract, are advantageous. By contrast, the polyphenols present in whole hops or pellets bring about more protein precipitation and elimination, but this loss is partially compensated by the proteins present in the hops.

The physiological condition of the yeast also plays a role in foam stability (cf. Yeast Technology and Fermentation). The influence of yeast vitality on the foam stability of the finished beer is shown in figure 1.

Figure 1: The influence of yeast vitality on foam stability [24]

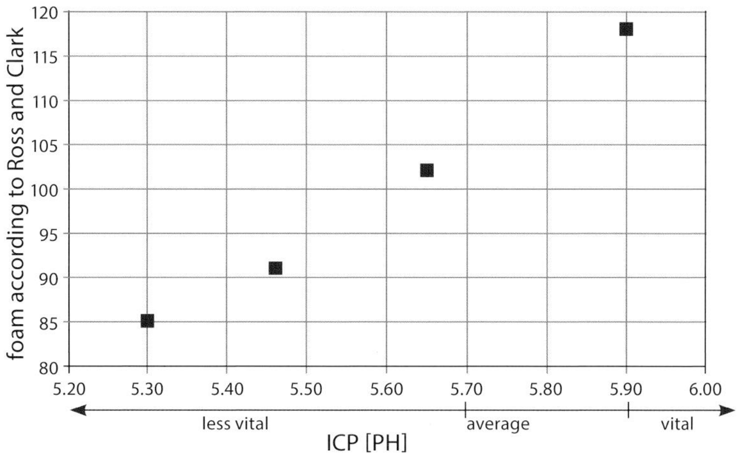

Above all, poor yeast vitality is manifested in the form of higher pH values and a stronger excretion of proteinase A into the beer. A higher pH value in the finished beer also leads to a stronger susceptibility to microbiological contamination. Moreover, a greater number of undesirable fermentation byproducts are created, such as medium chain fatty acids or fatty acid esters. These fatty acids and their corresponding esters influence the aroma of the finished beer. In particular, larger fatty acid esters, such as ethyl caprate, impart a sweaty note to the finished beer. For this reason, special care must be taken when recovering beer from surplus yeast (cf. Beer Recovery) [25].

Foam stability is reduced by protracted maturation and storage periods (yeast autolysis, proteinase A activity) if these processes are not carried out at sufficiently low temperatures. Equally negative are high temperatures during maturation. Under these conditions, the yeast excretes medium chain fatty acids as well as alkaline amino acids, of which arginine is especially detrimental with regard to foam [26].

The filtration process also has a marked effect on foam stability. Generally speaking, fine filtration should be avoided, since this removes a larger portion of the foam-enhancing proteins. When employing sterile filtration systems, an undue proportion of foam-positive substances are adsorbed onto the filter during the first hour of filtration, but then levels out as filtration progresses. For this reason, interim storage of these filtration charges in bright beer tanks is recommended. Protein stabilization treatment employing silica gel and/or especially bentonite also lowers the amount of foam-positive proteins. To achieve a good colloidal stability without sacrificing foam stability, polyphenol stabilization (PVPP) is preferable.

The uptake of metal ions (iron or aluminum) by the beer, e.g. in the past through defective steel cladding of brewing vessels, should be avoided. Although these metal ions improve beer foam, they also have a negative effect on flavor and flavor stability.

Reduced surface tension has been measured in the residual rinse water found in lager tanks and bright beer tanks which have been disinfected with surface-active substances (surfactants) and not rinsed properly (water-saving measures). The surface tension of residual rinse water should be approximately 70 mN/m (typical value for water: 72 mN/m).

Finally, only clean, intact bottles should be used in the bottling process. The bottles must be free of any detergent residues. It should be noted that when using new glass bottles, some manufacturers apply surfactants as part of their cold-end coating treatment to increase the strength of the bottles. These surfactants dissolve into the beer after filling and consequently reduce the surface tension of the beer, adversely affecting foam stability.

Beyond the reach of the *Reinheitsgebot*, foam stability can be improved through the addition of foam stabilizers. The use of propylene glycol alginates (PGA) for this application is widespread. Alginates can boost foam stability by approximately 15 % (fig. 2).

Figure 2: Relationship between foam stability and the concentration of alginate added to beer (confidence interval for P = 0.95)

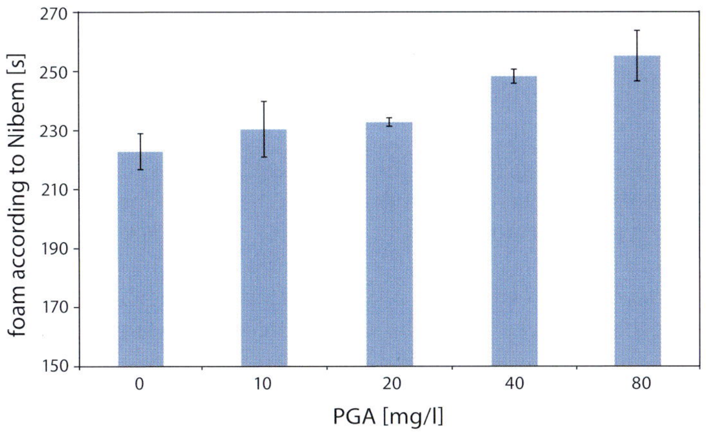

However, caution is advised when using alginate additives. These should only be dosed into the beer according to the range specified by the manufacturer, since PGA can result in unwanted turbidity and bitterness in the finished beer.

2.4 METHODS FOR DETERMINING FOAM STABILITY

There are a number of measurement methods for determining foam stability. In Germany, the Ross and Clark as well as the Klopper (Nibem) methods are widespread. In some cases, foam stability is measured using the method developed by Carlsberg LG (lg-Foamtester) [27, 28]. The DLG (German Agricultural Society) has adopted the lg-Foamtester as its official method for measuring beer foam. These methods are also used in other countries, but they do not receive the same level of recognition. At present, there is no uniformly recognized international method for characterizing foam stability, as is evident in the large number of methods available, some of which are specific to only one country. The wide range of measurement methods has hampered research into the factors influencing foam stability.

One of the oldest and simplest methods for determining foam stability is the manual pour test [29]. It is synonymous with the image of foam that the consumer sees. However, the disadvantage of this method is the lack of reproducibility. Nevertheless, the manual pour test can, under defined conditions, confirm the results of foam analyses or even be applied as a means to "identify outliers".

Measurement methods are commonly used today to assess the rate of foam degradation or collapse [30]. These methods either measure the amount of beer that accumulates as the foam collapses or they measure the level of artificially produced foam. Methods such as Ross and Clark [31], Klopper [32], Rasmussen [27] and Steinfurth [33] are all grouped in this category. Ambient conditions including air pressure, the temperature of the room and the temperature of the beer have an influence on the methods according to Ross and Clark and to Klopper. By contrast, the lg-Foamtester is a closed system, so variables like air pressure, relative humidity and air movement cease to be factors in the measurement.

3 SUMMARY

A consumer's first impression of the quality of a beer placed in front of them is the foam at the top of the glass. The layer of foam captures the carbon dioxide, thereby increasing the impression of carbonation. Simultaneously, foam protects against the loss of aroma. Therefore, the brewer is entrusted with the important task of producing beer with good quality foam. Individual stages of the brewing process that affect foam should not be considered independently from one another, but as a whole, since all aspects, from the malt to the beer in the glass, play a role in determining foam quality.

To achieve satisfactory beer foam, malt should be utilized which possesses sufficient amounts of foam-enhancing compounds and optimal modification properties. The protein fractions in malt containing protein Z and LTP-1 are known to increase the quality of beer foam. Modern brewhouse operations result in the extraction of higher amounts of polyphenols. This can be countered by optimizing brewing techniques to prevent the precipitation and loss of foam-positive proteins during wort boiling. This can be achieved through low-temperature heat transfer media or breaks in the movement of the wort during boiling. Proper yeast management is also necessary for adequate foam formation and head retention. Yeast populations exhibiting low vitality can release substances into beer (proteinase A, medium chain fatty acids, alkaline amino acids) that are very negative for foam. One must also understand that foam quality can be influenced throughout the entire beer production process and also when packaging the final product (surfactants). Furthermore, the brewery should ensure that the beer is served under optimal conditions (beer temperature, glass hygiene), in order to serve consumers a beer of the highest quality, crowned by a perfect head of foam.

4 OVERVIEW

Problems with malt quality	Cause	Suggested corrective measures/ reference values (RV)
raw protein content is too low (< 9.0 %)	harvest year and/or variety	alternative malt batches; RV 9.5–11 %
concentration of soluble protein (Kolbach index) is too high (> 45 %) or too low (< 38 %) based on Congress mash	variety/harvest year/ germination parameters	Kolbach index is too high: use a shorter mash program Kolbach index is too low: use a more intensive mash program (mash-in temperature: 50 °C)
FAN is too low (< 130 mg/100 g malt, d.m.) based on Congress mash	germination parameters	use a more intensive mash program, mash-in temperature < 50 °C, in order to form enough FAN for yeast nutrition, RV 135–155 mg/100 g malt, d.m. (20–22 % of the total nitrogen content)

Problems with mashing	Cause	Suggested corrective measures/ reference values (RV)
mash-in temperature is too low (< 60 °C)	especially with highly modified malts, too many foam-positive proteins are degraded by proteolytic enzymes	increase the mash-in temperature (> 60 °C)
zinc content is too low (sluggish fermentation)	mash pH is too high	lower the mash pH to 5.4–5.2 by acidification with natural lactic acid

Problems with lautering	Cause	Suggested corrective measures/ reference values (RV)
turbidity during lautering is too high (> 40 EBC)	lauter wort turbidity is caused by the malt	use malt with suitable specifications
	lauter wort turbidity is caused by incorrect lauter rake position and/or deep cuts with the rake	optimize the lautering process by adjusting the rake settings

Problems with wort boiling	Cause	Suggested corrective measures/ reference values (RV)
wort boiling is too intense (time, temperature)	conversion of DMS precursor (DMS-P) to compounds (DMS) that can be eliminated from the wort through evaporation	check the intensity of the wort boiling process; reduce boiling intensity; lower the temperature of the heat transfer medium RV for free DMS < 100 µg/l, coag. N 2–4 mg/100 ml
deposits of foam-positive proteins found on the surfaces in the heat exchanger	temperature of the heat transfer medium is too high	clean the surfaces in the heat exchanger after determining the presence of deposits; lower the temperature of the heat transfer medium (max. 15 °C higher than the boiling temperature)

Problems with wort handling	Cause	Suggested corrective measures/ reference values (RV)
yeast cells are covered with particulate matter	content of hot break material (trub) in the wort is too high	ensure that the maximum amount of hot break material is removed from the wort; RV < 0.5 g/l (membrane filtration method)

Problems with yeast/ fermentation/maturation/storage	Cause	Suggested corrective measures/ reference values (RV)
poor physiological condition of the yeast population	unfavorable storage conditions (time, temperature, over the weekend, breaks in production), incorrect yeast propagation techniques	improve yeast storage conditions (T < 4 °C, t < 1 week; store yeast in deaerated water); optimize the yeast propagation unit
sluggish fermentation	zinc content is too low (< 0.15 mg/l); FAN content is too low (< 200 mg/l or < 20–22 % of the total nitrogen); yeast cells are covered in hot break material (trub); poor physiological condition of the yeast population	lower mash pH (pH 5.5) by acidification with natural lactic acid; adjust mash-in temperature (see above); ensure that the maximum amount of hot break material is separated from the wort (see above); improve the physiological condition of the yeast population (RV for yeast vitality: ICP > 6.0)
yeast cell count is too high with yeast population exhibiting poor vitality		lower the yeast cell count for pitching; improve the physiological condition of the yeast population
proteinase A activity is too high (> 50 units x 10^{-5}/ml)	poor physiological condition of the yeast population; maturation phase is too intense (time, temperature); beer was stored on the yeast too long (especially applicable to horizontal lager tanks)	improve the physiological condition of the yeast population; shorten the maturation phase (RV for diacetyl < 0.1 mg/l); reduce the duration of lagering; flash pasteurize the beer after filtration (> 40 PU)

Problems with filtration	Cause	Suggested corrective measures/ reference values (RV)
proteinase A activity (> 50 units x 10-5/ml)	refer to previous point above	flash pasteurize the beer after filtration (> 40 PU)
content of foam-positive protein is too low	dosage of protein stabilizer is too high	reduce the use of protein stabilizers

Problems with filling	Cause	Suggested corrective measures/ reference values (RV)
reduction in the surface tension of water remaining on the bottle after exiting the bottle washer or rinser	residues of surfactants from cold-end treatment of new glass bottles at the manufacturer; cleaned bottles are not thoroughly rinsed	when purchasing new glass bottles, ask about the type of treatment during manufacturing; optimize the cleaning process in the bottle washer

Other problems	Cause	Suggested corrective measures/ reference values (RV)
surface tension of the residual water adhering to equipment surfaces after CIP procedures is too low	insufficient rinsing with clean water after cleaning or disinfection	ensure that the rinsing procedure is adequate after cleaning or disinfection (RV for surface tension of residual water adhering to surfaces ~ 70 mN/m)

5 REFERENCES

[1] Ronteltap, A. D., Hollemans, M., Bisperink, G. J., Prins, A.: Beer foam physics. In: Master Brewers Association of the Americas Technical Quarterly 28 (1991), no. 1. pp. 25–32.

[2] Aronson, M. P.: Influence of hydrophobic particles on the foaming of aqueous surfactant solutions. Langmuir 2 (1986). pp. 653–659.

[3] Simpson, W. J., Hughes, P. S.: Stabilization of foams by hop-derived bitter acids. Chemical interactions in beer foam. Cerevisia Biotechnology 19 (1994). pp. 39–44.

[4] Evans, D. E., and Sheehan, M. C.: Don't be fobbed off: The substance of beer foam – A review. Journal of American Society of Brewing Chemists 60, (2002), no. 2, pp. 47–57.

[5] Van Nierop, S. N. E., Evans, D. E., Axcell, B. C., Cantrell, I. C., Rautenbach, M.: Impact of different wort boiling temperatures on the beer foam stabilizing properties of Lipid-Transfer-Protein 1. In: Journal of Agricultural and Food Chemistry 52 (2004). pp. 3120–3129.

[6] Narziß, L., Reicheneder, E., Barth, D.: Über den Beitrag von hochmolekularen Proteinfraktionen und Glycoproteiden zum Bierschaum. Monatsschrift für Brauerei 35 (1982), no. 9, pp. 213–221.

[7] Smith, R.: Natural foam stabilizing and bittering compounds derived from hops. In: Journal of American Society of Brewing Chemists 56, no. 2 (1998). pp. 52–57.

[8] Bamforth, C.: Bringing matters to a head: The status of research on beer foam. In: European Brewery Convention Monograph, 1998. pp. 10–23.

[9] Lusk, L. T., Goldstein, H., and Ryder, D.: Independent role of beer proteins, melanoidins and polysaccharides in foam formation. In: Journal of American Society of Brewing Chemists 53, no. 3 (1995). pp. 93–103.

[10] Roberts, R., Keeney, P. J., and Wainwright, T.: The effects of lipids and related materials on beer foam. In: Journal of the Institute of Brewing 84 (1978). pp. 9–12.

[11] Wilde, P.: The physical mechanisms responsible for the destabilization of beer foam by fatty acids. Proceedings of the 29th European Brewery Convention Congress, Dublin 2003, pp. 875–882.

[12] Kondo, H., Yomo, H., Furukubo, S., Fukui, N., Nakatani, K., Kawasaki,Y.: Advanced Method for Measuring Proteinase A in beer and application to brewing. Journal of the Institute of Brewing 105 (1999), no. 5, pp. 293–300.

[13] Reicheneder E., Narziß, L.: Untersuchungen über die Schaumschädigung des Bieres durch Hefeproteinasen. Brauwelt 127 (1987), no. 21, pp. 956–962.

[14] Narziß, L., Reicheneder, E., Jogasuria, P., Eichhorn, P., Mayer, H., Gommelt, T.: Neueste Erkenntnisse zum Thema Bierschaum. Brauwelt 128 (1988), no. 3, pp. 1630–1642.

[15] Reicheneder, E., Narziß, L.: Untersuchungen über die Auswirkungen von Hefeproteinasen auf den Bierschaum. Proceedings of the 22nd European Brewery Convention Congress, Zurich 1989, pp. 577–584.

[16] Steiner, K., Lauplinger, U.: Hefeproteasen und Bierschaum. Proceedings of the 22nd European Brewery Convention Congress, Zurich 1989, pp. 585–592.

[17] Roettger, W.: Schaum und Molekülfraktionen. Dissertation, Technical University of Munich (Weihenstephan), 1973.

[18] Leisegang, R.: Proteinase A und Schaumstabilität – biochemische Untersuchungen der Wechselwirkungen von Proteinase A aus *Saccharomyces cerevisiae* und schaumaktivem Lipidtransferprotein. Dissertation, TU Berlin, 2003.

[19] Glas K., Scheuing, R.: Tenside und Bierschaum. Brauwelt 140 (2000), no. 46/47, pp. 2024–2025.

[20] Gromus, J.: Negative influences on beer foam. Monograph European Brewery Convention 27 (1998), pp. 69–83.

[21] Brierley, E. R., Wilde, P. J., Onishi, A., Hughes, P. S., Simpson, W. J., Clark, D. C.: The influences of ethanol on the foaming properties of beer protein fractions: A comparison of Rudin and micro conductivity methods of foam assessment. Journal of Scientific Food Agriculture 70 (1996), pp. 531–537.

[22] Evans, D. E., Nischwitz, R., Stewart, D. C., Cole, N., MacLeod, L. C.: The influence of malt foam-positive proteins and non-starch polysaccharides on beer foam quality. Monograph European Brewery Convention 27 (1999). pp. 114–128.

[23] Kobayashi, N., Segawa, S., Umemoto, S., Kuroda, H., Kaneda, H., Mitani, Y., Watari, J., Takashio, M.: A New Method for Evaluating Foam-Damaging Effect by Free Fatty Acids. Journal of American Society of Brewing Chemists 60 (2002), no. 1, pp. 37–41.

[24] Back, W., Imai, T., Forster, C., Narziß, L.: Hefevitalität und Bierqualität. Monatsschrift für Brauwissenschaft 51 (1998), 11/12, pp. 189–195.

[25] Schneeberger, M., Krottenthaler, M., Back, W.: Der Einfluss der Aufbewahrungsbedingungen von Hefesuspensionen auf die Qualität des darin enthaltenen, wiedergewinnbaren Hefebieres. Brauwelt 144 (2004), no. 37, pp. 1148–1152.

[26] Furukubo, S., Shobayashi, M., Fukui, N., Isoe, A., Nakatani, K.: A new factor which affects the foam adhesion of beer. Master Brewers Association of the Americas, Technical Quarterly 30 (1993), no. 4, pp. 155–158.

[27] Rasmussen, J. N.: Automated Analysis of foam stability. Carlsberg Research Communications 46 (1981), pp. 25–36.

[28] Anger, H. M., Glaro, L., Scholz, M.: Neues Verfahren zur Messung der Schaumstabilität. Brauwelt 150 (2001), no. 6/7, pp. 224–227.

[29] Weyh, H.: Die Schaumhaltbarkeit nach der Analyse und im Glas. Monatsschrift für Brauwissenschaft 40 (1987), no. 3, pp. 108–117.

[30] Yanagi, K., Ishibashi, Y., Kondo, H., Oka, K., Ushida, M.,: Neue Methoden zur Beurteilung von Geschmacksstabilität, Schaumeigenschaften und -stabilität von Bier. Anwendung in der Brauereipraxis. Brauwelt 137 (1997), no. 21/22, pp. 842–846, 857–859.

[31] Jacob, F. (ed.): Collection of Brewing Analysis Methods. Wort, Beer, Beer-based Beverages. Self published by MEBAK, Freising 2013.

[32] Klopper, W. J.: Foam stability and foam cling. In: Proceedings of the European Brewery Convention Congress, 1973. pp. 363-371.

[33] Jacob, F. (ed.): Collection of Brewing Analysis Methods. Wort, Beer, Beer-based Beverages. Self published by MEBAK, Freising 2013, method 2.18.4.

SENSORY ANALYSIS

1 INTRODUCTION

Sensory analysis plays a pivotal role in modern beer production with high demands being placed on quality, the wide reach of distribution channels and the extensive range of beer styles now on offer. The methods used in sensory analysis can be divided into three categories: those for distinguishing differences, those for describing characteristics and those designed to determine preferences [1]. Sensory analysis has gained in importance in recent years as a tool for quality assurance and product innovation. Methods for evaluating sensory quality have become well established in the brewing industry, and as a result, a wealth of information collected through sensory monitoring has been made available.

In contrast to wine, consumers tend to be less inclined to sensorially distinguish the numerous compounds present in beer. This is indeed puzzling since there are more than 800 different flavor and aroma components in beer. Compared to wine, beer actually exhibits more than twice the number of sensory characteristics accessible to drinkers [2]. Although beer is typically classified through the analysis of a given set of attributes, sensory analysis is the only way to evaluate its flavors and aroma characteristics.

Along with monitoring raw materials and the individual steps in the brewing process, sensory analysis is utilized in the evaluation of the final product and for quality control purposes. It is also an important instrument for determining flavor stability, for developing and optimizing products as well as for empirically assessing and processing complaints.

2 SENSORY ANALYSIS IN THE BREWING INDUSTRY

Sensory evaluation draws on visual, olfactory, gustatory and haptic impressions. Olfactory impressions comprise all odors that can be perceived through the nose. All of the characteristics identified by the tongue and within the mouth cavity include the five tastes: sweet, sour, salty, umami (glutamate-like or savory) and bitterness, as well as the perception of temperature. These are classified as gustatory impressions. Haptic characteristics of food are, for instance, consistency, structure and texture [3]. Figure 1 illustrates how a flavor diagram can be created for any given food product.

One unique feature of beer is its CO_2 content. Initially, upon pouring the beer, the foam hinders all aromas from escaping the surface of the beer. It is not until the bubbles have begun to collapse and some of the CO_2 has dissipated that aromas can be detected. The CO_2 content also influences how basic tastes are perceived on the tongue in various ways [4]. Non-volatile flavor compounds are perceived through the taste buds and are influenced by the further release of carbonation. When swallowing, retronasal olfaction of aroma compounds occurs, as they ascend out of the throat into the retronasal passage. Through this process, the olfactory stimuli are more intense than if they were perceived directly by the nose, due to the higher temperature of the throat [3].

Figure 1: An example of a flavor diagram [3]

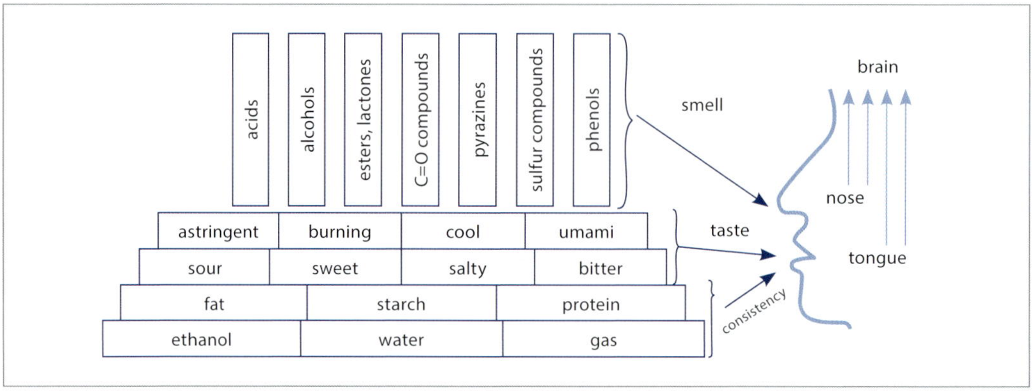

2.1 SENSORY EVALUATION OF BEER

Suitable sensory analysis methods are described in the various collections of methods published by MEBAK [5], the ASBC [6] and the EBC [7] as well as those from the German Institute for Standardization (DIN, Deutsche Institut für Normung) [8, 9, 10, 11, 12, 13, 14, 15, 16, 17, 18, 19], among others [20].

2.1.1 DISCRIMINATIVE TESTS

Discriminative tests involve comparing samples to one another. This method is also useful for revealing differences or similarities among beers that are very similar. It should be noted, however, that samples are not necessarily similar just because no significant difference among the samples is observed. Discriminative tests include the triangle test, the two-out-of-five test and the ranking test [5, 6, 7].

Figure 2: A "flavor wheel" showing the main categories and subcategories of descriptors [23]

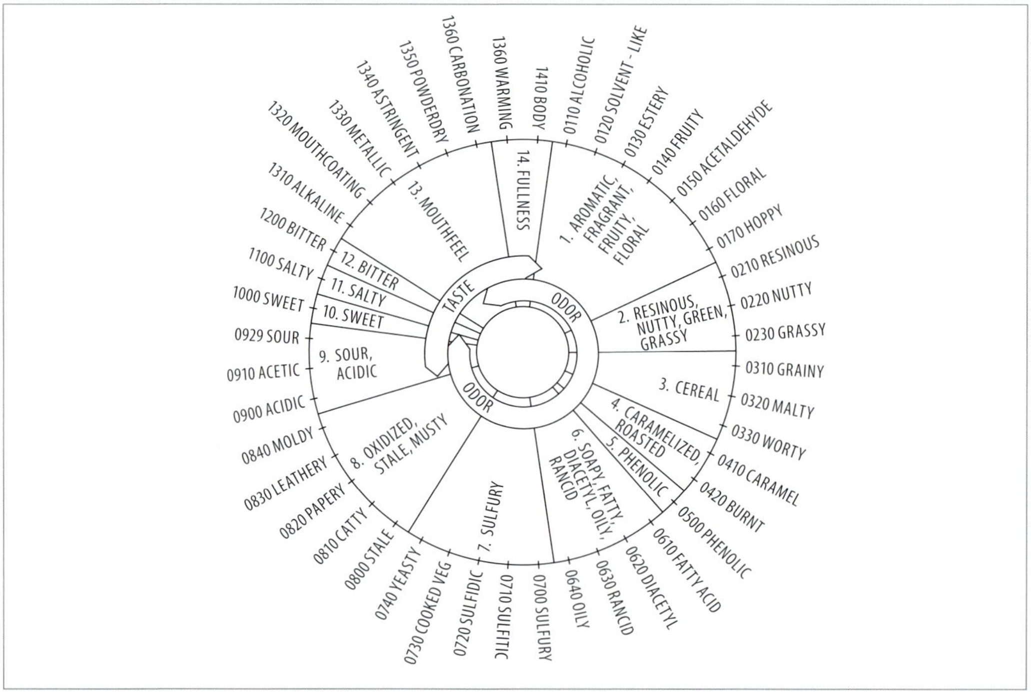

2.1.2 DESCRIPTIVE TESTS

As the name suggests, descriptive tests have been designed to describe samples qualitatively according to appearance, aroma, flavor and texture. In this kind of test, it is important that the evaluators are capable of describing the perceived impressions adequately. When performing a descriptive test at a beer tasting, one usually refers to the "flavor wheel" invented by Meilgaard (fig. 2). It encompasses 14 categories of terms to describe sensory impressions, with 44 main categories and 78 subcategories of descriptors. Categories 1–8 relate to the sense of smell, whereas categories 9–14 refer to taste [24]. In 1982, 27 reference substances were identified, which are associated with the 122 flavor descriptors defined by Meilgaard [25]. The flavor wheel provides assistance in selecting the terminology used in other sensory tests. In addition, it aids in determining which individual flavor impressions can be assigned to the same substance.

2.1.3 EVALUATION TESTS

When performing an evaluation test, qualitative impressions, such as malty, hoppy, floral, mouthfeel, drinkability and bitterness, are classified in a quantitative manner and assigned intensity values. This is carried out according to a scale, either with or without numbers, which is used to rate an impression from weak to strong. Two of the most established evaluation tests specifically designed for beer include the DLG test [5] and the aging test according to Eichhorn [26]. More recent work in the field of sensory analysis has resulted in the development of additional evaluation tests [27, 28].

2.1.3.1 Tasting scheme for determining aging according to Eichhorn

Eichhorn developed a testing scheme for the sensory evaluation of aged beers. In this scheme, only the degree of aging is evaluated. The criteria are derived from the DLG scheme, though only aroma, flavor and the quality of the bitterness are assessed. The degree of aging is appraised using a four-point scale ranging from 1 to 4, in which 1 represents an absolutely fresh sample and 4 an extremely aged sample. Furthermore, the taster evaluates the acceptance in percent (table 1). For example, an acceptance value of 20 % shows that only one in five tasters would find the flavor of this beer acceptable (cf. Flavor Stability). It is essential that the members of the tasting panel receive training on how to properly identify the flavors associated with aging prior to performing the analysis [26].

Table 1: Evaluation of aging characteristics in two beer samples (A, B)

	A		B	
	fresh	artificially aged	fresh	artificially aged
aroma	1.0	1.7	1.0	2.0
flavor	1.0	1.9	1.0	2.3
bitterness	1.0	1.5	1.0	1.8
weighted score $\left(\frac{2 \times aroma + 2 \times flavor + 1 \times bitterness}{5}\right)$	1.0	1.7	1.0	2.1
acceptance [%]	100	68	100	63

2.1.3.2 Tasting scheme according to Kaltner

This tasting scheme (fig. 3) was developed by Kaltner especially for beers with a pronounced hop profile. Scales are used to separately evaluate the intensity and quality of the hop aroma as well as the hop bitterness, as perceived by smell and taste [29].

Figure 3: Tasting scheme for beers with a pronounced hop profile

Tasting scheme for beers

Hop Aroma

		0	1	2	3	4	5	6	7	8	9	10
			slightly perceptible		somewhat perceptible		moderately perceptible		clearly perceptible		intense	
	intensity											
			unpleasant								pleasant	
	quality											
Aroma			slightly perceptible		somewhat perceptible		moderately perceptible		clearly perceptible		intense	
	intensity											
Flavor			unpleasant								pleasant	
	quality											

These terms refer **exclusively** to the hop aroma.

Description of the hop aroma

		0	1	2	3	4	5	6	7	8	9	10
			slightly perceptible		somewhat perceptible		moderately perceptible		clearly perceptible		intense	
Aroma	fruity											
	floral											
	citrus											
	green-grassy											
	hop-spicy/vegetal											
	other											
Flavor	fruity											
	floral											
	citrus											
	green-grassy											
	hop-spicy/vegetal											
	other											

If no aroma impression can be detected, mark the box under zero (0).

Perception of bitterness

	0	1	2	3	4	5	6	7	8	9	10
		mild								strong	
overall intensity											
initial impression											
flavor											
aftertaste											
harmony											

Harmony describes the personal overall impression for this beer style. 0 = not at all, 10 = completely harmonious

Bitterness Profile

normal (harmonius)

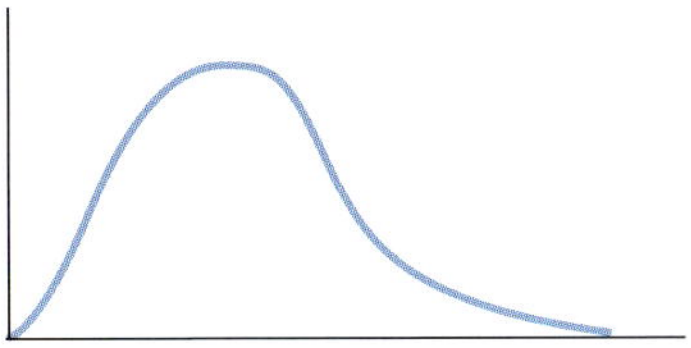

scratchy

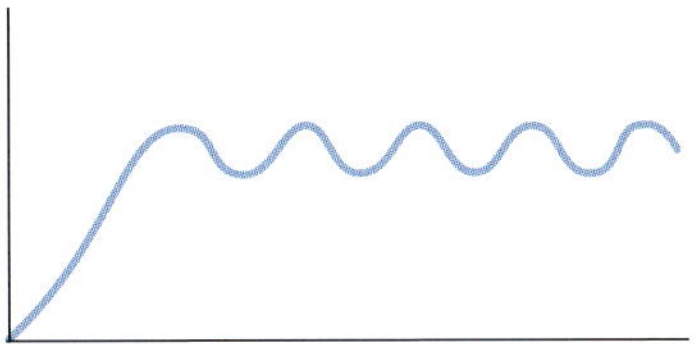

diminished

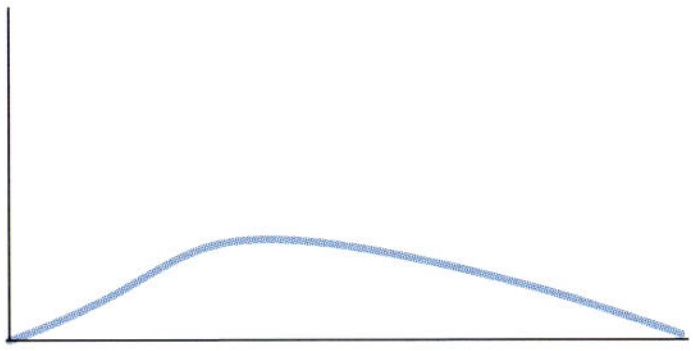

hard

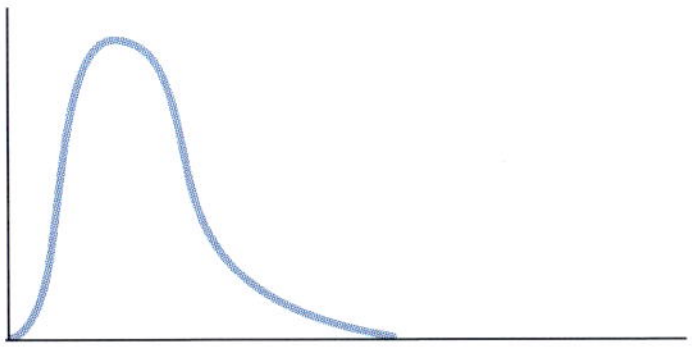

lingering and unpleasant

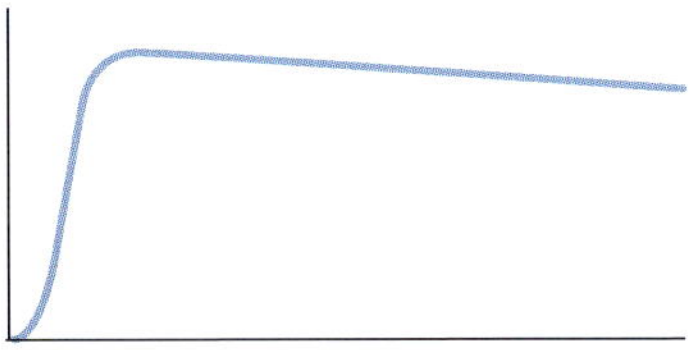

balanced, but with a lingering aftertaste

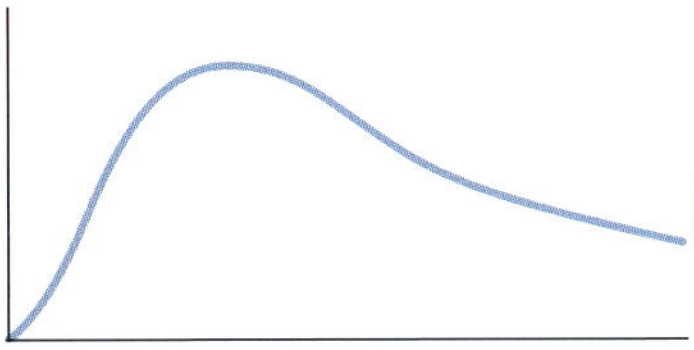

initial flavor is rounded, but lingers

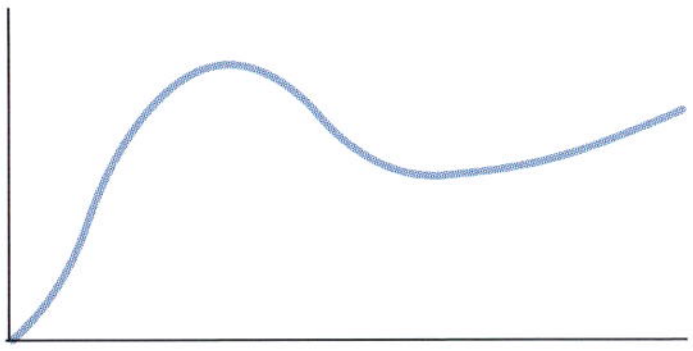

2.1.3.3 EBC Descriptive Analysis

Those trained in sensory analysis can utilize this method for describing the flavor and aroma of beer samples in a systematic manner. The applicable attributes are determined in advance. If possible, tasters can be trained to use the flavor wheel by practicing with reference standards to familiarize themselves with the relevant sensory attributes. The tasters then assign values for intensity to each of the selected attributes in the sample. The mean intensity value is then calculated for each attribute. The results can be illustrated as a graph or a spider diagram. This method is equally suitable for product comparisons and for purposes of internal quality control [7].

2.1.3.4 Tasting scheme for wheat beers

Southern German wheat beers can be divided into four main groups based on their aroma: estery, phenolic, yeasty and neutral. A specific tasting scheme has been developed to describe these criteria (fig. 4). In this scheme, the intensity as well as the quality of each primary aroma is assessed in addition to performing a standard DLG evaluation. A score of 1 is awarded if the intensity is imperceptible, whereas a score of 5 represents an easily discernible impression. In terms of quality, a score of 1 indicates that the product is considered unpleasant, while a score of 3 is neutral and a score of 5 is positive. Typical agents producing estery notes are 3-methylbutyl acetate (fruity, banana-like aroma) and ethyl acetate (solvent-like aroma). 4-Vinyl guaiacol primarily defines the phenolic note. Old yeast notes are usually associated with higher concentrations of short chain fatty acids and various higher alcohols. Neutral wheat beers are characterized by moderate concentrations of the relevant aroma compounds, which are also reflected in lower values for intensity [30].

Figure 4: Southern German wheat beer tasting scheme developed by Herrmann and Sacher

Taster: ____________________ **Sample:** ____________________

Date: ____________________

Aroma	Flavor	Fullness	Liveliness	Bitterness

Estery note					Phenolic note				
intensity					intensity				
1	2	3	4	5	1	2	3	4	5
quality					quality				
1	2	3	4	5	1	2	3	4	5
Yeasty note					**Malty note**				
intensity					intensity				
1	2	3	4	5	1	2	3	4	5
quality					quality				
1	2	3	4	5	1	2	3	4	5

Description

....

....

....

2.1.3.5 Modified "trueness-of-type" scheme

Schönberger created a modified test based on the "trueness-of-type" scheme developed by the Institute of Brewing. This scheme can be tailored specifically to the samples being evaluated. Therefore, it is important to define the sensory goals in advance and select the applicable positive and negative attributes by using a flavor wheel or other tool (see section 2.1.2). When doing so, one must determine which attributes must be fulfilled for the specific purpose of the evaluation along with the attributes considered to be faults. Positive attributes may include, for example, sweet, bitter, hoppy, malty, floral or fruity, whereas diacetyl, DMS, sourness and oxidation are considered negative.

Each of the positive attributes selected is evaluated according to a scale ranging from 0 to 3. A value of 3 represents an optimal intensity for the positive attribute; a value of 1 may describe an intensity that is either too weak or too strong. The negative attributes are evaluated on a scale from 0 to 3 and then totaled, which should be as low as possible. The final value is calculated by finding the difference between the sum of the points given for positive attributes and the sum of points given for negative attributes [28, 31].

The evaluation criteria can be selected individually, based on the product, the number of chosen attributes or the ways in which the scheme is applied (e. g. by tasting fresh and artificially aged samples for comparison or evaluating off-flavors). A suitable analysis can also be illustrated with the help of a spider diagram.

2.2 DESIGNING ROOMS FOR PREPARING AND PERFORMING SENSORY ANALYSIS TESTS

In order to perform the best sensory analysis possible, it is important that testers are not influenced in any way by their environment or their own state of well-being, so that objective evaluations can be made.

The minimum requirements for tasting rooms are defined in the DIN 10962 standard [21]. The tasting room must have at least 10 tasting stations of the same type. The table at each station should be 80 cm wide and at least 50 cm deep. If possible, tasting booths or carrels are preferable. These are essentially tables enclosed by three walls and equipped with a panel that can be opened and closed to allow the samples to be placed in the booth [3]. The tasting laboratory should be separated from the sample preparation area and should also be soundproof, so that the tasters are not influenced by external noises. Aside from the tasting laboratory, there should also be a group tasting table, where the tasters can sit to discuss and interpret their results. The preparation room should be designed in such a way that the samples can be prepared quickly. The work surfaces, tasting station surfaces and the floors must be easy to clean. They should be neutral in color and odorless as well. This ensures an undisturbed tasting atmosphere for all testers. Furthermore, the rooms must be equipped with a fully functional ventilation system. When designing the tasting rooms, it should be clear from the outset how often and how extensive the tastings are expected to be. When equipping the rooms, the option of collecting data electronically should be explored, as it facilitates subsequent data analysis [22].

2.3 SELECTING AND TRAINING A TASTING PANEL

Candidates who come under consideration as tasters for sensory analysis tests may be drawn from the following groups:

- untrained laymen (= consumers)
- trained tasters with the appropriate certification
- specialists or experts with specific knowledge of the product
- supervisors or those heading a tasting panel

2.3.1 SELECTING CANDIDATES TO SERVE AS MEMBERS OF THE TASTING PANEL

When first assembling a permanent tasting panel, a decision must be reached as to whether internal or external testers should be used. The advantage of an external tasting panel is that the testers are unbiased and motivated. However, higher costs as well as the need for more intensive training are among the most significant disadvantages [26].

A further factor in the decision is the size of the panel. This is relevant for the required statistical certainty. A group size of at least 10–14 testers is recommended; less than 7 does not allow for any statistical claims to be made. In order to secure the necessary number of testers, one must train at least two to three times as many testers as are needed.

One prerequisite for a tester is their ability to describe flavor and aroma attributes in a clear and understandable manner. It is highly beneficial if the tester possesses comprehensive experience in sensory analysis, enabling them to associate sensory impressions with past experiences and express their opinions. The testers must be in good health, should not suffer from allergies nor have any strong aversions to food ingredients. Moreover, age also plays a role when selecting the panel members. In particular, the sense of smell becomes compromised in individuals over the age of 50. The perception of bitterness and saltiness as well as localized perceptions for particular substances may also diminish with age. Sensory studies have shown that older testers have personal flavor thresholds 2 to 9 times higher than younger individuals [32, 33, 34]. Flavor perception may also be strongly affected by medication or pregnancy [35, 36].

The goals set for the tasting determine the extent to which the tasting panel must be trained and how sensitive the tasters must be. For most sensory tasks, average sensory abilities are normally sufficient. As a member of a tasting panel, it is important to possess the following abilities: good judgment, concentration, reliability, sensory memory and willingness to cooperate in a team.

2.3.2 TRAINING TO IDENTIFY THE BASIC TASTES

The basic tastes include: sour, salty, sweet, bitter and umami. The recommended concentrations for the reference substance employed in training sessions are listed in table 2. The solutions should first be prepared on the day of the tasting. Reference substances on the flavor wheel that induce the corresponding taste impressions are as follows: acetic acid (sour), sucrose (sweet), vanillin (sweet), sodium chloride (salty), isohumulone (bitter), iron sulfate (metallic) and γ-aminobutyric acid (astringent) [7]. Citric acid is also recommended for training the perception of sourness.

Table 2: Reference substances for the basic tastes and their recommended concentrations in water for training sensory analysis panelists

Taste	Reference substance	Concentration in g/l [8]
sweet	sucrose	6.0
sour	citric acid	0.4
salty	sodium chloride	1.3
bitter	isohumulone	0.012
umami	monosodium glutamate	0.5
astringent	γ-aminobutyric acid	0.051

Umami has only recently been recognized as a basic taste after it was discovered that its perception occurs through a unique path that is independent of the other four basic tastes [37]. Monosodium glutamate, which is often used as a flavor enhancer in foods, serves as the reference substance for umami. By contrast, astringent is not considered to be one of the basic tastes. Nevertheless, it is recommended that it be treated as such.

During training sessions, a water sample (tap water) should be provided in addition to the six reference standards and declared as a sample. A total of ten samples should be prepared for the test. These should include the water sample and the six reference standards, along with three duplicate samples (table 3).

The test should be conducted as follows: Diluted solutions containing sucrose, sodium chloride, acetic acid, isohumulone, monosodium glutamate and γ-aminobutyric acid should be placed on the table at each testing station. The tasters should taste the samples from left to right, entering an "x" in the corresponding column. Going back to previous samples and tasting them again should be avoided.

Table 3: Example of a testing sheet for identifying the basic tastes

Sample	Not detectable	Sweet	Salty	Sour	Bitter	Umami	Astringent	Correct	Incorrect
A				X				X	
B		X						X	
C			X					X	
D					X			X	
E						X		X	
F		X						X	
G							X	X	
H	X							X	
I					X			X	
J						X		X	

Tasters should be able to correctly identify eight of ten samples. This test should be repeated several times to judge the consistency of the tester. Practical experience has shown that bitter and sour samples in particular are more difficult to identify than sweet ones. The concentrations of the substances may be altered to determine the respective sensory thresholds of each tester.

2.3.3 TRAINING WITH REFERENCE SUBSTANCES

Aside from training the basic tastes, it is especially crucial for testers to gain experience in identifying key aromas by utilizing the relevant reference substances. There continues to be a discrepancy between the flavor and olfactory impressions found on the flavor wheel and the available substances used to train for their identification. Table 4 provides an overview of which reference standards are available for training sensory perception in a targeted manner and a list of the corresponding threshold values as individual substances in beer.

Table 4: Available reference substances for specific aromas found in beer

Number on the flavor wheel	Descriptor (according to EBC/ASBC)	Reference substance	Threshold in beer	Source
0131	isoamyl acetate (banana, estery)	3-methylbutyl acetate (isoamyl acetate)	1.6 mg/l	[40]
0132	ethyl hexanoate (apple, anise, fruity)	ethyl hexanoate	0.23 mg/l	[40]
0150	acetaldehyde (green apple)	acetaldehyde	10–25 mg/l	[40]
0161	floral, rose-like	2-phenylethanol	40–125 mg/l	[40]
0162	rose-like	geraniol	29 µg/l	[41]
	hoppy, citrusy	R-linalool	5 µg/l	[29]
0224	almond-like	benzaldehyde	2 mg/l	[40]
0231	fresh grass	hexanal	33 µg/l	
0320	malty	2-methylbutanal 3-methylbutanal	200 µg/l 89 µg/l	
0500	carbolic acid (phenolic, clove-like)	4-vinyl guaiacol	0.7 mg/l	[40]
0504	medicinal	chlorophenol	not specified	
0611	caprylic acid (goaty)	caprylic acid	15 mg/l	[40]
0613	cheesy (old hops)	isovaleric acid	1.3 mg/l	[42]
0620	diacetyl (buttery, rancid)	diacetyl	0.15 mg/l	[40]
0721	H_2S (rotten eggs)	hydrogen sulfide	4 µg/l	[7]
0722	mercaptan (sewer odor)	ethanethiol	1.7 µg/l	[40]
0724	lightstruck (skunky)	3-methyl-2-butene-1-thiol	4.4–35 µg/l	[43]
0732	DMS (cooked cabbage)	dimethyl sulfide	50–100 µg/l	[40]
0810	catty, ribes (blackcurrant)	p-menthane-8-thiol-3-one	50 ng/l	
0820	papery	5-methylfurfural	20 mg/l	[40]

It should be noted that the individual aroma impressions of different substances often have an additive effect in their overall impression. For example, it was found that the sensory threshold for 2-methylbutanal is 200 µg/l in beer, compared to 89 µg/l for 3-methylbutanal. These are typical aroma compounds associated with aging. When both were added to a beer sample, the sensory threshold for 3-methylbutanal was lowered to 39 µg/l, while that of 2-methylbutanal dropped to a mere 18 µg/l. Such additive effects are well known in aroma research [38, 39] and should be taken into account not only when training tasters, but also when determining the amounts to be added to unadulterated beer samples.

When working with volatile compounds, the solutions should be prepared immediately prior to the tasting session and kept at a temperature of around 20 °C. The reference substances can be offered as samples meant solely for the testers to smell or also as solutions to be tasted (in water or beer). Due to the absence of masking effects in the sample solutions, the concentrations required are much lower than the concentrations typically found in beer. The sensory thresholds for various compounds in beer are given in table 4, though the concentrations may vary depending on the tasting panel. It is important to be aware of the fact that in tasting sessions, the perception of aromas may change as the concentration increases (e.g. from fruity to solvent-like) [30].

First, the compounds are described and matched to the corresponding reference substance. In the event that the tasting is intended to focus on specific key aroma compounds, such as certain off-aromas or off-flavors, the sensory threshold for the taster's recognition of these compounds may also be determined. This aids in the selection of particularly sensitive tasters for specific sensory analyses [7].

2.3.4 CONSUMER TASTE TESTS – EXPERT TASTE TESTS

2.3.4.1 Consumers as tasters in sensory tests

In order to document consumer preferences when performing market surveys or comparing one's own products to those of competitors, large groups of untrained individuals (n > 50) are often asked to weigh in with an opinion. In such cases, it is prudent to keep the questions very simple. Consumers should only have to decide whether they can detect a difference between the products and to choose which one they prefer. This sort of feedback involves spontaneous reactions that only represent immediate sensory impressions experienced at a given moment. When selecting participants, it is advisable to ensure that the group represents an average cross-section of the population, unless the product has been designed for a specific target group. Drawing on a sufficiently high number of random samples can help offset any influences on consumers brought about by the environment or differences in testing situations.

2.3.4.2 Experts as tasters in sensory tests

Experts are those who have completed a certified training program and possess expert product knowledge. Thanks to their experience, experts are capable of describing the products they taste in great detail (e.g. aging). The test results provided by experts are not only relevant for quality assurance purposes but can also be quite valuable when modifying recipes as well as when making investment decisions. Involving experts requires their long-term commitment to participating in future tastings. Their willingness can be encouraged with the appropriate motivation.

2.4 STATISTICS IN SENSORY ANALYSIS

The integrity of sensory analysis results must be monitored. Statements made by individual tasters remain subjective conclusions until they are critically evaluated. This requires the application of statistics. The primary purpose of these methods is to generate an objective appraisal of the data, to evaluate statistical outliers and to determine statistical validity. A specific statistical evaluation (e.g. in tables) exists for most sensory analysis methods. However, certain key metrics for statistical evaluation are often not provided but they should be calculated and included in sensory tests to facilitate better interpretation of the data.

Table 5: Sensory analysis methods and their statistical evaluation

Test	Statistical evaluation	Description
paired comparison (directional, affective)	probability theory, binomial theorem (Bernoulli), χ^2 method, significance tables for one-tailed and two-tailed tests, variance analysis	which of the samples contains the higher concentration of the substance added? which of the samples does the taster prefer? [48]
triangle test, two-out-of-five test, duo-trio test, sequential test	probability theory, binomial-theorem (Bernoulli), χ^2 method, significance tables for one-tailed and two-tailed tests	determination of whether a difference exists between two samples
discriminative test	sequential analysis, method for testing a statistical hypothesis	determination of sensory threshold values
ranking test	ranking analysis method, multinomial distribution, rank sum test according to Kramer, rank correlation coefficient method according to Spearman	ranking of samples based on a certain characteristic (attribute or preference)
evaluation test	probability analysis, variance analysis, mean, standard deviation, confidence interval, repeatability, reproducibility, variances, creation of frequency distributions from individual findings	evaluation of multiple samples based on a specific attribute
evaluation test with scale	probability analysis, variance analysis, mean, standard deviation, confidence interval, repeatability, reproducibility, variances, creation of frequency distributions from individual findings	evaluation of multiple samples based on a specific attribute
simple descriptive test, profile test	frequency distribution, creation of frequency distributions from individual findings, principal components analysis, clustering systems, discriminant analysis	description of a sample

Calculating the key metrics necessary for statistical evaluation is reduced to a simple task with the relevant software. A number of programs have been developed to perform statistical analysis of sensory data.

2.4.1 UNIVARIATE EVALUATION METHODS

Among others, the following characteristics are of interest in the sensory analysis of beer:

- mean
- standard deviation
- variance
- coefficient of variation

The standard deviation is the measure of the variation or dispersion of a set of individual results, e.g. values for intensity or the like. The variance is a measure for the dispersion of the values around the mean. The variance enables differentiation of two samples with the same scope and the same mean.

The coefficient of variation shows the relative dispersion of the results in relation to the mean. It should only be utilized in order to compare distributions for different values for the mean.

A distribution according to Student (t-distribution) is a probability distribution for a limited number of individual values ($n \leq 30$).) The factor t can be calculated by referencing the appropriate table in conjunction with the requisite degrees of freedom ($f = n - 1$). If the value for the factor t is multiplied with the standard deviation s – while taking the number of tasters into account – this yields the distribution range around the mean T (= confidence interval), in which, for example, 95 % of the values fall [22].

2.4.2 MULTIVARIATE EVALUATION METHODS

In some cases, the results collected during sensory analyses cannot be fully described using univariate statistical methods. It is often the case, especially when making changes to recipes or process technology, for example, that sensory analysis data are combined with data from chemical-technical analysis. In this situation, it is important to find out whether specific chemical-technical analysis results (e.g. DMS content) correlate with a related sensory impression (e.g. vegetal). Such information can only be accessed through multivariate evaluation methods. Here, the focus is not on an isolated variable but rather the simultaneous interplay between multiple variables. Common multivariate methods include:

- principal components analysis [44, 45, 46]
- clustering systems [47]
- discriminant analysis [48]

Another field of application for multivariate evaluation methods is the reduction of data. This allows a large number of correlated measurement variables – resulting from a principal components analysis, for example – to be reduced to a smaller number of factors. In subsequent analyses, the measured values for the variables are replaced with the smaller number of defining factors.

3 SUMMARY

Sensory analysis has continued to become more relevant in recent years. Above all, sensory tests are performed as part of new product development, to inspect incoming goods, for process optimization and to monitor consumer acceptance. Humans, with their innate sensory abilities, serve as the instrument performing the analysis in this case. In order for a person to perform sensory tests effectively, they must possess at least an average level of sensory perception. It is equally important that tasters receive the proper training so that they meet the needs of the respective company and are able to become familiar with the tests for performing sensory analysis.

Panel members selected to participate in tastings can be sourced either in-house or externally. The size of the tasting panel depends on the demands of the company. However, the panel should consist of at least 7 members to enable a statistical evaluation of the results.

Different types of tests can be utilized within the scope of sensory analysis. Table 5 (p. 190) provides an overview of various testing methods and their statistical evaluation.

In addition to chemical-technical analysis, sensory analysis is a useful tool for evaluating raw materials as well as intermediate and finished products.

4 OVERVIEW

Sensory impression	Compound	Possible causes
banana-like, estery	3-methylbutyl acetate	product of yeast metabolism
apple-like, fruity	ethyl hexanoic acid	frequent reuse of the same yeast for fermentation in CCTs
sweet, malty	higher alcohols	product of yeast metabolism
solvent-like, estery	ethyl acetate (unpleasant at higher concentrations)	product of yeast metabolism, can also be formed by wild yeast strains
carbolic acid (phenolic)	4-vinyl guaiacol	is considered desirable in wheat beers, can also be formed by wild yeast strains
alcoholic	ethanol	product of yeast metabolism
rose-like	2-phenylethanol	product of yeast metabolism (indicator for fermentation temperature)

Sensory impression – off-flavor	Compound	Possible causes
medicinal	chlorophenol/cresol	this off-flavor is caused by chlorine compounds, especially those found in sources such as overly chlorinated water, unsuitable packaging materials, chlorophenols and disinfectant residues, but it also may occur as a result of yeast autolysis
goaty	octanoic acid	medium chain fatty acids are formed during fermentation; yeast cells excrete these fatty acids into beer that has been matured or stored for excessively long periods; this substance is especially perceptible in non-alcoholic beers
pungent, apple-like	acetaldehyde	metabolic product of the yeast (low vitality), green beer compound as a result of incomplete maturation, may also arise through microbial contamination
worty		dependent on wort boiling and fermentation (extremely pronounced in non-alcoholic beers or beer produced with the interrupted fermentation technique)
caramel-like	furaneol	Maillard product; the grain bill contains too much caramel malt for *Helles* and pilsner beer styles; wort boiling: overheating on the heating surfaces; pasteurization process is too long, excessively long pasteurization or storage, can also emerge as the beer ages
burnt		too much roasted malt in the grain bill or low quality malt; scorching of portions of the mash, dependent on dark or roasted malts

Sensory impression – off-flavor	Compound	Possible causes
cooked cabbage, vegetal	dimethyl sulfide (DMS)	wort boiling: insufficient degradation of DMS-P, insufficient evaporation of free DMS; the concentration of DMS-P in the raw materials is too high; contamination with wort bacteria
unpleasantly sour, rancid	butyric acid	contamination with microorganisms during wort production, especially when returning hot break material and the last runnings to the brewhouse to recover extract; also possible in packaged beer as a result of secondary contamination (*Megasphaera cerevisiae*), autolysis of the yeast
buttery, rancid	diacetyl	the maturation period was too short; low yeast vitality, contamination with *Pediococcus damnosus, Lactobacillus casei, Lactobacillus coryniformis* and members of the Enterobacteriaceae family
sulfurous, struck match	sulfite (SO_2)	SO_2 formation during fermentation is too strong; utilization of sulfite for preservation and to increase flavor stability; generation during fermentation
rotten eggs	hydrogen sulfide (H_2S)	formed during fermentation (low yeast vitality) or through contamination
lightstruck	3-methyl-2-butene-1-thiol	the effect of light is dependent on the packaging material and the color of the glass bottle when utilizing conventional hops and hop products
sewer odor	mercaptan	during fermentation or through autolysis of the yeast
onion-like		through the presence of amino acids containing sulfur, through wort boiling (shearing forces), hot wort centrifugation, cavitation on the hot side of wort production
moldy/musty		microbial contamination or unsuitable storage conditions, exposure of kieselguhr and stabilizing agents to moisture/mold or in less common cases, used filter sheets and those which have been inadequately rinsed.
sour	various acids	microbial contamination with lactic acid bacteria, acetic acid bacteria or wild yeast strains *(Brettanomyces* spp., *Dekkera* spp.)
metallic		corroded metal vessels and pipes, iron-rich water, lipid oxidation

Sensory impression – hops	Compound	Possible causes
floral	geraniol	the amount of hops and the timing of the addition; hop variety
hoppy aroma, citrusy	linalool	the amount of hops and the timing of the addition; hop variety
bitter	isohumulone, among others	the amount, the form and the method in which the hops are added can affect bitterness; a harsh bitterness can emerge as the volatile aroma compounds found in bitter hops undergo aging processes; the quality of the bitterness in bottled beer may deteriorate during storage
sweaty, cheesy	isovaleric acid	old hops; yeast autolysis

Sensory impression – aging	Compound	Possible causes
malty	2-methylpropanal, 2,3-methylbutanal	aging, incorrect storage of the finished beer, exposure to oxygen
cardboard (cardboard)	*(E)*-2-nonenal	aging, incorrect storage of the finished beer
almond-like	benzaldehyde	oxygen uptake, aging, incorrect storage of the finished beer
potato-like	methional	aging, incorrect storage of the finished beer
honey-like	phenyl acetaldehyde	oxygen uptake, aging, incorrect storage of the finished beer
ribes (blackcurrant), catty	3-methyl-3-mercap-to-butyl formate	aging, incorrect storage of the finished beer, exposure to oxygen

5 REFERENCES

[1] Carpenter, R. P., Lyon, D. H., Hasdell, T. A.: Guidelines for Sensory Analysis in Food Product Development and Quality Control. Aspen Publ., Gaithersburg 2000.

[2] Clarke, R. J., Bakker, J.: Wine flavour chemistry. Blackwell Publishing, Oxford 2004.

[3] Busch-Stockfisch, M. (ed.): Praxishandbuch Sensorik in der Produktentwicklung und Qualitätssicherung. Behr's Verlag, Hamburg 2002.

[4] Cowart, B. J.: The addition of CO_2 to traditional taste solutions alters taste quality. Chemical Senses 23 (1998), no. 4, pp. 397–402.

[5] Jacob, F. (ed.): Collection of Brewing Analysis Methods. Sensory Analysis. Self published by MEBAK, Freising 2014.

[6] ASBC: Methods of Analysis, ASBC, St. Paul 1992.

[7] Van Erde, P.: Analytica-EBC. https://analytica-ebc.com. Fachverlag Hans Carl, Nürnberg 1998 and 2018.

[8] Norm DIN 10961, Schulung von Prüfpersonen für sensorische Prüfungen, August 1996.

[9] Norm DIN 10964, Sensorische Prüfverfahren – Einfach beschreibende Prüfung, Februar 1996.

[10] Norm DIN 10966, Sensorische Prüfverfahren – Intensitätsprüfung, Dezember 1997.

[11] Norm DIN 10950-1, Sensorische Prüfung – Teil 1: Begriffe, April 1999.

[12] Norm DIN 10967-1, Sensorische Prüfverfahren – Profilprüfung – Teil 1: Konventionelles Profil, Oktober 1999.

[13] Norm DIN 10950-2, Sensorische Prüfung – Teil 2: Allgemeine Grundlagen, Oktober 2000.

[14] Norm DIN 10967 – Beiblatt 1, Sensorische Prüfverfahren – Profilprüfung – Beispiele zur statistischen Auswertung, Juli 2000.

[15] Norm DIN 10967-2, Sensorische Prüfverfahren – Profilprüfung – Teil 2: Konsensprofil, Oktober 2000.

[16] Norm DIN 10969, Sensorische Prüfverfahren – Beschreibende Prüfung mit anschließender Qualitätsbewertung, Mai 2001.

[17] Norm DIN 10967-3, Sensorische Prüfverfahren – Profilprüfung – Teil 3: Freies Auswahlprofil, August 2001.

[18] Norm DIN 10967-4, Sensorische Prüfverfahren – Profilprüfung – Teil 4: Prüfpläne, Mai 2002.

[19] Norm E DIN 10959, Sensorische Prüfverfahren – Bestimmung der Geschmacks-empfindlichkeit (Entwurf), Juli 2005.

[20] The Institute of Brewing (ed.): Methods of Analysis. London 1997.

[21] Norm DIN 10962, Prüfbereiche für sensorische Prüfungen – Anforderungen an Prüfräume, Oktober 1997.

[22] Stone, H., Sidel, J. L.: Sensory Evaluation Practices. Academic Press, New York 1993.

[23] Meilgaard, M. C., Dalgliesh, C. E., Clapperton, J. F.: Sammlung beschreibender Ausdrücke für die objektive sensorische Beurteilung von Bier (II). Monatsschrift für Brauwissenschaft 32 (1979), no. 2, pp. 33–40.

[24] Meilgaard, M. C.: Beer Flavour. Dissertation, Technical University of Denmark, Lyngby 1981.

[25] Meilgaard, M. C., Civile, G.V., Carr, B.T.: Sensory Evaluation Techniques. CRC Press, Boca Raton 1999.

[26] Eichhorn, P.: Untersuchungen zur Geschmacksstabilität. Dissertation, Technical University of Munich (Weihenstephan), 1991.

[27] Lawless, H. T., Heymann, H.: Sensory Evaluation of Food: Principles and Practices. Chapman & Hall, New York 1998.

[28] Schönberger, C., Krottenthaler, M., Back, W.: Formulation of an alternative tasting scheme for the sensory evaluation of beer. Monatsschrift für Brauwissenschaft 57 (2004), no. 4, pp. 16–24.

[29] Kaltner, D.: Untersuchungen zur Ausbildung des Hopfenaromas und technologische Maßnahmen zur Erzeugung hopfenaromatischer Biere. Dissertation, Technical University of Munich (Weihenstephan), 2000.

[30] Herrmann, M.: Entstehung und Beeinflussung qualitätsbestimmender Aromastoffe bei der Herstellung von Weißbier. Dissertation, Technical University of Munich (Weihenstephan), 2005.

[31] Schönberger, C.: Bedeutung nicht-flüchtiger Geschmacksstoffe in Bier. Books on Demand, Norderstedt 2003.

[32] Fukunaga, A., Uematsu, H., Sugimoto, K.: Influences of aging on taste perception and oral somatic sensation. The journals of gerontology. Series A, Biological sciences and medical sciences 60 (2005), no. 1, pp. 109–113.

[33] Kaneda, H., Maeshima, K., Goto, N., Kobayakawa, T., Ayabe-Kanamura, S., Saito, S.: Decline in taste and odor discrimination abilities with age, and relationship between gustation and olfaction. Chemical Senses 25 (2000), no. 3, pp. 331–337.

[34] Stevens, J. C., Cain, W. S.: Changes in taste and flavor in aging. Critical Reviews in Food Science and Nutrition 33 (1993), no. 1, pp. 27–37.

[35] Duffy, V. B., Bartoshuk, L. M., Striegel-Moore, R., Rodin, J.: Taste changes across pregnancy. In: Murphy, C. (ed.): Olfaction and Taste XII – Annals of the New York Academy of Sciences, New York Academy of Sciences, New York 1998, pp. 805–809.

[36] Schiffman, S. S., Graham, B. G., Suggs, M. S., Sattely-Miller, E. A.: Effect of psychotropic drugs on taste responses in young and elderly persons. In: Murphy, C. (ed.): Olfaction and Taste XII – Annals of the New York Academy of Sciences, New York Academy of Sciences, New York 1998, pp. 732–737.

[37] Kurihara, K., Kashiwayanagi, M.: Introductory remarks on umami taste. In: Murphy, C. (ed.): Olfaction and Taste XII – Annals of the New York Academy of Sciences, New York Academy of Sciences, New York 1998, pp. 393–397.

[38] Guadagni, D. G., Burr, H. K., Buttery, R. G., Okano, S.: Additive Effect of Sub-Threshold Concentrations of Some Organic Compounds Associated with Food Aromas. Nature 200 (1963), no. 491, pp. 1288–1289.

[39] Stevens, J. C.: Detection of very complex taste mixtures. In: Murphy, C. (ed.): Olfaction and Taste XII - Annals of the New York Academy of Sciences, New York Academy of Sciences, New York 1998, pp. 831–833.

[40] Meilgaard, M. C.: Flavor chemistry of beer. II. Flavor and threshold of 239 aroma volatiles. Master Brewers Association of the Americas Technical Quarterly 12 (1975), no. 3, pp. 151–168.

[41] Steinhaus, M., and Schieberle, P.: Transfer of potent hop odorants linalool, geraniol and 4methyl4sulfanyl2pentanone from hops into beer. Proceedings of the 31st European Brewery Convention Congress, Venice. Fachverlag Hans Carl, Nürnberg 2007, pp. 1004–1011.

[42] Schneeberger, M.: Verwertung von Prozessbieren in der Brauerei. Dissertation, Technical University of Munich (Weihenstephan), 2006.

[43] Irwin, A. J., Bordeleau, L., Barker, R. L.: Model Studies and Flavor Threshold Determination of 3Methyl2Butene1Thiol in Beer. Journal of the American Society of Brewing Chemists 51 (1993), no. 1, pp. 1–3.

[44] Clapperton, J. F., Piggott, J. R.: Differentiation of Ale and Lager Flavors by Principal Components-Analysis of Flavor Characterization Data. Journal of the Institute of Brewing 85 (1979), no. 5, pp. 271–274.

[45] Daems, V., Delvaux, F.: Multivariate analysis of descriptive sensory data on 40 commercial beers. In: Food Quality and Preference 8 (1997), no. 5/6, pp. 373–380.

[46] Moll, M., Flayeux, R., Muller, P.: Relationship between flavour and the composition of barley and malt. Master Brewers Association of the Americas Technical Quarterly 18 (1981), pp. 31–35.

[47] Siebert, K. J.: Chemometrics in brewing – a review. In: Journal of the American Society of Brewing Chemists 59 (2001), no. 4, pp. 147–156.

[48] Brown, D. G. W., Clapperton, J. F.: Discriminant analysis of sensory and instrumental data on beer. Journal of the Institute of Brewing 84 (1978), no. 6, pp. 318–323.

[49] Zarnkow, M., Rossmann, S.: Sensory – Critical Review on Different Discrimination Tasting Methods. MBAA 2013 Annual Conference, Austin, Texas, USA, 2013-10-26.

Additional sources on the internet: (accessed in July of 2019)

http://www.dlg.org

http://www.doemens.org

http://www.sensometric.org

http://www.compusense.com

http://www.sensorysims.com

http://www.camo.com

http://www.minitab.com

http://www.virtualstatistician.com

FLAVOR STABILITY

1 INTRODUCTION

The aroma and flavor of beer do not remain constant. As beer ages, positive flavor and aroma compounds are lost while other compounds associated with aging become more dominant. The aging process occurs principally through oxidation reactions. A very small amount of oxygen, such as that which is unavoidable even with an optimal filling process, is sufficient to fuel these reactions. Consequentially, stability is not really the correct term, but rather flavor instability [1]. Oxidation reactions counter the effects of various reducing compounds in beer, which can slow the formation of compounds associated with aging.

Flavor stability is a key quality attribute of beer. It can be enhanced through brewing techniques or the addition of antioxidants in countries not subject to the *Reinheitsgebot*. Important measures for increasing flavor stability from the standpoint of brewing technology include the selection of raw materials and proper production techniques, especially during the fermentation process.

2 ASPECTS OF FLAVOR STABILITY

2.1 THE FUNDAMENTALS

2.1.1 THE AGING PROCESS

After filling, beer is subject to a natural aging process. This is a dynamic process, characterized by the formation and degradation of substances, which constantly change the beer's aroma profile as it ages. Figure 1 chronicles the changes to the sensory profile of beer over its shelf life. If beer is stored for a longer period, the off-flavors formed during aging can lose some of their intensity; nonetheless, the flavor profile remains in a state of constant change.

2.1.2 REACTION PATHWAYS OF COMPOUNDS RELEVANT FOR AGING

Lustig describes aging in beer as a dynamic process, consisting of the formation and degradation of compounds in beer, which leads to a change in the aroma profile. Multiple reaction mechanisms are responsible for the formation of the compounds associated with aging. Some of these mechanisms function alone; others are linked to various pathways [3]. Oxidation reactions are the primary mechanism behind the formation of aging aromas. Their formation can be delayed by maintaining low concentrations of oxygen in the beer and a high concentration of reducing substances. Figure 2 shows the most important reaction pathways for aroma compounds during aging.

Figure 1: Sensory changes during the aging of beer [1, 2]

- a loss of body
- an increase in bitterness
- a change in the aroma profile
  - loss of hop character (e.g. hop bouquet in pilsner)
  - decline in the ester aroma
  - decline in the phenolic note in weissbier
  - loss in malt aroma in dark beers
  - formation of aging compounds/flavors

initial stages (after several days or weeks)

↓ blackcurrant-like (Ribes) or catty flavor
cardboard flavor
bread-like, cracker-like
caramel-like, honey-like, sweet
sherry-like, whisky-like

final stages (after several months or years)

Figure 2: The reaction mechanisms responsible for aging in beer [3, 4]

- Maillard reactions → furans, pyrones, → heterocyclic compounds
- Strecker degradation of amino acids → Strecker aldehydes
- oxidative degradation of isohumulones → aldehydes, ketones
- lipid degradation
  - enzymatic oxidation of free fatty acids → hydroxy acids, oxoacids, aldehydes
  - thermal-oxidative degradation → aldehydes, ketones, lactones, acids, esters
  - photo-oxidation of free fatty acids → hydroperoxides, aldehydes
- involvement of non-volatile substances in beer
  - metal-catalyzed reactions → radical mechanisms, $Fe^{2+/3+}$, $Cu^{+/2+}$
  - melanoidins as H^+ donors and acceptors → shifting of the redox potential (positive values)
- carbonyl sulfite complexes → aldehydes, ketones

Strecker aldehydes are substances with low aroma and flavor thresholds. The concentration of these aldehydes increases significantly as the beer ages and as such, they are considered to be indicator substances.

Many of these carbonyl compounds stem primarily from the Maillard reactions, particularly from the Strecker degradation of amino acids. Aldehydes are reduced to the corresponding alcohols through the reducing power of the yeast (fig. 3) [3].

Figure 3: Amino acids and their reaction products arising from Strecker degradation

Amino acids		Aldehydes		Alcohols
valine		2-methylpropanal		2-methylpropanol
isoleucine	→	2-methylbutanal	→	2-methylbutanol
leucine		3-methylbutanal		3-methylbutanol
phenylalanine		2-phenylethanal		2-phenylethanol

Lipid degradation also generates other substances associated with aging. Long chain fatty acids can be degraded through enzymatic reactions, autooxidation or photo-oxidation. Over the course of lipid degradation, different intermediate products are formed such as hydroxy and trihydroxy fatty acids. Eventually, the degradation process leads to the formation of a number of aging compounds including hexanal, (*E*)-2-nonenal and isomers of 2,4-decadienal [3, 5]. In addition to free fatty acids, ethyl esters of unsaturated fatty acids may serve as a source of aging carbonyls [3]. A common aging compound is (*E*)-2-nonenal, which is responsible for a cardboard flavor in aged beer (tab. 1).

Table 1: Carbonyl aging compounds formed by the degradation of long chain unsaturated fatty acids [6]

Oleic acid ($C_{18:1}$)	Linoleic acid ($C_{18:2}$)	Linolenic acid ($C_{18:3}$)
heptanal	pentanal	(*E,Z*)-2,4-heptadienal
(*E*)-2-nonenal	hexanal	(*E,E*)-2,4-heptadienal
octanal	(*E*)-2-octenal	(*Z*)-3-hexenal
decanal	(*E*)-2-nonenal	
(*E*)-2-decenal	(*E,Z*)-2,4-decadienal	
	(*E,E*)-2,4-decadienal	

Carbonyl compounds can also be formed through exposure to light and in the presence of photosensitizers (e.g. riboflavin). The best known product of this reaction is 3-methyl-2-butene-1-thiol (lightstruck flavor). In combination with iso-α-acids, this compound is formed at wavelengths between 250 and 500 nm (especially in beer packaged in clear or green glass bottles) and has an extremely low sensory threshold at 30 ng/l [7, 8].

Reactive forms of oxygen also have a significant effect on flavor stability. Despite the fact that molecular oxygen is slow to react, a variety of reactive oxygen species (ROS), ranging from singlet oxygen to superoxide radical anions and even hydrogen peroxide (H_2O_2), can arise through thermal and chemical reactions. The latter, catalyzed by metal, reacts further to form extremely reactive hydroxyl radicals (fig. 4).

Figure 4: Metal-catalyzed reactions which generate radicals in beer [3, 9]

Haber-Weiss reaction (copper):

$$Cu^{2+} + {}^{\cdot}O_2^- \rightarrow Cu^+ + O_2$$
$$Cu^+ + H_2O_2 \rightarrow Cu^{2+} + {}^{\cdot}OH + OH^-$$

reaction products: $${}^{\cdot}O_2^- + H_2O_2 \xrightarrow{Cu} O_2 + {}^{\cdot}OH + OH^-$$

Fenton reaction (iron):

$$Fe^{2+} + H_2O_2 \rightarrow Fe^{3+} + {}^{\cdot}OH + OH^-$$
$$Fe^{3+} + H_2O_2 \rightarrow Fe^{2+} + {}^{\cdot}O_2^- + 2H^+$$

reaction products: $$2H_2O_2 \xrightarrow{Fe} {}^{\cdot}OH + OH^- + {}^{\cdot}O_2^- + 2H^+$$

Oxidative disinfection agents (especially H_2O_2 and peracetic acid) are problematic in terms of flavor stability and, if for no other reason, all residues of these agents must be removed to the greatest degree possible from piping and tanks after disinfection.

Since metal ions act as catalysts in the formation of radicals, their introduction to the production process should be avoided to the greatest degree possible. Consequently, care should be taken to ensure that the concentration of heavy metal ions (iron and copper) in stabilizing and filter aids (e.g. kieselguhr) is low [10, 11].

2.1.3 EFFECTS OF ANTIOXIDANTS

Antioxidants are substances, which occur in considerably lower concentrations than those capable of being oxidized, and are characterized by their ability to significantly delay or prevent oxidation [12]. Typical reaction mechanisms include:

- direct capture of oxygen
- chelation: binding of metal ions (e.g. Fe^{2+})
- prevention of chain reactions through the capture of free radicals serving as initiators
- interruption of chain reactions involving free radicals through the capture of an intermediate radical

Many substances are either already present in the raw materials or are formed during malt or beer production, such as sulfur dioxide, phenolic compounds, Maillard products and vitamins. These substances have antioxidative effects and can cause the reactions listed above.

Sulfur dioxide (SO_2) is a strong reducing agent. In beer, it is generally bound to carbonyl compounds (especially to acetaldehyde). Because this bond is reversible, bound SO_2 exhibits the same antioxidative capacity as free SO_2 [13, 14, 44]. Sulfate is formed through the oxidation of sulfur dioxide. This antioxidative effect is primarily attributable to the fact that it inhibits reactions that give rise to free radicals. Here, SO_2 reacts with organic and inorganic peroxides, thus preventing the formation of free radicals [14, 15]. At normal storage temperatures, up to 3 mg/l SO_2 can be consumed per month in bottled beer. Oxidation reactions are suppressed as long as a sufficiently high concentration of SO_2 is present (over

3 mg/l based on the authors' experience). A strong correlation exists between lag time and the sulfur dioxide content in beer (fig. 5).

Figure 5: The relationship between lag time and the sulfur dioxide content of beer

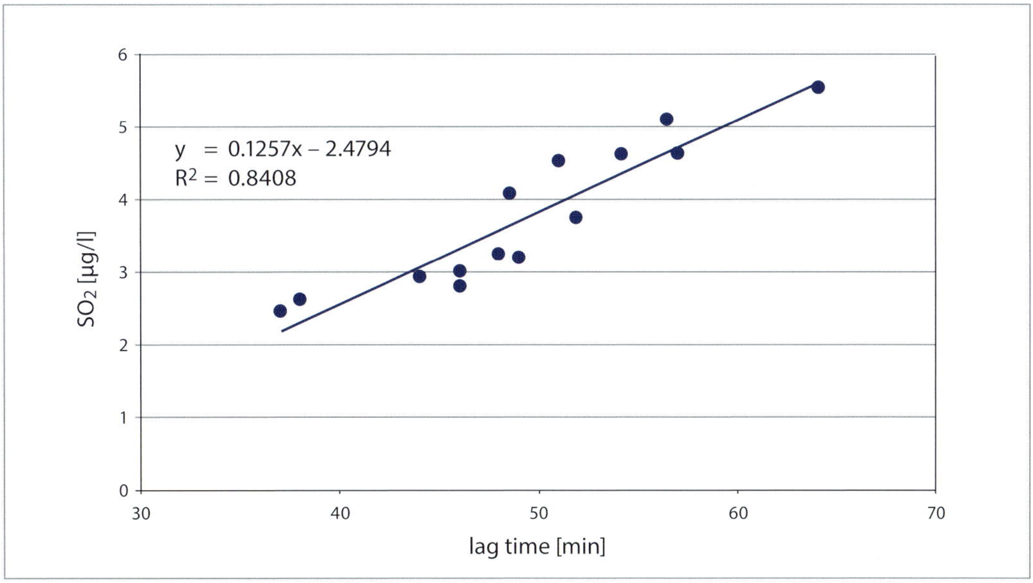

Phenolic compounds are known for their reducing and antioxidative effects. The following functional groups have been identified as being particularly advantageous. Figure 6 depicts the groups on a phenolic compound that are responsible for this antioxidative effect [16]:

- 5-OH group on the A ring → formation of a chelate complex with metal ions
- ortho-dihydroxy functional group on the B ring
- 2,3 double bond and a 4-oxo group on the C ring

Figure 6: Functional groups associated with antioxidative effects, depicted on a flavonoid compound

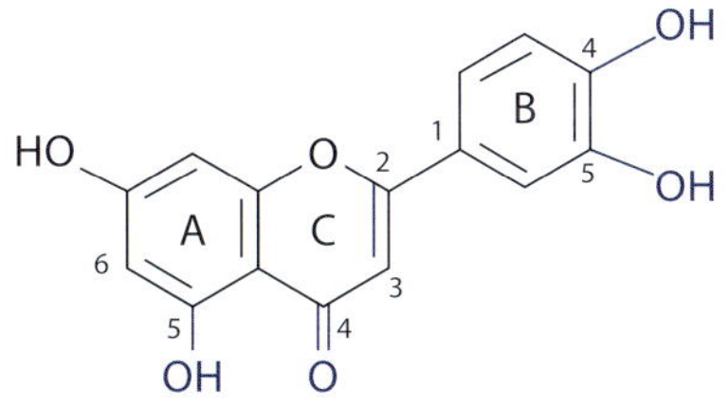

Figure 7 shows reactions involved in the generation of reactive oxygen species (ROS) [17], which can act as antioxidants in beer.

Figure 7: The antioxidative impact of sulfur dioxide and polyphenols on ROS [14, 15, 17, 18, 19]

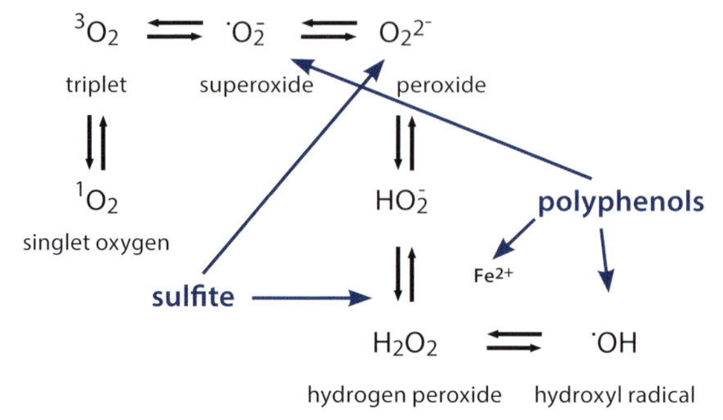

The Maillard reactions (non-enzymatic browning) result in pigments, which are known as melanoidins. They contain nitrogenous compounds of different molecular weights and degrees of solubility in water [6].

The vast majority of melanoidins are formed during the kilning process (section 2.2.1). In the brewing process, melanoidins assume the role of antioxidants. For instance, they suppress the oxidative degradation of isohumulone as well as the auto-oxidation of linoleic acid [20].

Some substances (e.g. Maillard products) possess a reductive capacity and thus are classified as reductones if they exhibit the structure shown in figure 8.

Figure 8: A reductone and its dehydroxy derivative [6]

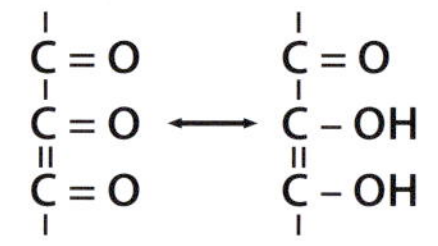

2.2 INFLUENCING FLAVOR STABILITY IN THE BREWING PROCESS

2.2.1 BARLEY AND MALTING

A sequence of events in the brewing process, both pro-oxidative as well as antioxidative, plays a definitive role in flavor stability. Table 2 lists the more important substances in barley and malt which can impact flavor stability [21, 22].

Table 2: Pro-oxidative and antioxidative substances in barley and malt

Pro-oxidative substances

- oxygenated fatty acids
- aldehyde products of lipid degradation
- pro-oxidative enzymes (lipoxygenase, polyphenol oxidase)
- heavy metals
- activated oxygen (e.g. from enzymatic reactions)

Antioxidative substances

- phenolic substances
- melanoidins and other Maillard products

The concentration of pro-oxidative and antioxidative compounds in malt, wort and beer can be influenced by the selection of raw materials and the appropriate parameters for each stage of the malting process. Certain antioxidative substances (e.g. melanoidins) created during the production of malt figure prominently in the aroma and flavor profiles of beer. In their function as reductones, these substances can positively impact flavor stability. Conversely, numerous products of the Maillard reactions (e.g. Strecker aldehydes) in addition to pro-oxidative substances (e.g. oxygenated fatty acids) have a negative effect on flavor stability.

Malt made from barley cultivated in moist and mild regions (e.g. maritime climates) exhibits a higher content of phenolic compounds as well as a lower polymerization index during mashing. Practical experience indicates that beers brewed with these malts have better flavor stability [10].

Flavor stability can be predicted through analysis of the malt. The aroma compounds present in the malt are measured using cold extraction or the Congress mash method. The expected flavor stability can be estimated based on these analysis results. The variety, the year of harvest and especially the cultivation region have an effect [3, 23]. The concentration of aging compounds in beer correlates very closely with the degree of proteolysis in the malt because a higher proportion of low molecular weight nitrogen results in the formation of more Strecker aldehydes. For this reason, any measures which promote proteolysis, e.g. higher moisture levels or intense modification processes during germination, have a negative effect on flavor stability. Similarly, an extended protein rest during mashing also enhances protein degradation (cf. Mashing).

The degradation of lipids is of great importance during malting since the formation of intermediate and final products have a direct influence not only on flavor stability but also on the flavor and the foam attributes of beer. Formation of aroma compounds derived from unsaturated fatty acids in barley occurs through enzymatic oxidation, particularly by lipoxygenases (LOX), as well as through autooxidation (section 2.1.2, p. 199). LOX activity in malt varies according to the variety and provenance [24, 26]. Hydroperoxides are generated from the fatty acids; however, these are unstable and degrade, forming short chain carbonyls and acids. Several of these degradation products, such as (*E*)-2-nonenal, are involved in the development of an aged flavor in beer [5, 24, 25].

The withering stage of kilning also plays a role in flavor stability. The concentration of aroma compounds generated through exposure to heat, such as Strecker aldehydes, furans and N-heterocyclic compounds, increases during withering. Naturally, higher temperatures (> 60 °C) at the beginning of the process bring about a correspondingly higher degree of thermal stress. This not only has a negative effect on the malt quality and that of the beer, but also on the flavor stability. By contrast, starting the withering process at lower temperatures (35–50 °C) and conducting it over a longer time interval leads to beers with higher flavor stability (fig. 9) [27, 28].

During the curing process, the temperature is not the deciding factor in the generation of Maillard products, but rather the amount of thermal stress to which the malt is subjected. Less thermal stress results from higher curing temperatures at shorter intervals compared to lower curing temperatures over longer periods. In addition, the rate at which the temperature is raised during curing is not a factor as long as the amount of thermal stress remains the same [27]. High levels of thermal stress during curing lead to an increase in the amount of aging compounds in beer subjected to forced aging (fig. 10). In kilning, low values for DMS precursor are equally as important as those for low thermal stress (TBI), making it important to work within a certain window with respect to time and temperature. The relevance of this relationship is explained in the chapter Malt.

Figure 9: Sum of aging compounds in beer as a function of the withering process [27]

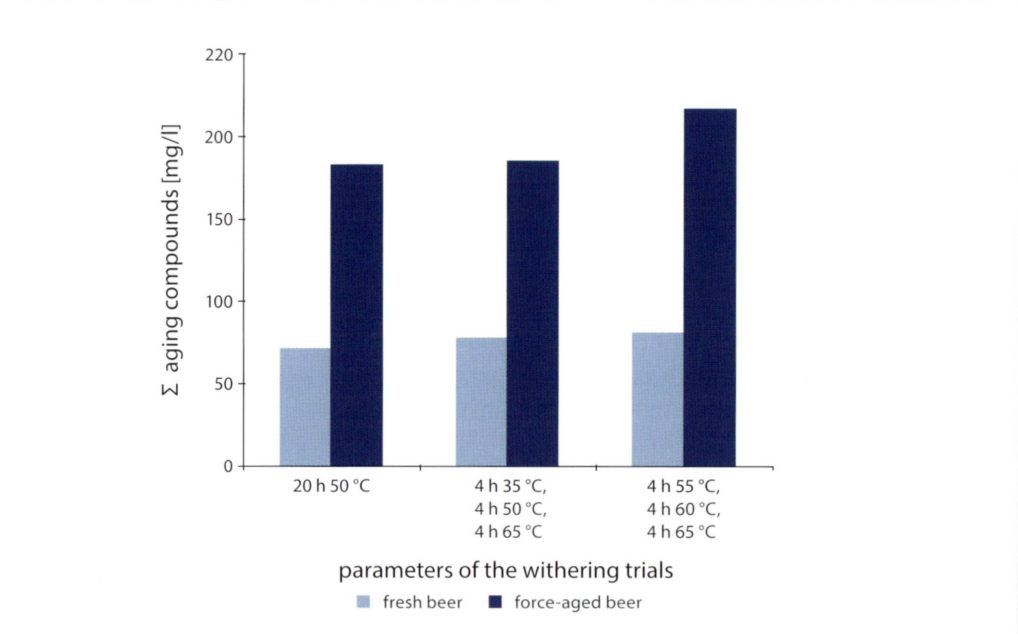

Figure 10: Sum of aging compounds in beer as a function of the curing temperature [28]

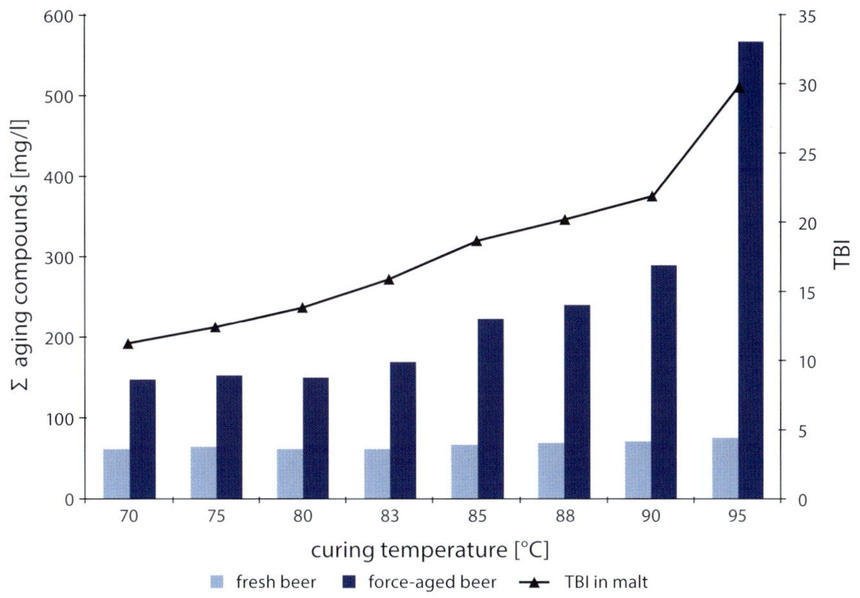

By varying the kilning parameters, darker malts can be created for the express purpose of imparting sweet, caramel-like, nutty or roasted aromas to dark beers. Darker malts result in beers with a high level of flavor stability, as they are produced with a more intense withering process (slower extraction of water) and are then added in large quantities (75 %) to the grist. Strecker aldehydes are considered true to style in the sensory profile of dark beers, generating more positive evaluations, while the same compounds are perceived as negative in paler beers [29, 30]. However, a great number of reductones are formed during the kilning process that play key roles in the color and aroma profile of the beer, not to mention in elevating the antioxidative capacity of the malt [31].

Color malts (e.g. caramel malts) or liquid coloring agents made from roasted malt are recommended for adjusting the color of paler beers rather than utilizing darker malts (cf. Mashing). If dark malts are employed for color correction, their inclusion introduces low amounts of Strecker aldehydes and thus aging compounds.

By contrast, dark beers brewed with roasted malt exhibit a less malty aroma and a lower flavor stability compared with beers produced with dark malts. Furthermore, they often possess an undesirable burnt character due to exceedingly high concentrations of pyrazines and other heterocyclic compounds found in roasted malt [27, 29].

2.2.2 MILLING AND MASHING IN

Under certain storage conditions, the lipids in malt tend to undergo degradation through autooxidation. This can be measured analytically. After the milling process in particular, the susceptibility of unsaturated fatty acids to peroxidation reactions increases dramatically [26]. However, a large number of other compounds, such as polyphenols or Maillard products are also affected through exposure to oxygen. Brewhouse operations should be planned so that grist is stored for the shortest time possible after milling.

The rate of enzymatic oxidation of lipids by LOX reaches a peak immediately after mashing in. For this reason, it is advantageous to avoid creating suitable reaction conditions for LOX in the first place, that is, when the grist initially comes into contact with the brewing liquor. Several systems have been developed to facilitate a rapid process for mashing in, free of clumps and with minimal exposure to oxygen (e.g. wet milling systems, pre-mashing technology, grist hydrators, etc.). The solubility, and therefore the effect of LOX during mashing, can be influenced by how finely the grist is milled and the corresponding size of the acrospire.

Milling systems (e.g. wet milling, conditioning) which preserve the structure of the acrospires, lead to incomplete extraction of the compounds found in them, thus minimizing the oxidative degradation of lipids in the brewing process [26].

Hammer mills designed for the new generation of mash filters produce grist particles of a very small size. Here, milling not only greatly increases the accessible surface area of the particles, it also raises the temperature, which may exceed 50 °C during the warmer seasons of the year. For this reason, modifying the conditions under which the grain is milled is recommended in order to accommodate an inert atmosphere (nitrogen).

Dropping the mash pH and mashing in at high temperatures (> 60 °C) bring about a significant reduction in LOX activity (section 2.2.3). Conditioning malt with hot water (> 80 °C; hot conditioning) results in extensive inactivation of LOX prior to wet milling. This technology can be readily utilized anywhere mashing in occurs at temperatures above 60 °C. The procedure is designed to minimize the enzymatic degradation of lipids during mashing, with the expectation that enhanced foam stability and flavor of the beer will result [26].

Oxygen uptake during mashing in leads to a significant amount of oxidation reactions. This oxygen is then transferred to the wort, measurable as a rise in the polymerization index. Furthermore, the color increases and poor lautering characteristics can be observed as a result of oxygen uptake [28, 32].

The oxidation processes occurring during mashing, especially during mashing in, can be monitored by means of detection with chemiluminescence (fig. 11). Even if the strike water is de-aerated prior to mashing in, the oxygen in the grist still presents a problem. Therefore, integrating an additional step to purge the oxygen from the grist (with CO_2 or N_2 gas) prior to mashing in would be advantageous.

Figure 11: Change in the chemiluminescence signal during the process of mashing in (60 °C, 1 h) [33]

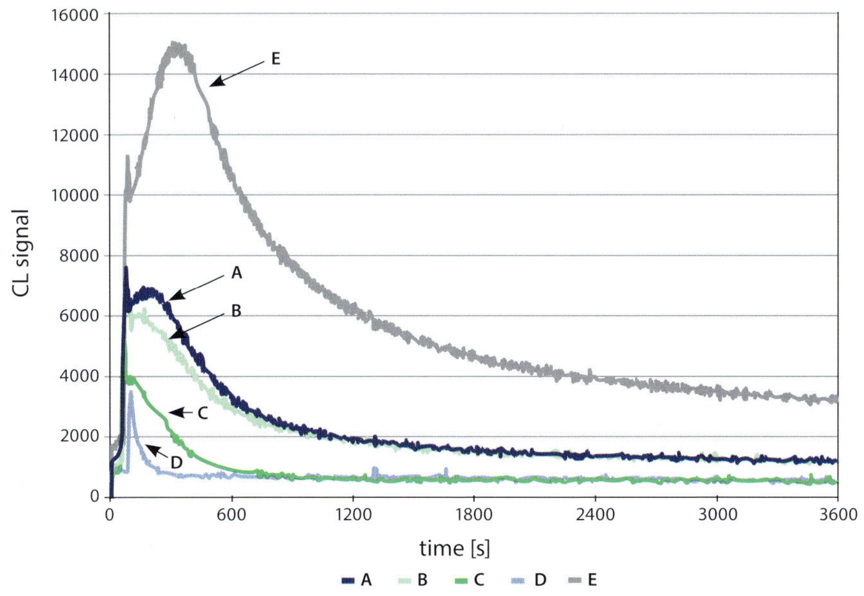

Legend:
A: normal conditions
B: usage of de-aerated water (N_2)
C: malt grist stored under an inert atmosphere (N_2) prior to mashing in
D: malt grist stored under an inert atmosphere (N_2) prior to mashing in and usage of de-aerated water
E: malt grist stored under an oxygen atmosphere prior to mashing in

2.2.3 MASHING

In order to minimize oxygen uptake during mashing, the following measures have proven to be advantageous in production:

- variable speed agitator
- variable speed pumps
- use of degassed water
- wet milling
- mashing in and out from the bottom of the mash vessel

Myriad opportunities for improving flavor stability during the mashing process exist. These include adapting the mash schedule to accommodate the unique properties of particular raw materials (malt quality, enzyme content of the malt) along with brewing techniques and equipment (e.g. mash acidification, mashing in at high temperatures, avoiding oxygen uptake while mashing in). Figure 12 provides an overview of factors, both positive and negative, which influence flavor stability during the mashing process and lists the relevant brewing techniques.

Figure 12: Factors stemming from raw materials and mashing techniques which affect flavor stability

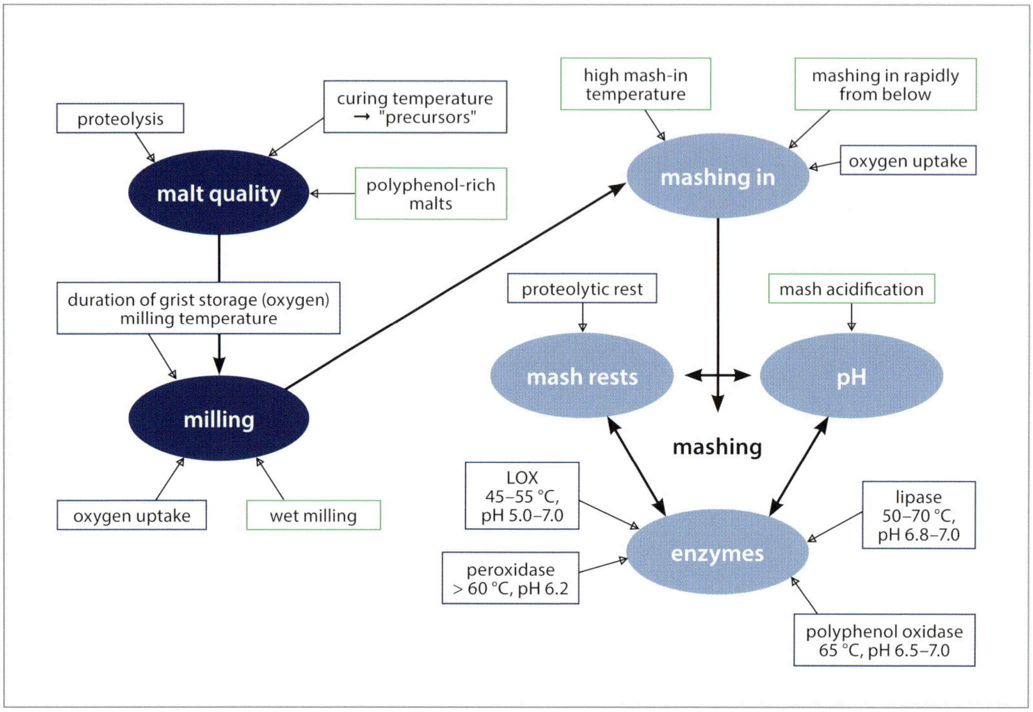

If the malt meets the requisite specifications for modification, mashing in is recommended at temperatures above 60 °C, in order to effectively prevent excessive proteolysis (cf. Mashing, Malt). This produces wort with less free amino nitrogen (FAN) and consequently fewer aging compounds (Strecker aldehydes) in beer.

Mashing in at higher temperatures also reduces LOX activity. The LOX activity is halved after 15 minutes at a mash temperature of 55 °C, while this is achieved after only five minutes at 60 °C. After 30 minutes of mashing, the LOX enzymes are completely inactivated. In this way, high mash-in temperatures also minimize the formation of fatty acid hydroperoxides [26].

Mash acidification with *Sauergut* (diluted first wort fermented by lactic acid bacteria) also exerts a positive influence on flavor stability. A pH adjustment of the mash to 5.4 produces a beer with lower concentrations of aging compounds (fig. 13). Mash acidification should only be carried out at temperatures above 60 °C with well modified malt. Alternatively, the mash should only be acidified to a pH that promotes proteolysis during mashing (cf. Acidification with Natural Lactic Acid).

Figure 13: The effects of mash PH on the sum of aging compounds in beer [28, 34]

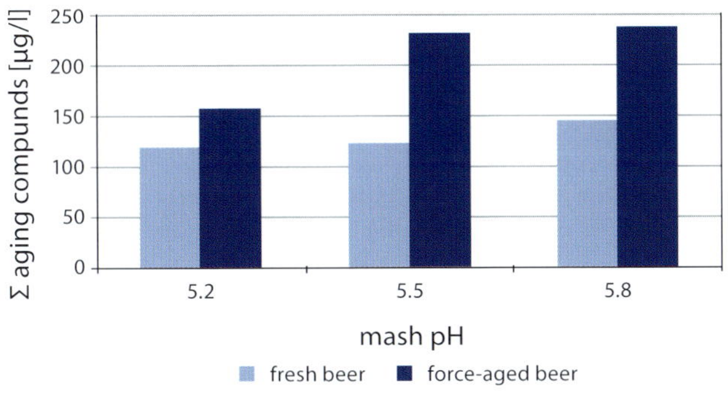

A particularly interesting variation on this theme is the *hoch-kurz* mash method acidified with *Sauergut* (pH 5.4–5.5). With this method, the effects of enzymes with a negative impact on flavor stability are limited: LOX, lipase, peroxidase, polyphenol oxidase and proteolytic enzymes. The presence of fewer lipid degradation products, a larger fraction of low molecular weight polyphenols and fewer protein degradation products (Maillard reactions, Strecker aldehydes) are the result. Moreover, the activity of phosphatases (stronger drop in pH) and β-glucan solubilase (lower viscosity, better filterability) is suppressed.

2.2.4 WORT PRODUCTION AND WORT HANDLING

Just as in mashing, oxygen uptake is to be avoided during the lautering and boiling processes. Thus, mashing out into the lauter tun should be carried out from the bottom; the surface of the wort should never fall below the top of the grain bed and de-aerated sparge water should be used. These measures are designed to prevent the oxidation reactions that lead to a higher polymerization index, an increase in the color of wort and beer, as well as an elevated concentration of aging compounds.

To produce beers with good flavor stability, the wort must be sufficiently clarified without overextraction of the husks. The lipids present in turbid lauter runoff as well as in the hot break material should be viewed critically as they contribute to the generation of aging compounds, especially those derived from carbonyl compounds (section 2.1.2, p. 199) [3, 5, 36]. However, one should note that the composition of particles comprising lauter turbidity is dependent on criteria associated with the brewing techniques, the equipment and the raw materials employed by the brewery and therefore must be evaluated on a case-by-case basis. According to the guidelines described in the DIN 8777 standard [35], the turbidity should remain below 40 EBC for more than 60 % of the total duration of lautering. Shorter boiling times and less thermal stress on the wort results in improved flavor stability (cf. Wort Boiling and Wort Boiling Systems). This also applies to the length of the whirlpool rest. Practical trials have demonstrated that curtailing the whirlpool rest leads to significant differences in the concentrations of aging compounds. Evidence of the positive effects of a shorter whirlpool rest is presented in figure 14. Practical trials have demonstrated that reducing the time in the whirlpool by 20 minutes through chilling more quickly, for instance from 50 to 30 minutes, lowers the quantity of aging compounds in beer subjected to forced aging [3].

Figure 14: The length of the whirlpool rest and its effect on the aging compounds in beer [3, 28]

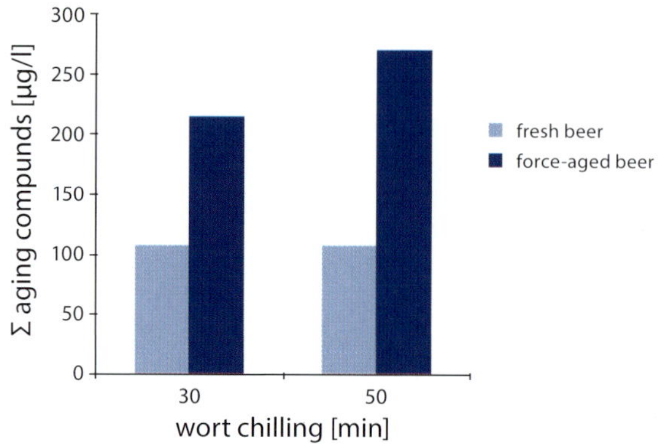

Moreover, a reduction in thermal stress creates beers that are lighter in color. In addition to shortening the whirlpool rest, pre-cooling the wort can also yield positive results (cf. Wort Boiling and Wort Boiling Systems).

2.2.5 FERMENTATION

The reductive capacity of the raw materials used in the brewing process plays a primary role in the flavor stability of the finished beer. The polyphenols, melanoidins and other reductones derived from the raw materials on the hot side as, well as the SO_2 generated by the yeast during fermentation on the cold side, contribute to the reductive capacity of the beer (section 2.1.3, p. 202).

Sulfur dioxide (SO_2) is produced during primary fermentation as an intermediate product in the biosynthesis of amino acids containing sulfur [37]. The cell initiates synthesis of these amino acids if a deficiency exists; this means that the cell no longer has access to them, neither from its own internal supplies nor from the medium (wort). The cells begins by taking up sulfate from the wort and then converting it to an active form. Subsequently, it is reduced to sulfite and later to sulfide, which is used in the synthesis of sulfur-containing amino acids. Toward the end of the growth phase, the amino acid requirements of the yeast cell decline sharply. Sulfur dioxide that is no longer needed is bound to acetaldehyde and excreted from the cell, accumulating in the green beer (figure 15) [37, 38, 39, 40].

Figure 15: Sulfur dioxide as an intermediate product in the synthesis of amino acids containing sulfur

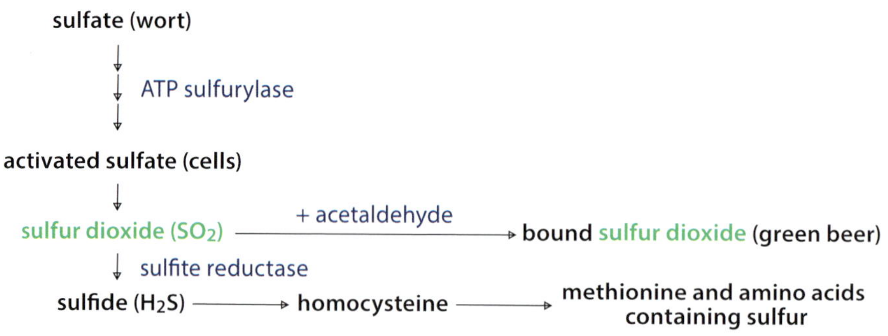

Furthermore, it is important that the brewing yeast culture exhibits a high degree of vitality with a correspondingly strong reductive capacity. This ensures not only the elimination of diacetyl but also of other carbonyl compounds in the wort. These relationships are discussed in detail in the chapter Yeast Technology and Fermentation.

SO_2 concentrations in beer of up to 10 mg/l are recommended for a satisfactory level of flavor stability. In Germany, SO_2 values above 10 mg/l must be declared on the label in order to comply with food laws [41].

2.2.6 LAGERING AND FILTRATION

As described in section 2.1.2, oxygen, heavy metal ions and residues from cleaning agents have a considerable effect on flavor stability. Therefore, due to its impact on beer stability, every effort should be made to prevent oxygen uptake during lagering and storage.

Problems such as oxygen in lager tanks and bright beer tanks, residual air in lines and piping, oxygen uptake during changeovers or through pressure fluctuations in the filter (ensure positive pressure!) are frequently encountered in the brewing process. This oxygen uptake is subsequently expressed in the high levels detected during lagering and filtration. For this reason, measures should be taken to ensure that tanks and piping are devoid of oxygen to the greatest extent possible. To prevent oxygen uptake, air should be purged from all vessels coming into contact with beer (lager tanks, bright beer tanks, buffer tanks) with CO_2 or N_2 gas, while dosing units for filter aids and stabilizers should be flooded with CO_2 [11, 42].

The oxygen concentration in beer can spike for a short time when performing changeovers (e.g. from water to beer) in a line outfitted with a swing pipe system. Air trapped in the swing pipe is the culprit. This can be purged to a large extent by opening the bleed valves (e.g. at the filter inlet or on the blending device). Utilizing de-aerated, carbonated water is essential at the beginning of filtration (starting the pump, coating the filter). In the production process, using degassed and carbonated water as a tool for reducing oxygen uptake is always recommended. Equally important is to avoid contamination of beer with heavy metal ions through contact with equipment or through raw materials and processing aids (e.g. kieselguhr, PVPP). Lastly, all vessels (tanks), piping and valves should be devoid of any residues of cleaning or disinfection agents (section 2.1.2).

2.2.7 FILLING

Exposure to oxygen during filling has a particularly negative effect on flavor stability. After achieving an adequate level of stability and maintaining high product quality throughout the brewing process, these efforts will all be for naught if mistakes are made during filling and oxygen enters the finished beer. This can be detected using oxygen indicators in GC analyses (section 2.3.2). High levels of oxygen in beer are manifested by a distinct oxidized flavor that is readily perceptible within a very short period of time (1–2 weeks) [3, 43].

A common practice for avoiding oxygen uptake involves the double evacuation of bottles and purging with CO_2 immediately prior to filling. However, eliminating all oxygen from the bottles is impossible and preventing any further uptake during filling and capping cannot be guaranteed. Therefore, it is necessary to employ high pressure jetting equipment between the filler and the capper. This causes the beer to foam over in a controlled manner, with the foam pushing the air out of the bottle as it exits. The correct balance must be found among numerous factors, including bottle position, pressure, temperature (microbiological safety) and the distance the bottle travels as it overflows, so that the desired quantity of foam is produced while avoiding imprecise fill levels due to beer loss.

Gassing the crown caps with CO_2 or N_2 is also beneficial. The barrier material in the liners of crown caps is also a factor in the oxygen uptake of beer. After oxygen has permeated the elastomer seal, aging processes commence and progress at different rates, depending on the barrier properties or the degree of saturation [10].

Figure 16: Example of a DLG scheme for conducting sensory analysis of beer and evaluation of aged beer according to Eichhorn [43, 45]

Tasting Scheme

Date:

Analysis series:

Taster:

DLG Tasting Scheme:

Evaluation in half-point steps from 1 to 5
Please record your impression of the aroma

No.	Sample	Aroma	Flavor	Fullness	Liveliness	Bitterness
A	A fresh	4.0	4.5	4.0	4.0	4.0
B	A forced	3.0	3.0	4.0	4.0	3.5
C						
D						
E	B fresh	4.0	4.0	4.0	4.0	4.0
F	B forced	2.5	2.5	3.5	4.0	3.5

Aged Beer Tasting Scheme

The acceptance is only in reference to the aging of beer.

1 = fresh
2 = slightly aged
3 = strongly aged
4 = extremely aged

No.	Sample	Aroma	Flavor	Bitterness	Acceptance [%]					
					100	80	60	40	20	0
A	A fresh	1.0	1.0	1.0	x					
B	A forced	2.0	2.0	3.0			x			
C										
D										
E	B fresh	1.0	1.0	1.0	x					
F	B forced	3.0	3.0	2.5					x	

2.3 ANALYTICAL EVALUATION OF FLAVOR STABILITY

2.3.1 SENSORY ANALYSIS

Sensory analysis is the key method by which the flavor stability of beer is evaluated (cf. Sensory Analysis). In order to provide insight into the flavor stability of packaged beer, beer samples are subjected to forced aging conditions and subsequently evaluated sensorially. The samples are placed on a laboratory shaker, shaken for 24 hours to simulate their movement through transport and distribution networks and then stored in the dark for four days at 40 °C. Fresh samples and force-aged samples are then tasted and compared.

Another more practice-oriented method for testing flavor stability is to store the samples at 28 °C for 30 days and use the scheme depicted in figure 16 to evaluate the beer (influence of bottle cap material on O_2 migration).

Sensory evaluation of flavor stability is carried out according to the five-point tasting scheme developed by the DLG [45]. In addition, a four-point scale is used to assess the aged condition of the beer [43]. The acceptance level of the force-aged samples is expressed as a percentage in order to integrate the subjective impression of the aged beer into the overall score (fig. 16).

2.3.2 DETECTION OF AGING INDICATORS USING GAS CHROMATOGRAPHY

Substances already present in beer change during storage to form aging compounds, also known as aging indicators. For the most part, these compounds are not directly associated with flavor but rather should be viewed as indicators with a strong correlation to the sensory impressions encountered when tasting aged beers (fig. 17).

These compounds increase significantly during forced aging. They are isolated through steam distillation for purposes of analysis followed by separation of the organic phase. The concentration of the compounds is determined through liquid extraction by GC-FID [3, 43].

Figure 17: Correlation between sum of aging compounds and the score obtained in the evaluation of aged beers according to Eichhorn [1–4]

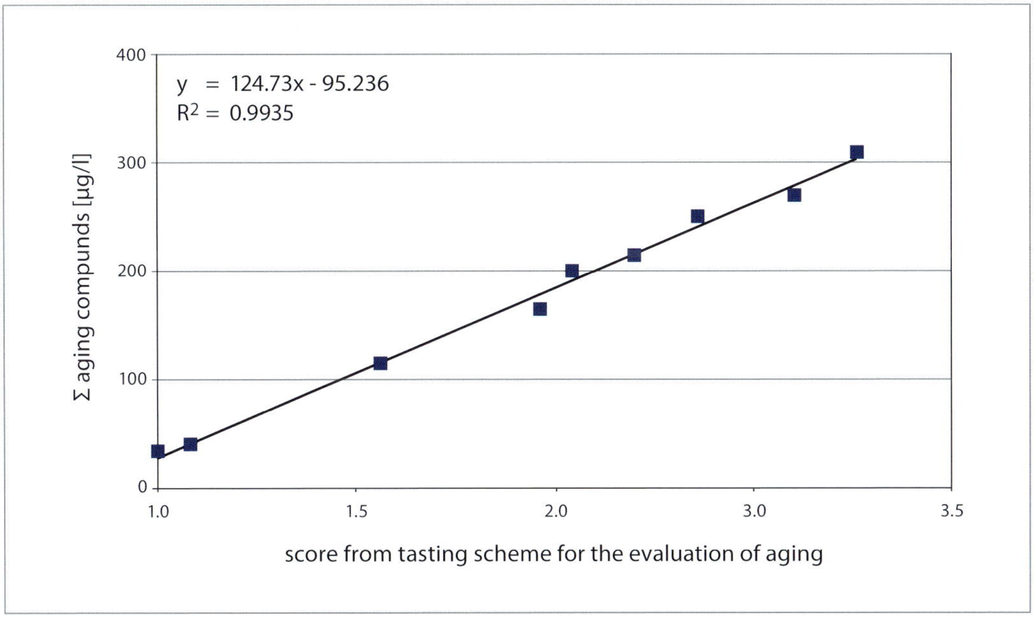

Table 3 lists the relevant aging compounds in beer. Several of these compounds are associated with thermal or oxidative processes.

Table 3: Classification of indicator compounds for aging [3, 43]

Compounds associated with aging (sum of aging indicators)
3-methylbutanal (O)
2-methylbutanal (O)
2-furfural (T)
5-methylfurfural
benzaldehyde (O)
succinic acid diethyl ester
2-phenyl acetic acid ethyl ester
nicotinic acid ethyl ester
2-acetylfuran
2-propionylfuran
γ-nonalactone (T)
(O): oxygen indicator compound
(T): thermal stress indicator compound

Thermal indicator compounds provide information about the degree of thermal stress during the production process or during storage of the packaged beer. Oxygen indicator compounds are primarily formed as a result of oxygen uptake during the filling process, but also reflect possible oxidation during wort production. However, oxygen uptake during the filling process can first be detected by analysis of force-aged beer samples. By contrast, exposure to oxygen during wort production can be identified in fresh beer due to the presence of elevated concentrations of oxygen indicator compounds [3, 43].

2.3.3 ANALYTICAL EVALUATION OF THE ANTIOXIDATIVE CAPACITY

From the standpoint of brewing technology, the term antioxidative activity is viewed as equivalent to chemical and sensory stability. This concept is not entirely correct, because the meaning of each one does not completely encompass the other. The term antioxidative activity describes the capacity of a substance to suppress oxidation reactions *in vitro* in simulated solutions (in food), or *in vivo* in a (different) system, thus imparting a degree of stability [46]. As a rule, this usually occurs through the capacity of the substance to capture free radicals, in particular its capacity to exert a reducing effect. Reducing substances have the capacity to neutralize oxidizing substances, thereby protecting other valuable substances from oxidation. A strong reductive capacity is synonymous with an overinclination to be oxidized. This extreme level of reactivity means highly reductive compounds are unstable. Though a substance with a strong reductive capacity can protect another system against oxidation, the substance itself is not stable. This is the basis for the effectiveness of many preservative agents which protect our food against oxidation, but for the very same reason, they may be easily oxidized.

This means that the stability of beer can be increased through the reductive capacity of individual substances which are not considered to be key contributors to the flavor and aroma profile of the beer (these substances are irrelevant from a sensorial, visual or nutritive standpoint). Sulfur dioxide comes closest to meeting this definition in beer. The total reductive capacity of beer can be analytically quantified as the sum of all compounds with the capacity to reduce oxidative compounds (such as melanoidins, polyphenols and proteins). However, even beer containing large quantities of substances with a high reductive capacity may not necessarily exhibit high overall stability. The reductive capacity is less of a protection mechanism and much more of a property or quality attribute to be protected and retained.

When evaluating the analysis results for antioxidative activity and chemical stability, one must distinguish between two fundamental approaches: the first involves measuring the reductive capacity of the product and comparing it to a test system (*in vitro*), while the other entails measuring the stability of the product in its own matrix (stability test, in food) [47, 48]. The effects of individual substances in a product can be measured using trials incorporating these substances as additives (measurement of the reductive capacity in foods).

2.3.4 DETERMINING THE REDUCTIVE CAPACITY

A strong reductive capacity does not automatically result in a high level of stability (section 2.3.3).

The determination of reductive capacity is generally performed *in vitro*. The reductive capacity of a sample can be determined directly in a redox indicator or a standardized measurement system (e.g. Fenton system, standardized food samples). In both cases, the results are expressed either as a percentage or reported in terms of the effect on a reference standard (e.g. Trolox or gallic acid) [47, 49]. Common redox indicators include DPI (2,6-dichlorophenolindophenol), DPPH (1,1-diphenyl-2-picrylhydrazyl) and ABTS (2,2'- azino-bis[3-ethylbenzothiazoline-6-sulfonic acid]). Analysis results vary according to the time interval used in the measurement (reaction kinetics), the affinity of the selected redox indicator and the designated measurement system (reaction environment).

Two methods for determining reductive capacity have become widely established in the brewing industry, one using DPI and the other, DPPH. Each is described below.

The analysis of reductive capacity with DPI is performed spectrophotometrically according to the MEBAK method [45]. In this analysis, the reduction of DPI is allowed to progress for 60s and is expressed as a percentage of the initial value (fig. 18). Calibration should be carried out with a phosphate/citrate buffer solution (pH 4.3) which corresponds to the average pH of beer.

Figure 18: Printout of the tannometer[1] analysis results for the reduction capacity of beer

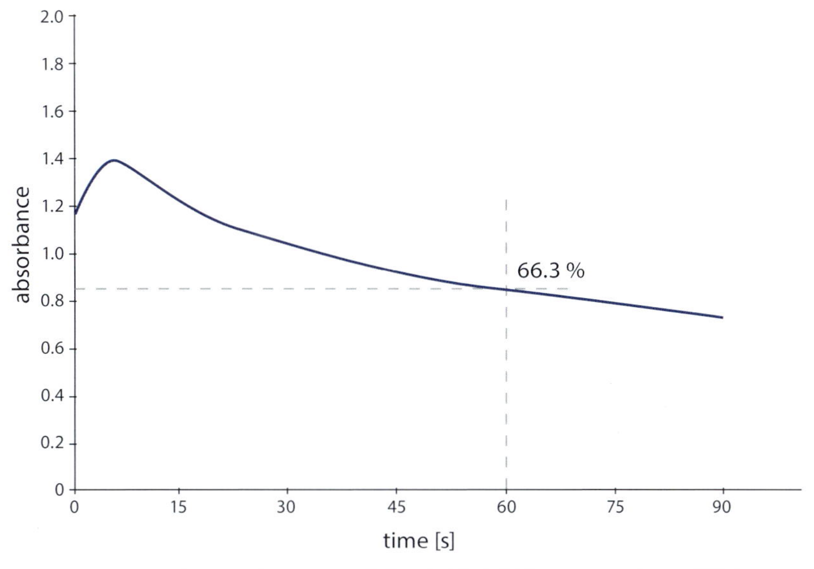

[1]source: company material from Pfeuffer GmbH, Kitzingen, Germany

Of the reducing substances detected in beer with this method, melanoidins, polyphenols (referred to as reductones), sulfhydryl compounds and hop resins are the best known [50, 51]. Among these substances, the melanoidins exercise a dominant influence. The reductive capacity of pale, bottom-fermented beers has been shown empirically to be a measure of melanoidins' propensity for reducing (activating) oxygen. Ultimately, however, this has a pro-oxidative effect (fig. 19) [52].

Figure 19: Influence on the activation of oxygen by reductones

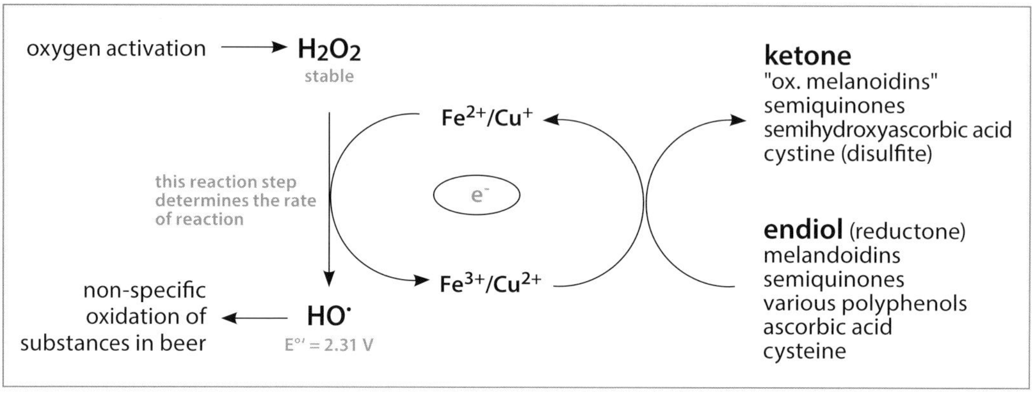

The determination of the reductive capacity with DPPH is carried out according to the antiradical potential method (ARP, modified method, originally according to Blois [53], application for beer according to Kaneda [54]). An electron spin resonance spectrometer or a spectrophotometer is employed for the measurement. The analysis involves measuring the reduction of the stable radical DPPH, a stable radical, is reduced by substances present in the beer over a period of 10 min. The reductive capacity is calculated based on the area under the curve and expressed as a percentage (fig. 20).

Figure 20: Evaluation of the antiradical potential [10, 46]

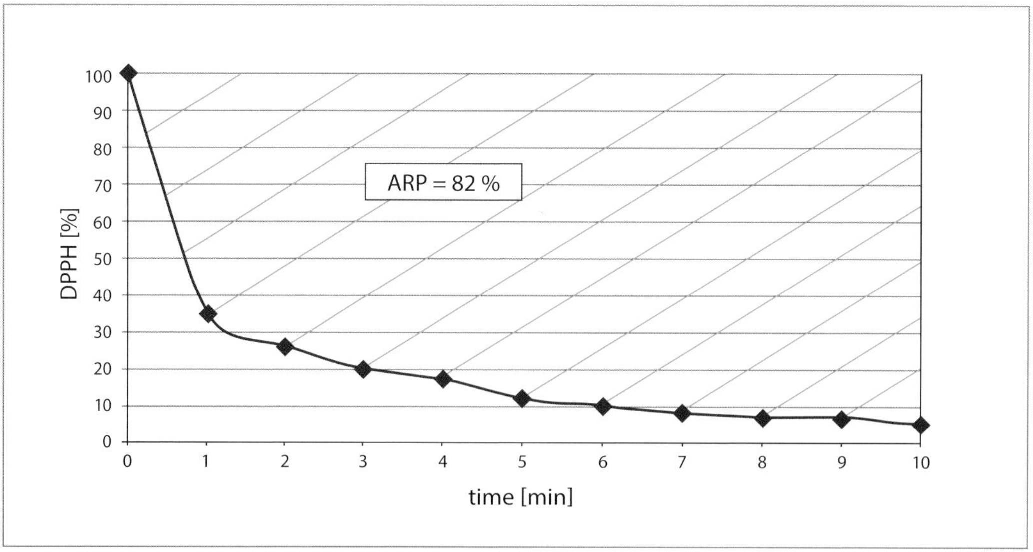

This method can be used in the analysis of wort, beer and other beverages and is not noticeably influenced by sulfur dioxide. Differences in the reductive capacity of pale beer samples (with comparable levels of thermal stress) determined with DPPH are strongly attributable to the positive effects of their polyphenol composition [54, 10]. However, since the antioxidative effect of polyphenols primarily occurs in the brewhouse [55] and their content in beer is subject to stabilization measures, comparison of packaged beer from different breweries is limited.

Malt can be characterized with regard to its reductive capacity using DPPH. For the analysis of malt, a solution must be prepared using a cold, aqueous extraction method [56].

2.3.5 STABILITY TEST

Stability tests are typically carried out in the matrix of the product itself (in food). Beer is stored at an elevated temperature for a defined time interval, either with or without exposure to oxygen. Oxidative stress can be induced through the addition of oxidizing reagents and/or analyzed in a measurement system (e.g. Fenton system, *in vitro*) [47].

The ensuing oxidation reactions can be quantified by measuring the final products of lipid oxidation and Strecker degradation as performed in a GC analysis of hexanal or other representative aldehydes or by means of a spectrophotometric determination of aldehydes with thiobarbituric acid (sections 2.3.2 and 2.3.7) [47]. For purposes of comparison, a fresh sample serves as the control.

Another possibility for quantifying the oxidation reactions in a stability test is to measure the amount of reactive oxygen species (ROS, free radicals), which appear at the beginning of the reaction, by means of electron spin resonance spectroscopy or by chemiluminescence.

The most common ESR method encountered in the brewing industry is measurement of the "lag time" of beer (otherwise known as endogenous antioxidative activity). This is conducted in a forced aging test in the presence of oxygen (stability test, in food) in which the induction phase of a progressive chain reaction of radical species is analyzed [14, 57, 58]. The lag time represents the capacity of a beer to resist oxidation. It is influenced, in particular, by the concentration of sulfur dioxide in the beer. In the stability test, the beer is exposed to oxygen and brought to a temperature of 60 °C. The free radicals generated in the subsequent oxidation reactions are stabilized by the addition of the free radical spin trap reagent N-*tert*-butyl-α-phenylnitrone (PBN), and are measured with ESR. Of course, it is possible to vary the reaction conditions (reaction environment, temperature) and use other spin trap reagents. With PBN, the rate of the forced aging process is even more accelerated due to the increase in pH [10, 59], although this does not negatively impact the measurements in beer.

In addition to the generation of free radicals, a second attribute of the free radical chain reaction can be determined with lag time analysis [60]. Accurately expressed, it is the trap rate that is measured, i.e. the proportion of the radicals that react with the spin trap reagent (in particular, hydroxyl ethyl radicals). To obtain the analysis results, the segment of the curve (starting at a point after 60 minutes) is calculated as an integral, as shown in figure 21. Due to this option for evaluating the results, the method has replaced the former, less sensitive antiradical behavior (ARB) method.

Figure 21: Lag time and trap rate as characteristics of the chain reaction of radical generation in the oxidative forced aging test of beer

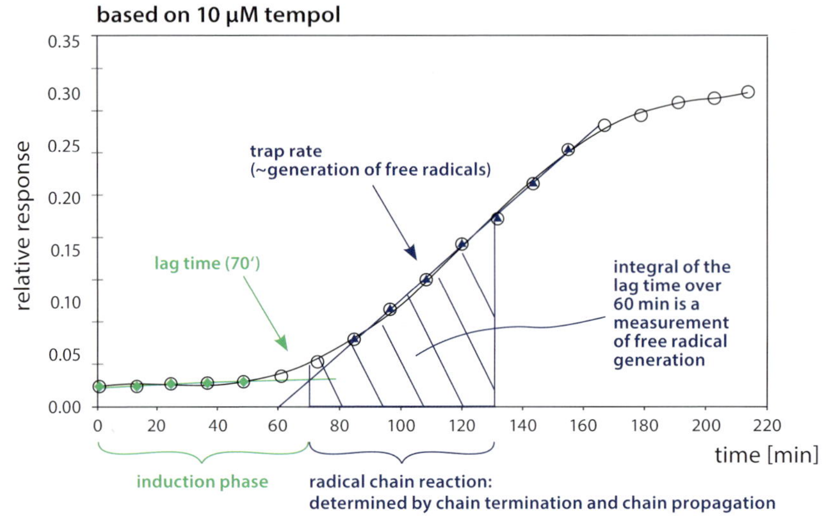

Since there is no visible lag time in wort, the integral is calculated for the time interval of 0 to 120 minutes. Prior to analysis, all samples must be adjusted to an ethanol concentration of 5 % by volume.

The stability index [61] has been revised in light of recent findings and new analysis methods [60]. The metric "reductive capacity" was retired for the aforementioned reasons. In the new stability index, only lag time and radical generation are included. The stability index is meant to be a simple number to help practical brewers evaluate the stability of their products without having to delve into the complexity of oxygen activation and antioxidative activity. Naturally, some information is lost in the calculation of the index, such as that which can be derived from the progression of the curves. However, the goal is to obtain a simple tool for classifying stability: poor – average – good – very good. The stability index (SI) is calculated with the following empirically derived equation:

$$SI = \frac{LT^2 - 4500 \times F + 45000}{500}$$

LT: lag time

F: radical generation (area)

good result: 40–60 (dimensionless)

The factors in the equation provide balance by weighting the two parameters, lag time and radical generation, and standardizing the results in a range from 0 to 100 for straightforward interpretation. By squaring the value for lag time, one can approximate the extent to which the forced aging test was accelerated based on the increase in pH [60]. As depicted in figure 22, this new measure is perfectly suited for analytically distinguishing beers with comparable lag times. The values for the new stability index (of fresh beers) reflect the sensory evaluation (acceptance [%]) of these beers after a natural aging process has taken place.

Figure 22: Stability index and sensory evaluation of two pale standard-strength beers from the same brewery (each determination was performed twice)

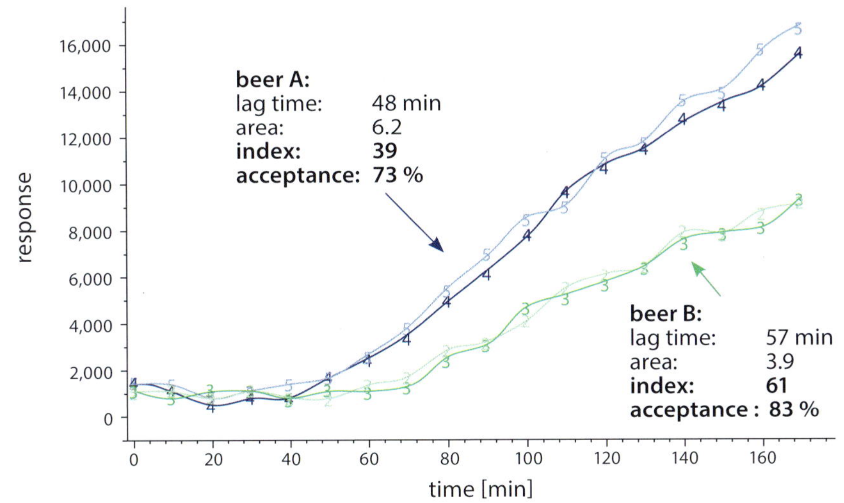

In some situations, good flavor stability in packaged beer can be exhibited by beer that only possesses average values for the stability index. This is attributable to the presence of higher concentrations of masking substances, such as esters (cf. Beer Recovery and Sensory Analysis) or hop aroma compounds (e.g. linalool). Yeast with a high level of vitality (cf. Yeast Technology and Fermentation) can produce beer with exceptional flavor stability, although the SO_2 concentrations and the corresponding antioxidative activity are usually rather low (fig. 23).

Figure 23: The relationship between the stability index and flavor stability

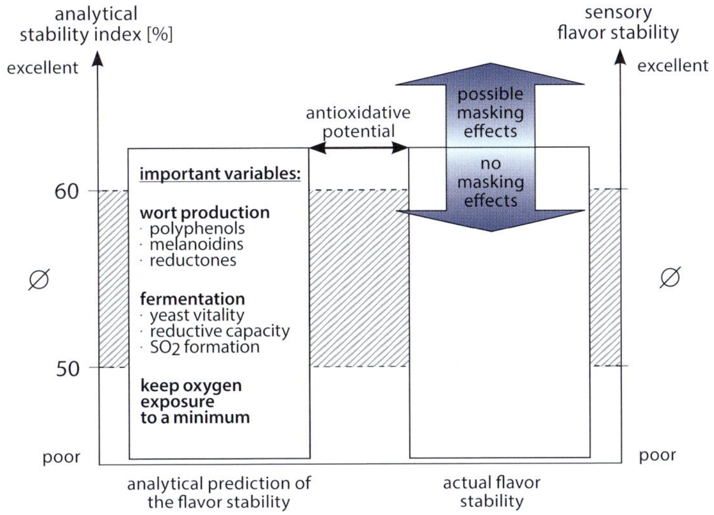

Determining the stability index only makes sense for freshly packaged beers that have not been exposed to excessive levels of oxygen. As beer ages, the influence of aging processes on the analysis is simply too complex to be described using a single metric.

However, to address this topic in a comprehensive manner, another option for evaluating flavor stability does exist. It is known as the T150 method. When performing the forced aging test, the height of the signal is recorded after 150 minutes have elapsed. The T150 method is widely used among English-speaking brewers [62]. This method also attempts to include radical generation as a quality attribute in the evaluation of flavor stability, but the weighted relationship between lag time and trap rate is obtained more arbitrarily.

If an effort is made to more fully understand the principles of ESR spectroscopy, this technique can be successfully used to monitor production processes in a brewery. It also allows inferences to be made regarding possible explanations for what may be causing lower antioxidative activity [60].

Table 4: Influence of thermal treatment on beer exposed to oxygen

	Lag time [min], fresh sample (2 days after sample collection)	Lag time [min] after thermal treatment
beer, exposed to oxygen	77	57
beer, not exposed to oxygen (control)	82	79

2.3.6 HELPFUL TIPS FOR SAMPLE COLLECTION AND STORAGE FOR STABILITY TESTS

When performing any analysis to measure antioxidative activity, it is immensely important that oxgyen uptake is minimized during sample collection and no changes occur during the storage of the samples. Samples of bottled beer should be stored at 0 °C until they are analyzed. Samples collected from sample ports on the cold side of production from sample ports should be analyzed as soon as possible after collection and stored at 0 °C until they are processed. Sample bottles should be flushed with N_2 or CO_2 gas prior to sample collection. Furthermore, during sample collection, the bottles should be filled completely and even be allowed to overflow briefly before closing them. This ensures that very little to no air remains in the headspace of the bottles.

According to the sampling principle, hot wort is filled in aluminum bottles and cooled immediately in cold water after collection. The aluminum foil liner in the lid should remain intact. Wort samples can be stored at 0 °C for several days or kept frozen for several weeks. In order to compensate for temperature-related fluctuations in volume and to ensure that the closure makes an airtight seal, creating a small indentation in the bottle prior to collecting the sample is recommended. If the wort sample is frozen, special care must be taken to protect the wort from contact with oxygen as it thaws (leave the bottle sealed). The analysis should be performed immediately after opening the bottle [60].

Furthermore, this method provides an opportunity for tracing oxygen uptake throughout the production process, assuming the samples are properly collected [60]. The analysis procedure is identical except for the step involving opening the bottle and exposing it to air, which is omitted. If a change is observed at this point, then it is an indication that the sample had been previously exposed to oxygen (table 4).

2.3.7 THIOBARBITURIC ACID INDEX (TBI) AND ANILINE NUMBER (AN)

The TBI is considered a key metric for measuring the sum of thermal stress in malt, wort and beer. It is a metric that encompasses numerous products from the Maillard reactions and other organic compounds, with the exception of 5-hydroxymethylfurfural (HMF). The TBI is determined according to the method developed by MEBAK [45] and is predominantly used for the analysis of malt and wort.

Standard values according to MEBAK [45]:
pale cast-out worts < 45
pale chilled worts (after wort chilling) < 60

Standard values for modern wort boiling systems:
pale cast-out worts < 35
pale chilled worts < 50

The aniline number is considered to be a better measure of the thermal stress in beer. It is based on the quantitative determination of 2-furfural as a typical indicator compound for thermal stress and aging in pale beers produced by means of bottom fermentation [63, 64, 65, 66]. The analysis is performed by distilling a beer sample with a rotary evaporator and adding the distillate to a reaction mixture (aniline/glacial acetic acid as reaction compounds). The reaction results in the formation of a red color, which is measured with a spectrophotometer at 518 nm [66]. Figure 24 shows the increase in the aniline number as a function of storage time and temperature.

Figure 24: Pilsner beer - increase in the aniline number (AN) as a function of storage time and temperature [63]

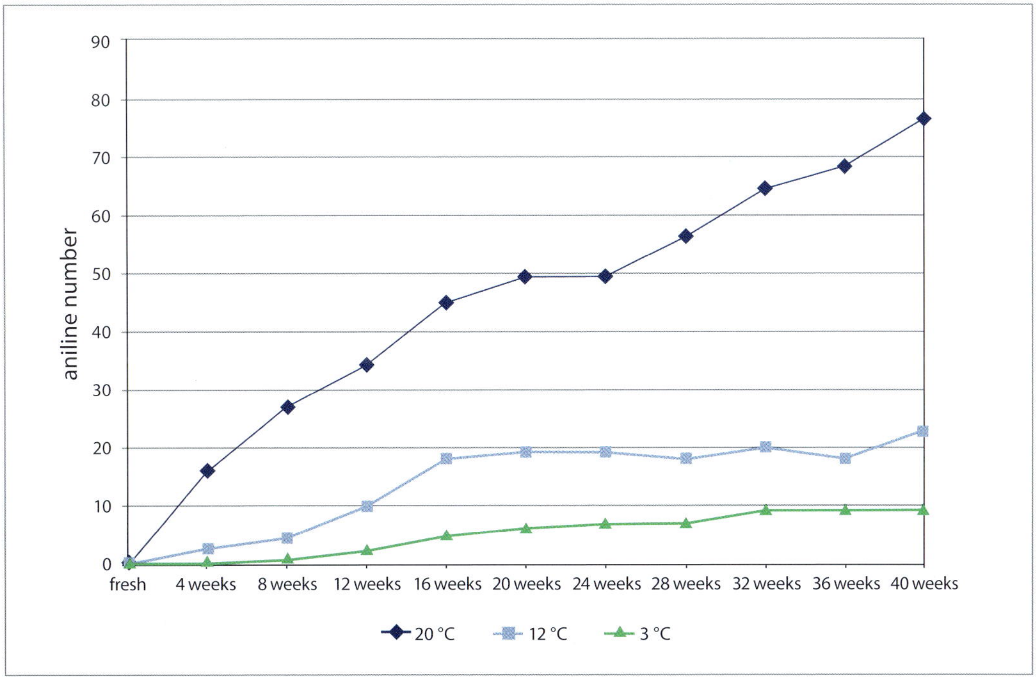

2.3.8 ABSORPTION INTEGRAL (AI)

Along with the aniline number, the absorption integral is another rapid determination method used for evaluating flavor stability. The absorption integral is primarily influenced by indicator compounds for thermal stress, such as 2-furfural. In this analysis, a sample of 200 ml of beer is steam-distilled. The UV spectrum is measured for the first 100 ml of distillate collected and the area between 240 nm and 310 nm is calculated [65, 66, 67, 68].

The flavor stability is evaluated according to the following scale [54]:

< 4.5 very stable	4.5–5.5 stable	5.5–6.5 somewhat stable	> 6.5 unstable

3 SUMMARY

Various aspects concerning flavor stability are explored in this chapter. It provides an overview of the fundamental reaction pathways involved in the formation of aroma compounds relevant for aging in beer. Furthermore, the positive and negative factors influencing flavor stability are examined along with options for enhancing it through technology and brewery operations.

Aging in beer is a dynamic process characterized by the ongoing formation and degradation of compounds responsible for aroma. This, in turn, evokes constant change in the sensory properties of beer. The goal of brewing technology should be to increase flavor stability while minimizing the occurrence of sensory changes within the shelf life of the product.

Perceptible improvements can only be achieved if the factors influencing flavor stability are rigorously monitored and controlled throughout the entire production process.

Certain measures can be implemented in beer production to improve flavor stability of the final product; these include:

1. minimizing oxygen uptake during mashing, lautering and subsequent process steps
2. producing lauter wort of a sufficient clarity without overextraction of the husks
3. minimizing the activity of enzymes and auto-oxidative reactions that impact flavor stability in a negative way
4. ensuring extensive, yet gentle evaporation of undesirable wort aroma compounds
5. utilizing vital yeast cells with a high reductive capacity
6. implementing measures to facilitate higher levels of SO_2 in beer (observe labeling regulations, concentrations > 10 mg/l, for example, must be declared in Germany)
7. avoiding oxygen uptake during the filling process
8. processing and packaging the beer in a gentle manner, so that the positive qualities of the beer are preserved

4 OVERVIEW

Sensory Analysis

Problem	Cause	Corrective measures/Prevention
oxidized note in beer soon after filling (approx. 2 weeks)	• excessive oxygen uptake in the bright beer tank or during filtration • problems during filling or capping: excessive oxygen uptake in the filler, high pressure jets, air in the head-space of the bottle • technical problems: effects caused by tunnel pasteurization or flash pasteurization → the beer is held too long at high temperatures (e.g. equipment disruptions at the filler or in the tunnel pasteurizer) • high pro-oxidant concentrations (Cu, Fe); low antioxidant concentrations	• check the high pressure jetting process: pressure, bottle position, distance bottle travels while foaming over • check pre-evacuation; identify any defective heads or valves on the filler, purge bottles with CO_2 • check the purity of the CO_2 if problems occur during filling (long downtimes, bottle accumulation on the conveyors) → transfer the remaining beer to the beer recovery tank; install a buffer tank downstream from the flash pasteurizer; if possible, run the tunnel pasteurizer until it is empty • use de-aerated water, especially in the filtration process
oxidized note in beer after a traditional lagering period (> 6 weeks)	• lower antioxidant content • piping or stabilizing agents were not purged to remove oxygen • oxygen uptake at the filler • problems with carbonation; impure CO_2 used for pressurizing tanks • quality of the elastomer material used to line the crown caps (permeation) • choice of container: green or clear bottles, shape of the bottle mouth and type of container; material (glass, PET), type of closure (screw cap)	• review the selection of raw materials • stepwise analysis of brewhouse operations (quality control) for oxygen uptake • use CO_2 to purge oxygen from kieselguhr or stabilizing agents during dosing • ensure that piping and tanks are devoid of oxygen; pressurize bright beer tanks and buffer tanks with CO_2 before transferring beer • check closures and containers • see above
"cardboard" off-flavor	• high concentration of lipid oxidation products, such as (*E*)-2-nonenal	• avoid oxygen uptake throughout the entire production process • produce lauter wort with a low solids content (~ 20–30 EBC), pitching wort should contain low amounts of break material

Sensory Analysis

Problem	Cause	Corrective measures/Prevention
"bread-like" and "worty" off-flavors	· malt with excessive amount of soluble nitrogen; high levels of thermal stress due to mistakes made during kilning or blended malt · excessive protein degradation during mashing · excessive thermal stress during boiling or the boiling process was too intense → high concentrations of Strecker aldehydes; fermentation problems due to FAN levels (either too high or too low) in the pitching wort · unsuitable lager conditions	· test raw materials and review their selection (TBI, color) · avoid excessive thermal stress → step-wise quality control (TBI, color) of lauter wort in the kettle prior to boiling, cast-out wort and chilled wort as well as green beer if necessary · check the progression of fermentation; yeast management
"lightstruck" off-flavor	· bottle color: brown < green < clear · clear or green bottles combined with excessively high hop rates	· use reduced iso-α-acid hop products (not in compliance with the *Reinheitsgebot*) · choose packaging that allows less light exposure (e.g. six-packs, cartons); bottles with UV protection

Analytics

Problem	Cause	Corrective measures/Prevention
high concentration of thermal stress indicator compounds in fresh beer	· due to the raw materials (e.g. high thermal stress during the kilning process) · high thermal stress during the brewing process: wort boiling was too long and too intense, long whirlpool rests, whirlpool temperature/insulated whirlpool · check the flash pasteurizer or tunnel pasteurizer for proper functioning · poor reductive capacity of the yeast	· avoid excessive thermal stress on the hot side of production · check the pasteurization units · use vital yeast with a high fermentative capacity
high concentration of oxygen indicator compounds in fresh beer	· exposure to oxygen during wort production: milling, mashing, wort boiling (e.g. external calandria) · exposure to oxygen in the bright beer tank or during lagering, filtration or filling	· do not mill ahead of time; purge oxygen from grist with CO_2 · mash in and mash out from the bottom of the vessel · utilize de-aerated water · check the settings for the agitator and the pumps (e.g. proper size) · avoid oxygen uptake during lagering, filtration and filling · see entry on oxidized notes in beer above
high concentration of thermal stress indicator compounds in force-aged beer (forced aging)	· due to the raw materials (e.g. high thermal stress during the kilning process) · high thermal stress during the brewing process: wort boiling was too long and too intense, insufficient evaporation during the boil, long whirlpool rests	· avoid excessive thermal stress on the hot side of production; check evaporative efficiency · see above
high concentration of oxygen indicator compounds in force-aged beer (forced aging)	· problems during filling: oxygen uptake at the filler, high pressure jetting, air in the headspace of the bottle	· check the high pressure jetting process: pressure, bottle position, distance bottle travels while foaming over · check pre-evacuation · purge bottles with CO_2

Analytics

Problem	Cause	Corrective measures/Prevention
high concentration of oxygen indicator compounds in beer that has undergone natural aging	· problems during filling or capping: very high oxygen uptake in the filler, high pressure jets, air in the headspace of bottle · quality of the elastomer material used to line the crown caps (permeation) · choice of container or closure: shape of the bottle, material, type of closure · high pro-oxidant concentrations (Cu, Fe); low antioxidant concentrations	· check the high pressure jetting process: pressure, bottle position, distance bottle travels while foaming over (fine-tune operations between filler and capper) · check pre-evacuation · pressurize tanks with CO_2 prior to transferring beer · review choice of container (if purchasing new or different equipment) · test filter aids for the presence of heavy metal ions; excessive treatment with PVPP; preserve the reductones in beer · see entry on oxidized notes in beer above
decrease in the concentration of sulfur dioxide in the lager cellar	· oxidation reactions, e.g. through exposure to O_2 or H_2O_2 from disinfectant residues · pressurizing tanks with air prior to transferring beer	· avoid leaving residues of cleaning agents in vessels, piping, valves and pumps (check for proper concentrations of the cleaning agents, the CIP system, the spray balls, or any increase in the intensity of individual cleaning stages) · pressurize tanks with CO_2 prior to transferring beer
very low concentration of sulfur dioxide in fresh beer, short lag time	· yeast strain · problems with yeast management or pitching techniques · unsuitable procedure for fermenting multiple batches in a single tank · excessive aeration	· use propagation yeast with a high fermentative capacity (20 %) together with cropped yeast of a high vitality (80 %) · reduce the degree of aeration for yeast with a high vitality · use a different yeast strain · review the timing for the addition of subsequent batches to a single tank · consider switching yeast strains
rapid decline in the concentration of sulfur dioxide during lagering or natural aging of the beer	· problems during filling: excessive oxygen uptake in the filler, incorrect settings for high pressure jets, air in the headspace of the bottle · quality of the elastomer material used to line the crown caps (permeation) · choice of container: shape of the bottle, material (PET) · high concentration of pro-oxidants	· check the high pressure jetting process: pressure, bottle position, distance bottle travels while foaming over (fine-tune operations between filler and capper) · check pre-evacuation · pressurize tanks with CO_2 before transferring beer · see entry on oxidized notes in beer above

5 REFERENCES

[1] Dalgliesh, C. E.: Flavour Stability. Proceedings of the 16th European Brewery Convention Congress, Amsterdam (1977). pp. 623–659.

[2] Miedaner, H., Narziß, L., Eichhorn, P.: Einige Faktoren der Geschmacksstabilität – sensorische und analytische Bewertung. Proceedings of the 23rd European Brewery Convention Congress, Lisbon (1991), pp. 401–408.

[3] Lustig, S.: Das Verhalten flüchtiger Aromastoffe bei der Lagerung von Flaschenbier und deren technologische Beeinflussung beim Brauprozess. Dissertation, Technical University of Munich (Weihenstephan), 1994.

[4] Narziß, L., Back, W., Gastl, M., Zarnkow, M.: Abriß der Bierbrauerei, 8th ed., WILEY-VCH Verlag GmbH, Weinheim 2017.

[5] Tressl, R., Bahri, D., Silwar, R.: Bildung von Aldehyden durch Lipidoxidation und deren Bedeutung als "Off-Flavor-Komponenten" in Bier. Proceedings of the 17th European Brewery Convention Congress, Berlin (1979), pp. 27–41.

[6] Belitz, H.-D., Grosch, W., Schieberle, P.: Lehrbuch der Lebensmittelchemie. 5th ed., Springer-Verlag, Berlin Heidelberg 2001.

[7] Komarek, D.: Key Odorants in Beer – Influence of Storage on the Flavour Stability. Dissertation, Technical University of Munich (Weihenstephan), 2001.

[8] Hill, P. G., Lustig, S., Barklage H.-J.: Seeing the light – using innovative instrumental analysis to assess new packaging. Proceedings of the 28th European Brewery Convention Congress, Budapest (2001), pp. 910–915.

[9] Kaneda, H., Kano, Y., Osawa,T., Kawakishi, S., Kushino, S.: Role of Active Oxygens on Deterioration of Beer Flavor. Proceedings of the 23rd European Brewery Convention Congress, Lisbon (1991), pp. 433–440.

[10] Franz, O.: Systematische Untersuchungen zur endogenen antioxidativen Aktivität von hellem, untergärigem Bier unter besonderer Berücksichtigung technologischer Maßnahmen beim Brauprozess. Dissertation, Technical University of Munich (Weihenstephan), 2004.

[11] European Brewery Convention (ed.): Beer Filtration, Stabilisation and Sterilisation (Manual of Good Practice). Fachverlag Hans Carl GmbH, Nürnberg 1999.

[12] Halliwell, B., Gutteridge, J.: Free Radicals in Biology and Medicine. 2nd ed., Clarendon Press, Oxford 1989.

[13] Wedzicha, B. L.: Chemistry of Sulphur Dioxide in Foods. Elsevier, London 1984.

[14] Andersen, M. L., Outtrup, H., Skibsted, L. H.: Potential Antioxidants in Beer assessed by ESR Spin Trapping. J. Agric. Food Chem. 48 (2000), no. 8, pp. 3106–3111.

[15] Uchida, M., Ono, M.: Determination of Hydrogen Peroxide in Beer and its Role in Beer Oxidation. J. Am. Soc. Brew. Chem. 57 (1999), no. 4, pp. 145–150.

[16] Rice-Evans, C., Miller, N., Paganga, G.: Structure-Antioxidant Activity Relationships of Flavanoids and Phenolic Acids. Free Rad. Biol. & Med. 20 (1996), no. 7, pp. 933–956.

[17] Uchida, M., Ono, M.: Improvement of Oxidative Flavour Stability of Beer – Role of OH-Radical in Beer Oxidation. J. Am. Soc. Brew. Chem. 54 (1996), no. 4, pp. 198–204.

[18] Kaneda, H., Kobayashi, N., Furusho, S., Sahara, H., Koshino, S.: Reducing Activity and Flavor Stability of Beer. MBAA Technical Quarterly 32 (1995), no. 2, pp. 90–94.

[19] Ueda, J.-I., Saito, N., Shimazu, Y., Ozawa, T.: A Comparison of Scavenging Abilities of Antioxidants against Hydroxyl Radicals. Arch. Biochem. Biophys. 333 (1996), no. 2, pp. 377–384.

[20] Hashimoto, N., Kuroiwa, Y.: Proposed Pathways for the Formation of volatile Aldehydes during Storage of Bottled Beer. ASBC Proceedings 33 (1975), no. 3, pp. 104–111.

[21] Boivin, P.: A review of pro- and anti-oxidant of malt. European Brewery Convention Symposium Flavour and Flavour Stability, Nancy (France). Monograph 31. Fachverlag Hans Carl, Nürnberg 2001.

[22] Van Waesberghe, J.: Anti-Oxidants and Pro-Oxidants in Preprocessed Brewing Ingredients. MBAA Technical Quarterly 33 (1996), no. 2, pp. 96–101.

[23] Mader, F.: Ermittlung geschmacklich relevanter Faktoren zur frühen Bewertung von Braugersten-Neuzüchtungen. Dissertation, TU München, 1998.

[24] Kretschmer, H.: Der Einfluss physiologischer Vorgänge beim Mälzen auf das Verhalten der Lipide und deren Einfluss auf Bieraromastoffe und Bieralterung. Dissertation, Technical University of Munich (Weihenstephan), 1996.

[25] Ketterer, M.: Untersuchungen zum Fettstoffwechsel bei der Keimung und der hieraus resultierenden Metabolismen. Dissertation, Technical University of Munich (Weihenstephan), 1994.

[26] Zürcher, A.: Der Einfluss des Blattkeims von Gerstenmalz auf die Geschmacksstabilität und weitere Qualitätsmerkmale von Bier. Dissertation, Technical University of Munich (Weihenstephan), 2003.

[27] Forster, C.: Der Einfluss der Darrtechnologie auf die Malz- und Bierqualität. Dissertation, Technical University of Munich (Weihenstephan), 1996.

[28] Back,W., Forster, C., Krottenthaler, M., Lehmann, J., Sacher, B., Thum, B.: Neue Erkenntnisse zur Verbesserung der Geschmacksstabilität. Brauwelt 38 (1997), no. 38, pp. 1677–1692.

[29] Forster, C., Narziß, L., and Back, W.: Untersuchungen zum Bieraroma und der Geschmacksstabilität mit verschiedenen Malzschüttungen hergestellter dunkler Biere. Proceedings of the 26th European Brewery Convention Congress, Maastricht (1997), pp. 561–568.

[30] Preuß, T.: Technologische Maßnahmen zur Erzielung malzaromatischer dunkler Biere hoher Geschmacksstabilität nach Charakterisierung der Schlüsselaromastoffe in ausgewählten dunklen Bieren. Dissertation, Technical University of Munich (Weihenstephan), 2001.

[31] Cantrell, I. C., Griggs, D. L.: Malt: Its Role In Oxidation. MBAA Technical Quarterly, 33 (1996), no. 2, pp. 82–86.

[32] Muts, G. C. J., Pesman, L.: European Brewery Convention Symposium Wort production, Maffliers (France). Monograph XI. Fachverlag Hans Carl, Nürnberg 1986.

[33] Franz, O., Back, W.: Continous Measurement of Oxidation Reactions During Mashing by Chemiluminescence Detection. Proceedings of the 29th European Brewery Convention Congress, Dublin (2003), pp. 941–951.

[34] Takahashi, Y.: Pilotversuche über den Einfluss der verschiedenen Maischparameter auf die Eigenschaften der Würze und des Bieres unter besonderer Berücksichtigung der Geschmacksstabilität. Dissertation, Technical University of Munich (Weihenstephan), 1996.

[35] Deutsche Norm Sudhausanlagen in Brauereien (ed.): DIN 8777. Normenausschuss Maschinenbau (NAM) im DIN Deutsches Institut für Normung e.V. Berlin, Beuth Verlag GmbH, 1996.

[36] Drost, B. W., Van Eerde, P., Hoekstra, S. F., Strating, J.: Fatty Acids and Staling of Beer. Proceedings of the 13th European Brewery Convention Congress, Estoril (1971), pp. 451–458.

[37] Ilett, D. R.: Aspects of the Analysis, Role and Fate of Sulphur Dioxide in Beer – a Review. MBAA Technical Quarterly 32 (1995), no. 4, pp. 213–221.

[38] Nordlöv, H.: Formation of sulphur dioxide during beer fermentation. Proceedings of the 20th European Brewery Convention Congress, Helsinki (1985), pp. 291–298.

[39] Gyllang, H., Winge, M., Korch, Chr.: Regulation of SO_2 Formation During Fermentation. Proceedings of the 22nd European Brewery Convention Congress, Zurich (1989), pp. 347–354.

[40] Dufour, J. P.: Influence of Industrial Brewing and Fermentation Working Conditions on Beer SO_2 Level and Flavour Stability. Proceedings of the 23rd European Brewery Convention Congress, Lisbon (1991), pp. 209–216.

[41] Verordnung über die Kennzeichnung von Lebensmitteln – LMKV (Fassung vom 15. Dezember 1999), Anlage 3 (zu § 3 Abs.1 Nr. 3, § 5 Abs. 3 und 6), 16.01.2004.

[42] European Brewery Convention (ed.): Fermentation & Maturation (Manual of Good Practice). Fachverlag Hans Carl, Nürnberg 2000.

[43] Eichhorn, P.: Untersuchungen zur Geschmacksstabilität des Bieres. Dissertation, Technical University of Munich (Weihenstephan), 1991.

[44] Kaneda, H., Osawa, T., Kawakishi, S., Munekata, M., Koshino, S.: Contribution of Carbonyl-bisulfite Adducts to Beer Stability. J. Agric. Food Chem. 42 (1994), no. 11, pp. 2428–2432.

[45] Pfenninger, H. (ed.): Brautechnische Analysenmethoden. Methodensammlung der Mitteleuropäischen Analysenkommission (MEBAK). Band II, 4th ed., Weihenstephan 2002.

[46] Halliwell, B., Gutteridge, J.M.C.: Free Radicals in Biology and Medicine, Oxford University Press, Oxford 2007.

[47] Antolovich, M., Prenzler, P. D., Patsalides, E., McDonald, S., Robards, K.: Methods for Testing Antioxidant Activity. The Analyst 127 (2002), pp. 183–198.

[48] Halliwell, B., Aeschbach, R., Lölinger, J., Aruoma, O. J.: The Characterization of Anti-Oxidants. Food and Chemical Toxicology 33 (1995), no. 7, pp. 601–617.

[49] Nenadis, N., Lazaridou, O., Tsimidou, M. Z.: Use of Reference Compounds in Antioxidant Activity Assessment. Journal of Agricultural and Food Chemistry 55 (2007), no. 14, pp. 5452–5460.

[50] Reich, H., Bock, R. M.: Dichlorophenolindophenol-Reducing Substances in Beer. ASBC Proceedings-Annual Meeting 18 (1960), pp. 78–84.

[51] Bellmer, H.-G.: Studie über die Polyphenole und deren Polymerisationsindex in den Rohstoffen des Bieres und ihre Veränderungen während der Bierbereitung. Dissertation, TU München, 1976.

[52] Wurzbacher, M.: Bedeutung von Reduktionskraft und Antioxidativer Aktivität im Brauprozess. 41. Technologisches Seminar, Weihenstephan (2008), pp. 5/1–5/9.

[53] Blois, M. S., Antioxidant Determinations by the Use of a Stable free Radical. Nature (London, UK), 181 (1958) pp. 1199–200.

[54] Kaneda, H., Kobayashi, N., Furusho, S., Sahara, H., Koshino, S.: Reducing Activity and Flavor Stability of Beer. Technical Quarterly – Master Brewers Association of the Americas 32 (1995), no. 2, pp. 90–94.

[55] Wurzbacher, M., Kinitz, C., Back, W. Galensa, R.: Endogene antioxidative Aktivität von Bier unter besonderer Berücksichtigung der Polyphenole. Brauwelt 146 (2006), no. 33, pp. 964–967.

[56] Wurzbacher, M.: Untersuchungen zum Einfluss antioxidativer Substanzen auf die Geschmacksstabilität des Bieres. Dissertation, Technical University of Munich (Weihenstephan), 2011.

[57] Uchida, M., Suga, S., Ono, M.: Improvement for Oxidative Flavor Stability of Beer – Rapid Prediction Method for Beer Flavor Stability by Electron Spin Resonance Spectroscopy. J. Am. Soc. Brew. Chem. 54 (1996), no. 4, pp. 205–211.

[58] Uchida, M., Ono, M.: Technological Approach to improve Beer Flavor Stability: Analysis of the Effect of Brewing Processes on Beer Flavor Stability by the Electron Spin Resonance Method. Journal of the American Society of Brewing Chemists 58 (2000), no. 1, pp. 8–13.

[59] Bishop, D. F., Stern, G., Fleischman, M., Marshall, L. S.: Hydrogen Peroxide Catalytic Oxidation of Refractory Organics in Municipal Waste Waters. Industrial & Engineering Chemistry Process Design and Development, 1968. 7(1): pp. 110–17.

[60] Wurzbacher, M., Back, W.: Einsatzmöglichkeiten der ESR-Analytik zur Brauereiprozesskontrolle im Hinblick auf die Geschmacksstabilität. Brauwelt 147 (2007), no. 22–23, pp. 594–599.

[61] Franz, O., Back, W.: Stability Index – A new Approach to Measure the Flavor Stability of Beer. MBAA Technical Quarterly 40 (2003), no. 1. pp. 20–24.

[62] Foster, R.T., Samp, E. J., Giarratano, C. E., Fletcher, S., Miller, M., Quilliam,W.: Electron Paramagnetic Resonance Studies Comparing Wort Boiling Temperatures and Various Levels of Sulfur Dioxide in Packaged beer. Technical Quarterly 42 (2005), no. 3, pp. 209–213.

[63] Bößendörfer, G., Birkenstock, B., Thalacker, R.: Ein Beitrag zur Geschmacksstabilität des Bieres. Brauwelt, 141 (2001), no. 27/28, pp. 1042–1047.

[64] Thalacker, R; Bößendörfer, G., und Birkenstock, B.: Eine neue Kennzahl in der brautechnischen Analyse – die Anilinzahl (AZ). Brauwelt 138 (1998), no. 10/11, pp. 421–425.

[65] Urban, A., Aichinger, K., Klein, H., Huber, C.: Absorptionsintegral und Anilinzahl. Brauwelt 142 (2002), no. 49, pp. 1880–1883.

[66] Urban, A., Aichinger, K., Klein, H., Huber, C.: Verhalten von Absorptionsintegral und Anilinzahl während der Bierherstellung. Brauwelt 142 (2002), no. 17, pp. 598–602.

[67] Klein, H., Krammer, R., Natter, M.: Schnellbestimmung zur Vorhersage der Geschmacks-stabilität von Bieren. Proceedings of the 26th European Brewery Convention Congress, Maastricht (1997), pp. 553–560.

[68] Klein, H., Glinsner, M., Natter, M., Steiner, I.: Einflussfaktoren des Absorptionsintegrals im Hinblick auf die Geschmacksstabilität. Monatsschrift für Brauwissenschaft 53 (2000), no. 11/12, pp. 217–222.

SOUTHERN GERMAN WHEAT BEER

1 INTRODUCTION

Southern German wheat beer, also known as weissbier or *Hefeweizen* (referred to hereafter as wheat beer) is a popular style of beer that has managed to defy the general decline in beer consumption over a number of decades. In fact, production increases have even been noted in some cases. At the beginning of the 60's, the Bavarian wheat beer production volume did not even reach 500,000 hl. Whereas northern Germany experienced a turnaround in the trend from export beer to pilsner, constant gains were being seen in Bavaria in the production volumes of wheat beer. The growth observed in recent years is primarily due to its growing popularity in other German states as well as in sales outside of the country. In 2014, wheat beer turnover from the 33 largest German breweries rose from 9.811 million in the previous year to 10.238 million hl – an increase of 427,000 hl, or 4.3 %. The growth in output represents a volume of 100,000 hl destined for the German market while the rest is attributable to exports [1].

2 BREWING TECHNIQUES IN WHEAT BEER PRODUCTION

2.1 THE BASICS

According to the current beer laws in Germany, malt from all types of malted grains except rice, corn and dari (sorghum) are permitted in the production of top-fermented beers. In contrast, bottom-fermented beers may only be produced using malted barley. Cane, beet or invert sugar may also be used when brewing top-fermented beers, as well as sugar derived from starch and coloring agents made from the aforementioned types of sugar. Top-fermented, low-alcohol beers may be supplemented with sweeteners in Germany in accordance with the relevant version of the regulation governing additives (*Zusatzstoff-Zulassungsverordnung*) [2]. However, such additives are not permitted in Bavaria and Baden-Württemberg. The use of wheat malt is permitted in the production of Southern German-style wheat beer, provided that its proportion of the grain bill is at least 50 % [3].

The German market alone boasts more than 1000 wheat beers [4]. The wide range of processes that can be applied in wheat beer production contribute to this rich beer landscape. Table 1 provides an overview of the various processes; the figures are in reference to the breweries rather than production volumes. As can be seen from the information provided, many of the high-volume breweries utilize cylindroconical fermentation tanks and/or lager tanks.

Table 1: Variations in production techniques for Bavarian-style wheat beers

Parameters	Range/distribution/frequency reported
portion of grain bill (wheat)	50–100 %
malt modification (Kolbach index)	35–43 %
mash-in temperature	30–57 °C
intensity of the protein rest	45–58 °C 10–26 min
intensity of the maltose rest	59–69 °C 10–120 min
infusion/decoction	60 % perform infusion
mash acidification	18 % perform mash acidification
turbidity of the first wort	23 % don´t care for lautering bright wort
duration of the boil	50–210 min
removal of cold break material	25 % do not remove the cold break material
fermentation vessel	25 % use traditional vats 19 % use horizontal tanks 25 % use vertical tanks 31 % use cylindroconical tanks
fermentation temperature	18–25.5 °C
fermentation time	2.0–7.5 days
addition of wort or kräusen	42 % add cast-out wort 8 % add first wort 23 % add bottom-fermented kräusen 7 % add a mixture of top-fermented and bottom-fermented kräusen
maturation in bottles or tanks	75 % in bottles 25 % in tanks
warm lagering phase	16–22 °C 3–7 days
cold lagering phase	4–10 °C 6–18 days
yeast addition for maturation	40 % do not add additional yeast 7 % add top-fermenting yeast 25 % add bottom-fermenting yeast 7 % add a mixture of top-fermenting and bottom-fermenting yeast
flash pasteurization	47 % do flash pasteurize* 53 % do not flash pasteurize

*if yeast with a low vitality is utilized → cell lysis → release of medium chain fatty aids → problems with flavor/foam

The effects of these very different methods of production are reflected in the great variety of aroma compounds found in the beer (see tab. 2). As a result, wheat beers not only contain a higher concentration of these compounds, especially those generated as a by-product of fermentation, but they also exhibit a wider range of variation.

Table 2: Aroma compounds in wheat beers and pale, standard-strength lager beers

Beer aroma compounds/ higher alcohols		Conc. in wheat beer	Avg. in wheat beer	Conc. in pale lager	Avg. in pale lager
propanol-1	mg/l	15–30	21	5–20	10
2-methylpropanol-1	mg/l	20–60	40	10–20	8.8
2-methylbutanol-1	mg/l	15–30	18	10–20	15
3-methylbutanol-1	mg/l	40–100	61	30–50	41
hexanol-1	µg/l	15–50	27	10–30	24
heptanol-1	µg/l	10–50	32	10–30	12
octanol-1	µg/l	10–40	15	20–40	20
decanol-1	µg/l	5–20	7.5	5–20	3.8
2-phenylethanol	mg/l	15–45	31	10–30	20
ethyl acetate	mg/l	10–50	29	5–20	9.7
acetic acid isobutyl ester	mg/l	0.05–0.80	0.25	0.05–0.10	0.05
3-methylbutyl acetate (isoamyl acetate)	mg/l	0.5–8.0	3.6	0.5–2.0	0.7
acetic acid hexyl ester	µg/l	3.0–15.0	6.5	3.0–15.0	3.3
acetic acid heptyl ester	µg/l	2.0–10.0	9.4	2.0–6.0	2.2
acetic acid octyl ester	µg/l	1.5–5.0	2.6	1.0–5.0	1.5
acetic acid furfuryl ester	µg/l	3.0–20.0	8.0	10.0–30.0	10.0
acetic acid-2-phenylethyl ester	mg/l	0.2–2.0	1.0	0.1–0.3	0.25
butyric acid ethyl ester	mg/l	0.05–0.20	0.18	0.05–0.15	0.05
hexanoic acid ethyl ester	mg/l	0.05–0.30	0.14	0.05–0.30	0.06
octanoic acid ethyl ester	mg/l	0.10–0.40	0.21	0.10–0.50	0.15
decanoic acid ethyl ester	mg/l	0.01–0.05	0.01	0.01–0.05	0.01
dodecanoic acid ethyl ester	µg/l	1.0–5.0	2.4	2.0–10.0	2.1
isovaleric acid	mg/l	0.20–1.00	0.35	0.20–1.00	0.30
hexanoic acid	mg/l	1.0–4.0	1.7	0.5–2.0	1.5
octanoic acid	mg/l	2.0–10.0	4.0	2.0–10.0	4.0
nonanoic acid	mg/l	0.01–0.05	0.01	0.01–0.07	0.01
decanoic acid	mg/l	0.10–2.00	0.65	0.10–2.00	0.46
9-decenoic acid	mg/l	0.05–0.30	0.20	0.05–0.20	0.09
dodecanoic acid	mg/l	0.01–0.10	0.04	0.01–0.07	0.03
γ-nonalactone	µg/l	20.0–50.0	36.0	20.0–40.0	13.0
4-vinyl guaiacol	mg/l	0.5–4.0	2.0	0.1–1.0	0.3

Wheat beers can be classified into four basic types based on the array of aroma compounds present [5]: estery, phenolic, yeasty and neutral.

Estery wheat beers exhibit a dominant fruity character and, of the broad range of esters, isoamyl acetate (3-methylbutyl acetate) with its banana-like character, is undoubtedly the most common. Further positive impressions are a result of 2-methylbutyl acetate, while hexanoic acid ethyl ester with its apple note and ethyl acetate with its solvent-like character tend to induce more negative impressions.
Wheat beers with a phenolic character are usually described as having a clove-like note. The compound 4-vinyl guaiacol is the key aroma compound in these beers, while 4-vinyl phenol is also said to contribute a positive effect. However, concentrations of 4-vinyl guaiacol higher than 4 mg/l are often described as unpleasant or medicinal.

The yeasty aroma is a pleasant interplay of numerous fermentation by-products, although the precise composition and their relative concentrations are unknown. In any case, it is produced by a variety of esters and sulfur compounds. The primary constituents responsible for the old yeast character are medium chain fatty acids (rancid, goaty) and short chain fatty acids (cheesy, sweaty), as well as certain higher alcohols (e.g. heptanol, earthy).

Wheat beers are considered neutral whenever they exhibit a low overall concentration of aroma compounds and do not exhibit any dominating aroma notes [5, 6, 7, 8].

Figure 1: The influence of 3-methylbutyl acetate and ethyl acetate on the sensory profile of wheat beer

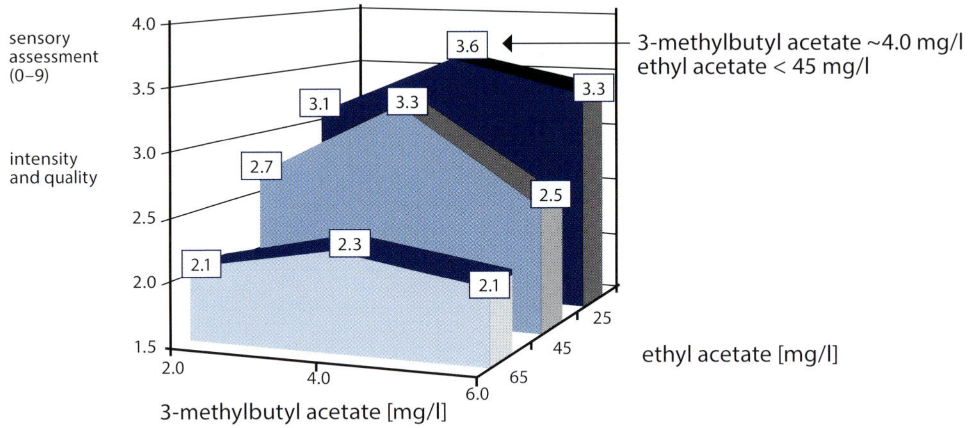

It should be noted that higher concentrations of aroma compounds do not inherently lead to a stronger aroma impression. This is especially true for esters, where an isoamyl acetate concentration between 4.5 and 6.0 mg/l is no longer perceived as being pleasantly fruity. Due to its solvent-like character, ethyl acetate can conceal such unpleasant attributes starting at a concentration of 45 mg/l and can even lead to a bland flavor impression [9]. Esters and phenolic compounds can mutually veil or otherwise influence one another. Higher alcohols with a total concentration of 100 mg/l or more can also have a negative impact on the beer character.

2.2 WHEAT AS A RAW MATERIAL AND IN THE GRAIN BILL

Wheat used for brewing does not have equal standing with malting barley, because unlike barley, no effort has been made to breed different varieties of wheat to improve their characteristics for brewing. At present, primarily wheat lots with poor baking qualities are being used as the raw material for producing malted wheat for brewing applications. The processability of malted wheat is classified according to the work done by Sacher [10].

Table 3: The classification of wheat malts according to their processability

Groups according to degree of protein modification	Variety	Extract	Viscosity	Final attenuation	Raw protein	Protein modification	Malting technique
group 1: low to medium viscosity with low to medium protein modification	Herrmann	++	(+)	(+)	+	(+)	every malting method is possible in principle
	Dekan	+	(+)	(+)	+	(–)	
	Skater	(+)	(+)	+	+	0	
group 2: low to medium viscosity with high protein modification	Anthus	+++	(+)	0	++	++	increasing temperatures with low moisture content after steeping
	Maltop	+++	+	0	+	+	
group 3: high viscosity with low to medium protein modification	Boheme	0	–	0	–	+	decreasing temperatures with high moisture content after steeping
	Piko	+	– –	++	0	0	
	Greif	++	– –	0	0	0	
group 4: high viscosity with high protein modification	Ephoros	+	(-)	(-)	+	++	no reasonable technique for compensation exists

In addition, the insights gained from the analysis of wheat malt (cf. Malt) require further research. As such, the quality attributes for evaluating barley malt are drawn upon for assessing wheat malt, although they are based on an entirely different material. Many values regarding cytolytic modification are related to β-glucans, although in actuality the pentosans in wheat play a larger role. It is important to remember that the key metrics for amylolysis do not take into account that higher extract values are attained due to the lack of husks on wheat, whereas the final attenuation is generally lower due to less α-amylase activity. Only the proteolytic values remain unchanged for the most part when evaluating wheat malt [11]. Generally speaking, the proportion of high molecular weight nitrogen (35–40 %) is particularly distinctive to wheat malt. This comes at the expense of the low molecular weight, easily assimilable fraction.

Wheat has only a small influence on the formation of esters and other fermentation by-products and ultimately on the aroma [10]. Comparatively, the proteolytic modification has a far greater impact. In many cases, lower proteolytic values result in much livelier and more estery beers [5]. Yet, the 4-vinyl guaiacol content can be heavily influenced by wheat malt, which contains very high levels of ferulic acid, among other things, and serves as the precursor to 4-vinyl guaiacol. However, higher concentrations of ferulic acid can usually be found in barley malt. As such, the intensity of the phenolic character can be regulated by controlling the amount of wheat malt in the grain bill, but even more so through the mashing procedure (cf. section 2.3) [9, 13].

An addition of 3 to 5 % light caramel malt or 0.5 to 1.0 % dark caramel malt is recommended in order to increase the body. Pronounced dark malt notes can also be found in wheat beers, which can be traced back to the addition of the respective special malts [3, 12].

2.3 MASHING

When producing yeasty wheat beers, water possessing a broad range of hardness – from soft up to medium or even hard water (10–12 °dH) – can be utilized for brewing liquor [3, 12]. Utilizing harder water may, however, lead to increased gushing problems in bottle-conditioned beers. In addition, gushing may also be observed in beers that have not been subjected to a sufficiently long cold phase prior to separation, due to the slow precipitation of calcium oxalate (cf. section 2.8). The ratio of grist to brewing liquor is around 1 : 3 to 1 : 4. A rest at 45 °C during mashing, along with ratio of barley malt to wheat malt influences the quantity of ferulic acid released and ultimately the phenolic character of the beer. A rest of at least 10–15 minutes must be performed in order to achieve a distinctly phenolic character in the beer. The pH optimum for ferulic acid hydrolase is between 5.9 and 6.1. As such, acidification of the mash suppresses the release of ferulic acid [5, 13, 14].

Table 4: The impact of a 45 °C rest on the intensity of primary aroma notes

	Length of the 45 °C rest [min]		
Aroma impression	0	10	> 25
phenolic	1.2	2.1	3.3
estery	4.1	3.4	2.6
yeasty	1.8	2.6	2.8

On the other hand, a prolonged 45 °C rest reduces the intensity of the estery character and promotes a more yeasty character. The release of ferulic acid is suppressed by mashing in at 50 °C followed by a protein rest. Even at temperatures below 40 °C, the activity of ferulic acid hydrolase is reduced. As wheat malts have a lower FAN content (cf. section 2.2), it is often necessary to ensure a more extensive degradation of the protein, which is usually achieved through a rest designed to release ferulic acid. A proportion of FAN amounting to about 18 % of the total nitrogen is desirable. In the event that the proteolysis of the malt is insufficient, progressive protein rests (45/48/52 °C) may be necessary.

The impact of the amylolytic rests on fermentation by-products is rather negligible, so that standard rest periods at 60–62 °C and 70–74 °C can be implemented (cf. Mashing). As with bottom-fermenting beers, a 65 °C rest can bring about a synergistic effect between α-amylase and β-amylase.

A comparison of infusion and decoction processes showed no significant change in flavor. One advantage of the decoction process is that the beers tend to exhibit better colloidal stability [9, 15]. Improved iodine test results for conversion have been observed in the double decoction process.

Mash and wort acidification are uncommon in wheat beer production because it suppresses the expression of estery, phenolic and yeasty characteristics in the beer [5]. Although biological acidification does lead to an increase in the ester content, the strong formation of higher alcohols and ethyl acetate occurring simultaneously masks the positive aromas. Therefore, mash and wort acidification usually results in wheat beers with a more neutral character [9]. This may be desirable, within measure, for very pale, filtered wheat beers.

Figure 2: The influence of mash acidification on sensory properties (MA = mash acidification)

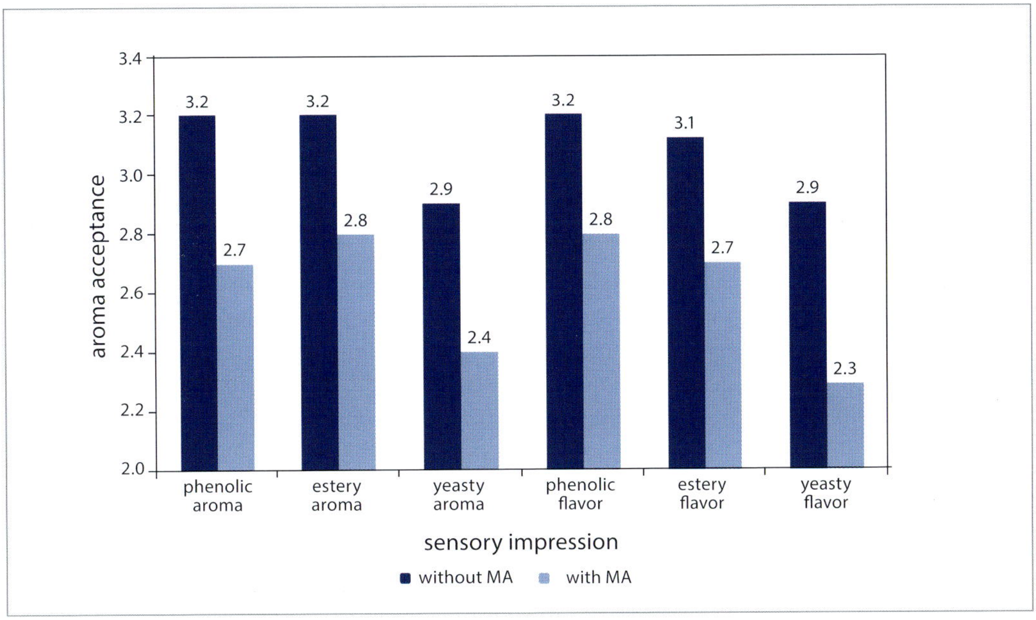

By contrast, the production of esters is more intensive when brewing to a higher original gravity [5]. Also, when using the special maltase mashing program (cf. Mashing), which serves to elevate the glucose concentration at the expense of maltose concentration, the production of ester compounds increases significantly. The increase in the concentration of 3-methylbutyl acetate, compared to that of ethyl acetate, can be seen as a very positive aspect of this process. In particular, yeast strains which do not form a large quantity of esters, tend to show a greater response to changes in the sugar spectrum [16].

This somewhat more complicated mash program is not necessary for breweries with traditional fermentation vessels – and to a certain degree for breweries using bottle conditioning – as a more pronounced estery character can also be achieved within the parameters of classical fermentation and aging (cf. sections 2.5 and 2.6).

2.4 WORT BOILING PROCEDURES

When producing hazy wheat beers, a moderate boil is recommended in order to retain a sufficient amount of proteins that contribute to turbidity. On the other hand, a longer boil is recommended for filtered wheat beers (*Kristallweizen*) to assist in their clarification [3, 12]. Hops additions at rates of up to 30–40 mg/l of α-acids usually take place at the beginning of the boil. Hop aroma is only desired in a few specialty wheat beers and is achieved by adding aroma hops (e.g. 50 % of the EBC bittering units) late in the boiling process (cf. Hops). However, hopping wheat beer generally plays a subordinate role, as bitterness in wheat beer is usually limited to between 8 and 15 EBC bittering units. Hop aromas are not common in wheat beers due to the presence of other intense aromas.

2.5 FERMENTATION

Yeast is usually pitched at a rate of 2–4 million cells/ml. The temperature during primary fermentation ranges from 15–22 °C, although temperatures of 18–22 °C may be reached if the yeast is initially pitched at 15 °C [3].

Higher fermentation temperatures promote the production of an estery aroma [17]. Yet, an excessively high concentration of esters can also lead to the negative synergetic effects previously mentioned. In any case, temperatures above 24 °C should be avoided, as this promotes yeast autolysis.

A higher pitching rate often leads to greater ester production. Low as well as extremely high aeration rates may also lead to increased ester production. The individual yeast strain also plays a role [5, 8].

The phenolic character is affected by the individual yeast strain as well as by the content of ferulic acid, although the fermentation parameters have minimal influence [5, 8].

It follows that the choice of yeast strain (fig. 3) is a determining factor in the development of the aroma; however, yeast management and the condition of the yeast (viability and vitality) have an equally large impact (cf. Yeast Technology, Fermentation).

Figure 3: Yeast strain and aroma formation [8]

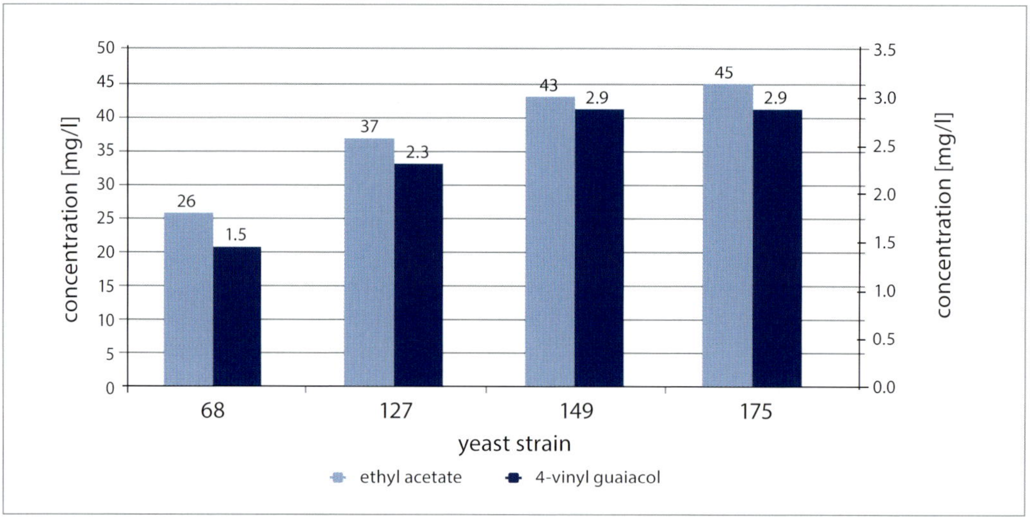

Tank geometry and the various heights of fermentation tanks have a strong impact on the development of aromas. As such, aromas develop most intensively in open fermentation vessels and progressively decline with horizontal and vertical tanks – especially cylindroconical tanks (fig. 4). The reduction in aroma compounds is most likely caused by the increased partial pressure of CO_2 in the lower sections of tall tanks. CO_2 directly inhibits pyruvate decarboxylase thus leading to lower concentrations of acetyl-CoA, resulting in fewer acetate esters. Indirectly, it may also reduce the production of alcohol acetyl transferases – and therefore acetate esters – due to inhibited metabolism of fatty acids.

The phenolic character is also diminished in cylindroconical fermentation tanks, although only to a small degree [18, 19, 20, 21]. When using propagated yeast cultures in cylindroconical fermentors, the first fermentation produces decent levels of higher alcohols and esters, as well as 4-vinyl guaiacol. Starting with the second fermentation using this yeast, the concentrations of these substances begin to decrease. By the third and all subsequent fermentations, the beers lose character, taking on an empty, harsh flavor with an aroma defined by apple esters.

When using traditional fermentation vats, harvesting the yeast is usually performed by repeatedly removing the yeast layer floating on the top or installing a channel to enable the yeast to flow off of the side of the vat into a collection vessel. Yeast is cropped from cylindroconical fermentors by pulling it off of the cone [3]. The amount of yeast required for maturation is established by separating the green beer with a centrifuge (with a bypass, in some cases), or through a partial filtration.

Figure 4: The influence of fermentation vessel geometry on the content of 4-vinyl guaiacol and 3-methylbutyl acetate

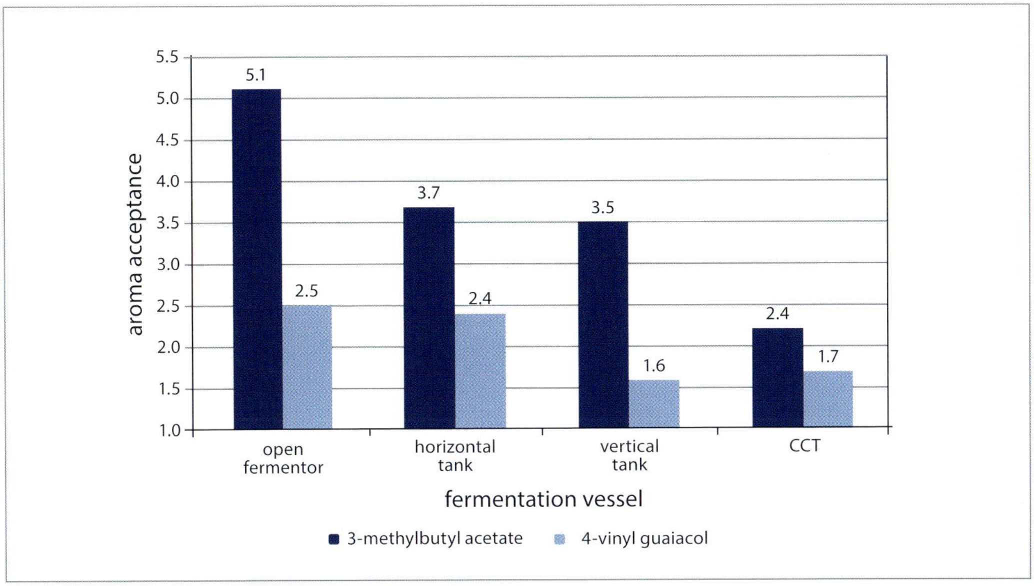

2.6 MATURATION

Traditional bottle conditioning involves a higher temperature (15–20 °C, 1 week) initially, followed by a cold phase of about 2 weeks at temperatures ranging from 10–12 °C or below. A similar process is implemented in tank maturation processes, although cold maturation occurs at significantly lower temperatures (to –1 °C). In order for the necessary volume of CO_2 to be generated during maturation, the required amount of extract is either added in the form of unfermented wort or kräusen or by allowing some residual extract to remain in the beer. The yeast used for conditioning either stems from primary fermentation (concentration set by centrifugation and bypass) or through the addition of fresh brewing yeast or kräusen [3]. Occasionally, bottom-fermenting yeast or mixtures of yeast are utilized for maturation, although the aroma that develops is less intense in such cases. As in fermentation, higher temperatures during the warm phase of maturation lead to an increase in the ester content. The length of the warm phase also promotes the development of aromas [5], although keeping the beer at high temperatures bears a stronger risk of yeast autolysis. For this reason, a warm phase of more than 7 days is not recommended.

In general, bottle conditioning results in a greater aroma yield compared to maturation in a tank, although high levels can also be achieved in tanks with the aid of appropriate temperature settings and pitching the correct amount of yeast [5].

The form of extract (wort, kräusen) added for maturation only has a limited impact on the development of an estery character in the finished beer. However, the addition of unfermented wort does increase the concentration of 4-vinyl guaiacol slightly, as it brings in more unconverted ferulic acid [9].

2.7 THE AGING PROCESS IN WHEAT BEER

Aging in wheat beer is associated with three main processes:
The breakdown of aroma-positive substances: The esters [22], the phenolic compounds and the fresh yeast aroma, along with malt aroma in some instances, can mask aging aromas that have begun to develop. The higher the concentration of aroma-positive substances, the longer a wheat beer may appear to be fresh. Degradation or the conversion of these substances is accelerated by oxygen, temperature as well as the presence of live yeast cells or damaged yeast in particular.

Autolyzed yeast or old yeast flavor: This is the most commonly found off-flavor in wheat beers [23]. At first, this is manifested as a dull, sweaty-cheesy or earthy off-flavor (short chain fatty acids like 3-methylbutyric acid, but also higher alcohols such as heptanol), which, depending on the initial situation, can reach a maximum after just three months [24]. The autolyzed yeast aroma usually diminishes as the process continues with an unpleasant taste (yeast bitterness) becoming more dominant as time progresses. The following measures must be taken in order to counteract this phenomenon: Yeast with an extremely high level of vitality should be utilized for tank maturation and bottle conditioning. A maximum of 1 million cells/ml should be added to the beer to be bottled first, and the beer should be then be flash pasteurized (only in combination with vital yeast cells!). Any oxygen uptake should be avoided. Oxygen uptake should be avoided (high pressure jetting is recommended when packaging beer destined for bottle conditioning).

The production of aromas typical in aged beers: Oxidation, cardboard, blackcurrant and sherry notes, all aromas commonly associated with bottom-fermented beers, are also found in wheat beers. Though present, they are less pronounced in wheat beers, since they are usually masked by aroma-positive substances or overpowered by the autolyzed or old yeast flavor. They may, however, be more noticeable in neutral wheat beers. Determining the concentration of aging compounds in wheat beer remains an important criterion for estimating its condition with regard to aging. It should be noted that during bottle conditioning, living yeast cells initially bring about a reduction in the carbonyl compounds associated with aging, prior to the normal increase in aging compounds that occurs approximately 1–2 months later (cf. Flavor Stability) [28].

2.8 UNIQUE ASPECTS OF WHEAT BEER PRODUCTION

Increasing water hardness: in order to prevent gushing, the brewing liquor used is often treated with calcium chloride (usually about 50 g/hl) to increase hardness, causing excess oxalic acid to precipitate as calcium oxalate over the course of the production process. This is only practical if the cold phase prior to separation is sufficiently long. In other cases, the precipitation and the corresponding increase in the risk of gushing occurs at a later stage for bottled beers in cold storage.

Flash pasteurization: This type of pasteurization has a stabilizing effect on flavor stability and foam stability, because it deactivates autolysis enzymes as well as proteases. In this case, it is essential to use vital yeast, since older yeast cells tend to burst due to the stress of flash pasteurization, causing their contents to be released into the medium.

2.9 HAZE STABILITY IN WHEAT BEER

Turbidity or haze plays a special role in wheat beers and serves as a visual sign of quality, even though fluctuating levels of haziness can be especially problematic in draft beers.
The stability of haze depends primarily on the particle size distribution, particularly for particles < 1 μm in size. Within this context, the proportion of particles < 1 μm in the finished beer should reach a value of 1.2 μl/100 ml, or better yet 2 μl/100 ml, in order to maintain sufficient haze (> 30 EBC) for periods exceeding three months [25].

During the production process, an evaluation of the haze stability can be performed by means of a scattered-light measurement (angles of 90° and 25°) in yeast-free beer samples (visual examination with a microscope). Wheat beers with stable haze show very similar values obtained through measurement at 90° and 25° [26], whereby high values for the scattered light measurement (> 60 EBC at 12 °C) also signifies good haze stability within this context. In contrast, a ratio of 1:2 (90°/25° value) indicates rapid clarification and low haze stability.

From the standpoint of brewing technology, haze stability can be positively impacted through flash pasteurization (with a simultaneous reduction in coagulable nitrogen) as well as through the addition of beer recovered from sedimented yeast or tank bottoms. Beer recovered in this manner can be added in quantities of up to 10 % by volume on the cold side without significantly affecting the sensory characteristics of the resultant beer [27].

Measures such as storing wheat beers for long periods at low temperatures or treatment with modern, high-performance separators to separate the yeast and clarify the beer remove the particles relevant for haze. As such, they have a negative impact on haze stability.

3 SUMMARY

Wheat beers are divided into four main categories based on their characteristic aromas: estery, phenolic, yeasty and neutral. Various fermentation parameters, the type of fermentation vessel and the yeast strain are the primary factors influencing the production of esters. In this context, higher ester contents are achieved by means of higher fermentation temperatures, whereas the height of the fermentation tank (cylindroconical fermentation tanks) has a dampening effect. Brewhouse operations also play a role in the development of esters. When it comes to a phenolic character in wheat beer, the proportion of wheat malt in the grain bill, the length of the rest at 45 °C, the yeast strain, and to a lesser degree, the geometry of the fermentation vessel, have all been proven to play a role. The number of cells in the bottle, the yeast strain and the temperature during primary fermentation are mainly responsible for a yeasty character in beer. Neutral wheat beers are characterized by a moderate level of aroma compounds.

Three main processes occur in the aging of wheat beers: the degradation of positive aroma components, the development of an old yeast or autolyzed flavor and the production of aroma compounds related to aging. Changes which can be perceived through the senses and are part of the aging process can be managed effectively by avoiding oxygen intake throughout the entire production process, using vital yeast, employing flash pasteurization and reducing the number of yeast cells in the bottle.

4 OVERVIEW

Figure 5 provides an overview of the primary factors influencing the aroma of wheat beer.

Figure 5: The primary factors influencing the aroma of wheat beer

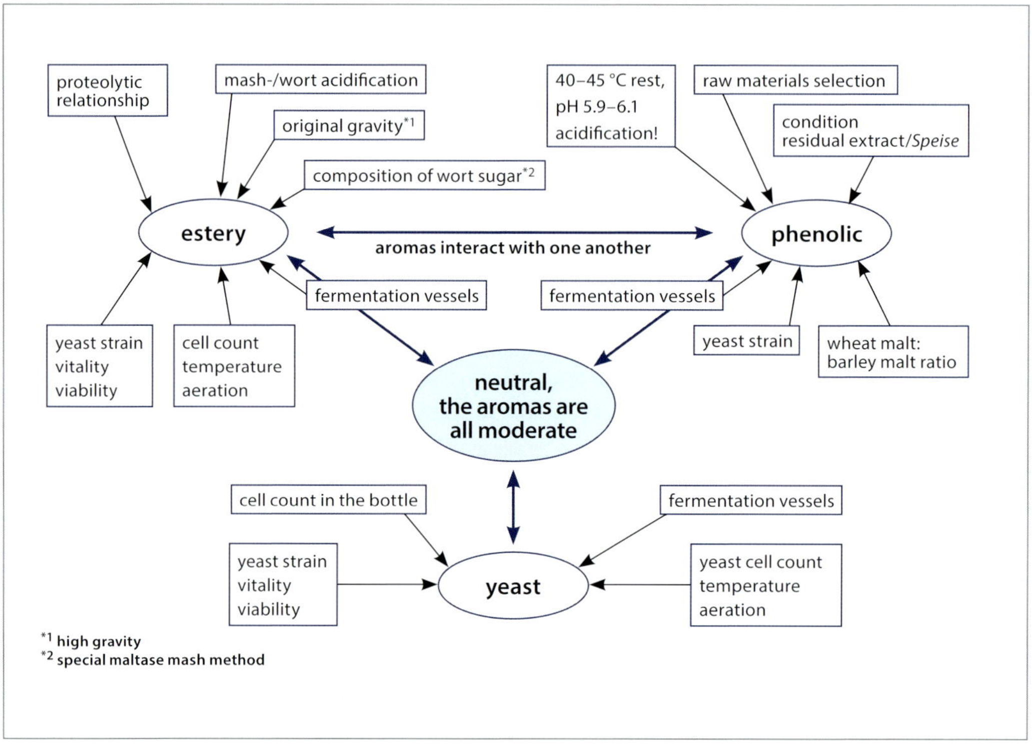

Tables 5–7 provide further explanation with regard to figure 5.

Table 5: Estery notes

Parameter	Cause	Effect
proteolytic ratios	influence of FAN on yeast metabolism	concentrations up to approx. 20–22 mg/100 ml FAN cause a reduction in estery notes
mash/wort acidification	addition of yeast nutrients (e.g. zinc)	disproportionally high formation of ethyl acetate and higher alcohols causing a reduction in estery notes
original gravity	influence on yeast metabolism	disproportionally high formation of esters due to high original gravity
sugar composition	shift in the diauxic growth pattern	increased ester formation with an elevated proportion of glucose
fermentation vessels	influence on yeast metabolism	suppression of ester formation with increasing tank height and with cropped yeast that has been used in multiple fermentations
yeast cell count	change in the oxygen/cell factor	elevated formation of esters is frequently encountered with high yeast pitching rates
temperature	influence on yeast metabolism	elevated formation of esters with higher fermentation temperatures
aeration	change in the oxygen/cell factor	elevated formation of esters is frequently encountered with very low and very high aeration rates
viability and vitality	influence on yeast metabolism, change in the oxygen/cell factor	change in the actual number of yeast cells pitched (see above) due to low yeast viability; elevated formation of esters is frequently encountered in later fermentations with cropped yeast that has been used multiple times
yeast strain	different reactions to fermentation parameters	elevated or reduced ester formation, depending on the individual yeast strain

Table 6: Phenolic notes

Parameter	Cause	Effect
rest at 40–45 °C pH 5.9–6.1	influence on ferulic acid hydrolase	long rests at the pH and temperature optima for ferulic acid hydrolase increase the concentration of ferulic acid and therefore also increases the concentration of 4-vinyl guaiacol
selection of raw materials/ratio of wheat to barley malt	different levels of ferulic acid in the raw materials	an increase in the proportion of barley malt generally leads to an increase in the concentration of 4-vinyl guaiacol
geometry of the fermentation vessel	influence on yeast metabolism	increasing tank height and multiple runs with cropped yeast results in a slight decrease in the formation of 4-vinyl guaiacol
yeast strain	influence on yeast metabolism	elevated or reduced formation of 4-vinyl guaiacol, depending on the yeast strain

Table 7: Yeasty notes

Parameter	Cause	Effect
cell count in the bottle	number of yeast cells	the more yeast cells in the bottle, the stronger the yeasty aroma is in the beer
yeast strain, vitality, viability	selection of individual yeast strain, condition of the yeast	the intensity of yeast aromas varies according to the individual yeast strain, cell vitality and cell viability; low yeast vitality contributes to more problems with yeast autolysis
fermentation vessels	influence on yeast metabolism	increasing tank height brings about a slight reduction in yeasty notes
duration of maturation	excessively long maturation times with slow degradation of 2-acetolactate	presence of an unpleasant yeasty note, sometimes even a soapy note
temperature and aeration	influence on yeast metabolism	the higher the temperature and degree of aeration is, the more intensive the fermentation process is and the more pronounced the yeast aroma is

5 REFERENCES

[1] Kelch, K.: Weißbier weiterhin beliebt. Brauwelt 155 (2015), no. 29, pp. 826–827.

[2] BierStG (i. d. F. vom 29.07.1993) § 9 Abs. 1–8.

[3] Narziß, L., Back, W., Gastl, M., Zarnkow, M.: Abriss der Bierbrauerei, 8th ed., Wiley-VCH 2017.

[4] Deutscher Brauerbund: http://www.brauer-bund.de, 2004.

[5] Back, W., Diener, C., Sacher, B.: Hefeweizenbier – Geschmacksvarianten und Technologie. Brauwelt 137 (1998), no. 28/29, pp. 1279–1284.

[6] Schieberle, P.: Primary Odorants in pale lager beer. Zeitschrift für Lebensmitteluntersuchung und Forschung, (1991), no. 193, pp. 558–565.

[7] Hecht, S.: Über die Veränderung wertbestimmender Stoffgruppen bei der Herstellung von bayerischen Weizenbieren. Dissertation, Technical University of Munich (Weihenstephan), 1983.

[8] Narziß, L.: Über den Biergeschmack und seine Beeinflussung durch Rohstoffe und technologische Faktoren. Brauwelt 134 (1995), no. 45, pp. 2286–2300.

[9] Herrmann, M.: Entstehung und Beeinflussung qualitätsbestimmender Aromastoffe bei der Herstellung bayerischer Weizenbiere. Dissertation, Technical University of Munich (Weihenstephan), 2005.

[10] Sacher, B.: Über den Einfluß von Sorte, Umwelt, agronomischen Maßnahmen und Mälzungstechnologie auf die wertbestimmenden Eigenschaften von Winterweizenmalzen. Dissertation, Technical University of Munich (Weihenstephan), 1997.

[11] Kreisz, S., Hartmann, K., Zarnkow, M., Herrmann, M., Back, W., Sacher, B.: Verschiedenheit von Gerste und Weizen. Aussagekraft von Weizenmalzmerkmalen. Brauindustrie 90 (2005), no. 1, pp. 12–15.

[12] Narziß, L., Back, W.: Die Technologie der Würzebereitung, 8th ed., Wiley-VCH 2009.

[13] Nitzsche, W.: Untersuchungen zur Bildung phenolischer Komponenten bei der Herstellung von bayerischem Weizenbier. Dissertation, Technical University of Munich (Weihenstephan), 1992.

[14] McMurrough, I., Madigan, D., Donnelly, D., Hurley, J., Doyle, A.-M., Hennigan, G., McNulty, N., Smyth, M.: Control of Ferulic Acid and 4-Vinyl Guaiacol in Brewing. Journal of the Institute of Brewing 102 (1996), pp. 327– 332.

[15] Köhler, P., Krottenthaler, M., Herrmann, M., Keßler, M., Kühbeck, F.: Untersuchungen zu Infusions- und Dekoktionsverfahren. Brauwelt 145 (2005), no. 6, pp. 143–150.

[16] Herrmann M., Back, W., Sacher, B., Krottenthaler, M.: Möglichkeiten zur technologischen Beeinflussung der Bieraromastoffe. Monatsschrift für Brauwissenschaft 56 (2003), no. 5/6, pp. 99–106.

[17] Verstrepen, K., Derdelinckx, G., Dufour, J.-P., Winderickx, J., Thevelein, J. M., Pretorius, I. S., Delvaux, F.: Flavor-Active Esters: Adding Fruitiness in Beer. Journal of Bioscience and Bioengineering 96 (2003), no. 2, pp. 110–118.

[18] Kumara, H. M. C. Shanta, H., Fukui, N., Kojima, K., Nakatani, K.: Regulation Mechanism of Ester Formation by Dissolved Carbon Dioxide During Beer Fermentation. Technical Quarterly 32 (1995), no. 3, pp. 159–162.

[19] Renger, R., van Hateren, S., Luyben, K.: The Formation of Esters and Higher Alcohols during Brewery Fermentation. The Effect of Carbon Dioxide Pressure. Journal of the Institute of Brewing 98 (1992), no. 6, pp. 509–523.

[20] Mason, A., and Dufour, J.-P.: Alcohol Acetyltransferases and the Significance of Ester Synthesis. Yeast 16 (2000), pp. 1287–1298.

[21] Dufour, J.-P., Weaver, A., Mason, B.: Towards our Understanding of the Physiological Role of Ester Synthesis. European Brewery Convention, Symposium Yeast Physiology, Nutfield 1999. Monograph 28, pp. 83–91.

[22] Neven, H., Delvaux, F., Derdelinckx, G.: Flavour Evolution of Top-fermented Beers. Technical Quarterly 34 (1997). pp. 115–118.

[23] Gerstenberg, H., Uhlig, R.: Über sensorische Qualitätsabweichungen bei Bier. Brauwelt 137 (1998), no. 51/52 pp. 2512–2514.

[24] Herrmann M., Back, W.: Erkenntnisse zur Alterung von Hefeweißbieren. Der Weihenstephaner, 72 (2004), no. 1, pp. 16–20.

[25] Schwarz, C., Kreisz, S., Back, W.: A Measurement System for Haze Stability of Turbid Beers. Proceedings of the 31st of the European Brewery Convention Congress, Venice, Fachverlag Hans Carl, Nürnberg 2007.

[26] Englmann, J.: In: Lehrstuhl für Technologie der Brauerei I, TU München-Weihenstephan: 35. Technologisches Seminar, Weihenstephan 2002.

[27] Schneeberger, M.: Verwertung von Prozessbieren in der Brauerei. Dissertation, Technical University of Munich (Weihenstephan), 2006.

[28] Schneiderbanger, H.: Entstehung, Abbau und potentielle Verlustquellen wertgebender Aromen während der Produktion von Weizenbieren. Dissertation, Technical University of Munich (Weihenstephan), 2016.

NON-ALCOHOLIC BEER

1 INTRODUCTION

Non-alcoholic beers, with their beer-like properties and low alcohol content, enjoy substantial popularity as a foodstuff as well as a social beverage around the world, and they are becoming more popular as time passes. In 2006, their percentage of the total beer production in Germany was 2.6 % [1]. Beers containing less than 0.5 % alcohol by volume are considered to be non-alcoholic in Germany. In other countries, especially in the Arab world, the alcohol content must be less than 0.05 % by volume. Non-alcoholic beers are not subject to the beer tax in Germany. Moreover, in many countries, the labeling requirements for non-alcoholic beer are much more liberally regulated than those for beer containing alcohol [2, 3].

Consumer interest in non-alcoholic beer is primarily rooted in calorie consciousness, restrictive laws governing the operation of vehicles and safety rules forbidding alcohol in the workplace. A series of investigations has shown that the residual content of alcohol present in non-alcoholic beers has no demonstrable physiological or pharmacological effect [4]. Non-alcoholic beer produced using interrupted fermentation is usually isotonic or slightly hypotonic and therefore can be marketed as a functional drink, e.g. for athletes [5, 6, 7, 8]. Non-alcoholic beer is the official beverage of the German national soccer team and other selected German professional sports teams.

Different techniques have been developed for producing non-alcoholic beer. The underlying problem with each of these is that alcohol makes a significant contribution to the flavor of beer. For this reason, a non-alcoholic beer can never be completely equivalent to the flavor of the original beer. Therefore, when producing non-alcoholic beer, it is advantageous to use a different recipe than that of the regular beer and to take additional measures to adapt the process to the kind of beer being brewed. The methods described below are also employed in the production of low-alcohol beer and some are also used to make special beer suitable for diabetics.

2 PRODUCTION METHODS FOR NON-ALCOHOLIC BEERS

The production methods for non-alcoholic beers can be divided into two main groups: physical and biological methods. With physical methods, the alcohol produced must be removed from the beer after fermentation, while the formation of alcohol is suppressed or limited to certain concentrations with biological methods. The choice of a particular method has a significant influence on the characteristics of the finished product as well as on the cost of production [9, 10].

2.1 PHYSICAL REMOVAL OF ALCOHOL

Thermal methods are primarily used in the removal of alcohol from fermented beer. For example, this includes vacuum distillation or membrane separation methods, such as reverse osmosis or dialysis. From a legal perspective in Germany, operating distillation equipment puts a brewery in the position of running a distillery without the proper licensing. The brewery must register with the relevant tax authority in order to install and operate this type of equipment [11].

2.1.1 THERMAL REMOVAL OF ALCOHOL

Distillation and rectification of alcohol/water mixtures are thermal separation methods based on the unit operations of evaporation and condensation. In these operations, components with lower boiling points become more concentrated in the gas phase while those with higher boiling points accumulate in the liquid phase. Placing the contents under a vacuum lowers the boiling point of the liquid mixture (fig. 1). This vastly reduces the amount of thermal stress on the beer. Vacuum distillation equipment is used almost exclusively in the thermal removal of alcohol and is generally operated at 60–200 mbar of absolute pressure within a temperature range of 37 to 60 °C [12, 13].

Figure 1: The vapor pressure of water

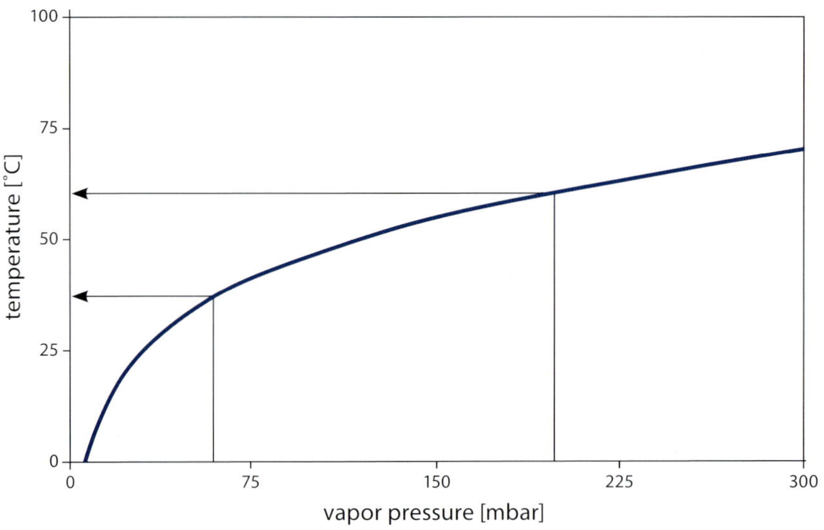

As the intensity of the evaporation process increases, the amount of residual alcohol in the beer decreases. Separation of alcohol from water is not only dependent on the equilibrium of the mixture undergoing evaporation but also on the kinetics of the component transport to the phase interface. Thin film evaporators with mechanical generation of thin films or falling film evaporators are primarily employed for vacuum evaporation. The construction of the evaporator influences how the vapors form and therefore the requisite rate of evaporation, as well [12, 14]. Evaporation rates between 40 and 60 % are generally required for simple distillation to reduce the alcohol in beer to a concentration below 0.5 % by volume with thin film evaporators or falling film evaporators. If the evaporators are used in conjunction with a rectification column, evaporation rates between 6 and 10 % can be achieved to lower the concentration of alcohol to below 0.05 % by volume in the finished product. The significantly higher separation capacity of the rectification process results from greater enrichment and separation compared to simple distillation (multiple phase changes occurring countercurrent to the flow direction of the steam/liquid mixture) [14, 15].

In addition to the primary fermentation products of ethanol and CO_2, there are a number of other volatile compounds in beer that are removed with the distillate or exit through the vacuum pump. Although most volatile aroma compounds possess a higher boiling point than ethanol, they have rectification quotients greater than 1, as a result of fermentation. This means that they accumulate more strongly in the gas phase than ethanol does [13, 14, 17, 18]. The evaporation behavior exhibited by some important aroma compounds in beer during vacuum distillation (simple distillation) is depicted in figure 2 below.

Figure 2: Concentrations of aroma compounds during the vacuum evaporation of beer

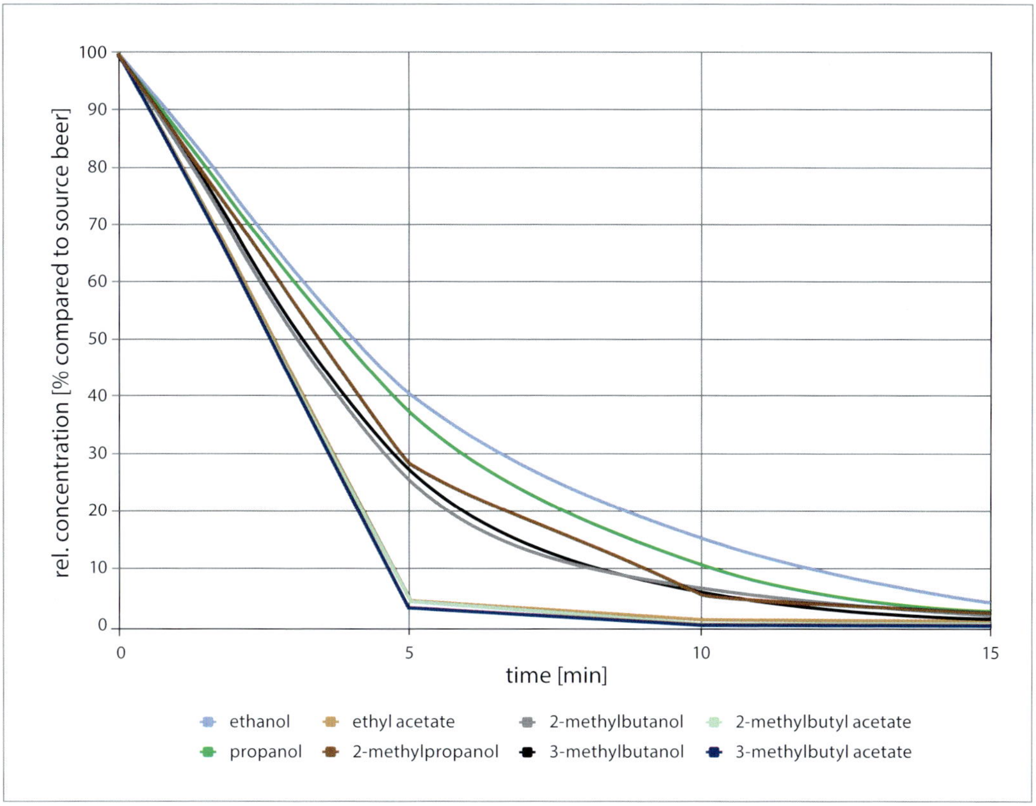

It is evident from their behavior that the highest concentrations of aroma compounds are achieved in vacuum-evaporated beers when the alcohol is removed to the greatest extent possible. This beer is subsequently blended with the regular, untreated beer to a final alcohol concentration of 0.5 % by volume.

A modern dealcoholyzer featuring a falling film evaporator and a rectification column is pictured in figure 3. The beer is first heated to a temperature near the boiling point (at regular operating pressure) in a plate heat exchanger (W1) and the majority of the CO_2 is liberated in a degasser (E). Subsequently, the beer is sent through the stripping section of a rectification column (R). The product flows from plate to plate at a temperature of 40 to 50 °C, while the vapors move in a countercurrent direction, constantly evaporating and condensing, causing alcohol to be selectively separated.

Figure 3: Process diagram for a Sigmatec dealcoholization system (source: API Schmidt-Bretten GmbH & Co KG, Bretten, Germany [19])

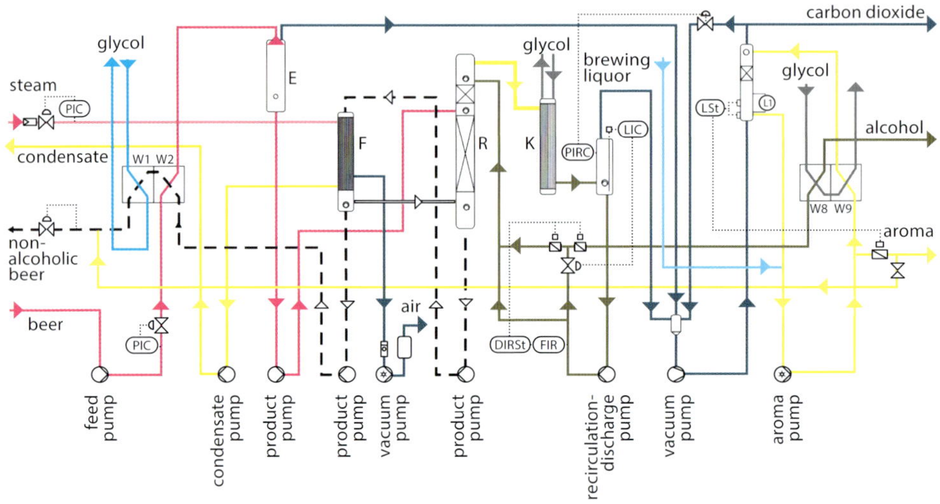

E = degasser F = falling film evaporator K = tubular surface condenser R = rectification column
W1 = beer pre-warmer W2 = beer chiller W8 = alcohol chiller W9 = aroma chiller

The reduced-alcohol beer collects at the bottom of the column and is pumped into a falling film evaporator (F). There, the vapors required for the rectification process are generated and sent back into the column. The dealcoholized beer is chilled in plate heat exchangers (W1 + W2) and pumped out of the system. The alcohol-enriched vapors pass through the stripping section of the column and enter the dephlegmator portion of the column (R) where they are concentrated to an alcohol content of more than 80 % by volume. The gases leaving the vacuum pump are washed with brewing liquor in an aroma recovery system. This aroma extract is low in alcohol and can be used to dilute the dealcoholized beer, while simultaneously adding the aroma back into the final product.

Thin film evaporation can be carried out with a falling film evaporator, a centrifuge (Centritherm®, Flavourtech, Griffith, Australia) or a spinning cone column. All of these systems operate under a vacuum.

Many of the aroma compounds are virtually eliminated in processes that remove almost all of the alcohol present (> 99 %). Thus, the concentrations of most esters and volatile higher alcohols along with DMS are reduced to undetectable levels (tab. 1). Less volatile substances, such as diacetyl, medium chain fatty acids and higher alcohols like phenylethanol and furfuryl alcohol remain unaffected for the most part. Furthermore, although it is assumed that a certain amount of aging compounds is formed during vacuum evaporation, a reduction in the carbonyl compounds associated with aging also occurs. After vacuum evaporation, the beers exhibit a slight increase in color (0.5–1.5 EBC) and bitterness (1–2 EBC). The real extract concentration increases somewhat depending on the evaporation rate (around 7 %). The pH of the beer after the process is around 0.1–0.2 units higher than the source beer which is attributable to the loss of volatile organic acids and CO_2. The foam characteristics remain largely unchanged, while the increase in turbidity is generally less than 1 EBC. However, to achieve these values, it is necessary to start with a well-stabilized source beer and to conduct all of the process steps in such a way that oxygen uptake is kept to a minimum.

Blending fermented beer with unfermented wort prior to dealcoholization can be problematic. The beer foams vigorously during the degassing step, causing protein and polyphenol compounds to accumulate on the bubbles. These substances react to form the particles responsible for turbidity. It is preferable to only degas the fermented beer, remove the alcohol and subsequently carry out the blending process. Alternatively, the alcohol can be removed from the unfiltered beer prior to filtration.

The sensory characteristics of beer are changed significantly by the thermal removal of alcohol. Compared to the source beer, a strong decrease in aroma, mouthfeel and drinkability is detected. The aroma profile is different, and less desirable aromas come to the fore, such as bread-like, worty and caramel notes. Additionally, the beer may, in some cases, exhibit the flavor of fatty acids (e.g. goaty flavor). The sensory quality of vacuum-evaporated beer is usually considered to be acceptable for non-alcoholic beer [20].

Table 1: Aroma compounds in beer with alcohol removed using a Sigmatec dealcoholizer (API Schmidt-Bretten GmbH & Co KG. Bretten, Germany)

Aroma compounds		Source beer	Concentrate	Non-alcoholic beer
alcohol	% v/v	5.5	0.03	0.45
acetaldehyde	mg/l	10.2	2.1	3.3
propanol-1	mg/l	25.5	traces	3.6
ethyl acetate	mg/l	21.7	traces	1.7
isobutanol	mg/l	25.5	traces	3.6
isoamyl acetate	mg/l	2.7	n. d.	0.2
3-methylbutanol	mg/l	82.3	0.3	12.1
2-methylbutanol	mg/l	25.7	0.1	3.2
phenylethanol	mg/l	39.7	42.0	40.1
furfuryl alcohol	mg/l	3.5	2.8	3.0
diacetyl	mg/l	0.1	0.1	0.09
DMS	µg/l	86	n. d.	10
hexanoic acid	µg/l	1089	812	835
octanoic acid	µg/l	2578	1987	2086
decanoic acid	µg/l	423	173	186
dodecanoic acid	µg/l	21	16	18
aging compounds	µg/l	67	32	36

n. d. = not detectable

Obviously, the flavor and aroma of this beer is improved when it is blended with untreated beer to a maximum alcohol content of 0.5 % by volume in the final product. In some breweries, the aroma extract collected from the dealcoholization unit is blended back into the beer. Other post-treatment procedures entail adding separately fermented beer rich in aroma to the dealcoholized beer. The addition of fresh yeast (yeast contact) and especially the kräusening process have a noticeably positive impact on the aroma and flavor of non-alcoholic beer. Modifying the recipe of the source beer can bring about other quality improvements, while the degree of attenuation can be lowered through utilization of special malts or mashing methods. Stripping the dealcoholized beer with CO_2 can also be beneficial.

2.1.2 EXTRACTION WITH CO_2

Extraction with carbon dioxide is generally carried out in two steps [22]. In the first step, an aroma extract is collected using liquid CO_2 at a pressure of 125 bar. Subsequently, the ethanol is removed with supercritical CO_2 at 250 bar. During the continuous extraction process, the ethanol must be separated from CO_2 in the gas phase. This method is particularly widespread in applications involving the removal of alcohol from wine.

2.1.3 MEMBRANE SEPARATION METHOD

The technique of diffusion through a suitable membrane for removing alcohol from beer is widely established. Of these, reverse osmosis and dialysis are of particular importance. With reverse osmosis, the driving force behind the separation process is to raise the pressure across the membrane to above the osmotic pressure. In dialysis, the substances are separated on the basis of a concentration gradient (diffusion). The composition of the dealcoholized beer is dependent on the composition of the source beer, the selectivity of the membrane to certain substances in the beer (permeability) and the various process parameters.

Membrane separation processes are characterized by a lack of thermal stress on the product. However, due to a tendency to become blocked and the fact that they are quite delicate, membrane separation systems pose special demands on the cleaning and sterilizing procedures (they require their own CIP equipment) [21, 23, 24].

2.1.3.1 Reverse osmosis

Reverse osmosis exploits the separation effect of a semi-permeable membrane between two fluids with different solute concentrations. Osmosis occurs when the solvent, in this case water, is transported through the membrane into the fluid containing more solute. The direction of water transport can be reversed if enough hydrostatic pressure is applied to the side with the higher solute concentration (reverse osmosis). In this instance, parts of the water, the alcohol and other compounds found in beer traverse the membrane, entering the permeate [25].

A reverse osmosis system essentially consists of a high-pressure pump and a membrane module block. Separation of the alcoholic permeate and the alcohol-reduced concentrate (retentate) is performed at pressures ranging from 20 to 60 bar. Substances in the extract incapable of diffusion collect in the retentate, which is later diluted with treated brewing liquor to the original extract concentration. In order to achieve an alcohol content of less than 0.5 % by volume, a permeate runoff volume of 2–3 times the volume of the source beer is to be expected. Figure 4 depicts a flow diagram of a reverse osmosis system in continuous operation.

Another option for diluting the concentrated beer is to continuously add brewing liquor by means of diafiltration. Here, the flux is stabilized at a high value across the membrane.

Figure 4: Schematic flow diagram of a reverse osmosis unit

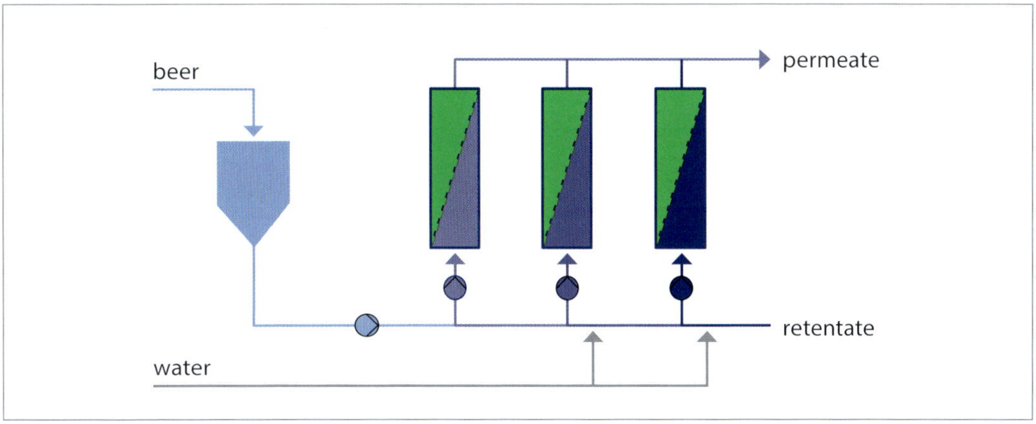

Membranes composed of various materials, such as cellulose derivatives, polysulfones, polymethylsiloxane or other plastics, are readily available on the market. The membranes differ according to their separation performance which is characterized by their capacity to retain specific substances (e.g. sucrose, NaCl). Two different separation models are used to describe the selectivity of membranes: with ideal solubility membranes, separation occurs through permeation (diffusion quotient ~ Δp), whereas with ideal pore membranes, separation is dependent on molecular size (diffusion quotient ~ r^{-1}). Further parameters determining the performance of a reverse osmosis system include working pressure, the volumetric ratio of retentate to permeate, temperature and the concentration of dissolved and undissolved substances [26]. Stringent demands are placed on the water used for dilution with regard to oxygen concentration, pH and salt content. While the latter two characteristics should conform to those typical of beer, the oxygen concentration must be kept as low as possible.

The behavior of the aroma compounds in beer treated with reverse osmosis can be found in table 2. Compared to thermal methods, several beer aroma compounds are present in somewhat higher concentrations in non-alcoholic beer produced using reverse osmosis. Proportionally, many esters and higher alcohols are detected in much higher concentrations with respect to the residual alcohol concentration in the finished product. The reduction in the amount of medium chain fatty acids is average to strong depending on their solubility in water. Primarily, it is the thermal stress indicators that are reduced among the aging compounds. In total, the aging compounds are lowered by 42 %.

Beer with alcohol removed by means of reverse osmosis exhibits a somewhat fuller flavor than those subjected to thermal methods. Here as well, bready and worty aromas tend to emerge and dominate. Occasionally, these are accompanied by the flavor of fatty acids. Frequently, the treated beer exhibits a slightly sour impression compared to the source beer. If yeast with a strong vitality was used to ferment the beer, the concentration of medium chain fatty acids is lower in the dealcoholized beer, and therefore the quality of the resultant non-alcoholic beer is considered to be higher. The loss of bitter compounds is greater in thermal dealcoholization processes. The increase in turbidity in the retentate is generally less than 1 EBC.

Table 2: Concentrations of aroma compounds in beer with alcohol removed by reverse osmosis

Aroma compounds (see Appendix)	Source beer	Dealcoholized beer (retentate)	Relative to content in source beer [%]
ethanol [% v/v]	4.7	0.46	10
DMS [µg/l]	64	n. d	n. d.
Esters [µg/l]			
acetic acid isobutyl ester	23	4.2	18
acetic acid hexyl ester	3.7	0.7	19
acetic acid 2-phenyl ethyl ester	192	24	13
butyric acid ethyl ester	69	11	16
hexanoic acid ethyl ester	122	15	12
octanoic ethyl ester	253	24	9
decanoic acid ethyl ester	62	3	5
Higher alcohols [mg/l]			
propanol	14	1.9	14
2-methylpropanol	10	2.2	22
3-methylbutanol	46	9.9	22
2-methylbutanol	12	2.6	22
2-phenylethanol	28	4.1	15
Fatty acids [µg/l]			
hexanoic acid	2593	273	11
octanoic acid	5630	871	15
nonanoic acid	27	6.7	25
decanoic acid	1070	316	30
dodecanoic acid	25	22	87
Aging compounds [µg/l]			
thermal stress indicators	28	3.8	14
oxygen indicators	31	30	97
Σ aging compounds	84	49	58

n. d. = not detectable

2.1.3.2 Dialysis

In contrast to reverse osmosis, no pressure is applied in dialysis as the substances are transported through spontaneous diffusion. The driving force behind diffusion is the difference in concentration of substances in the retentate and the dialysate. In the case of the dialysis of beer against water, all of the substances in the beer, except for water, diffuse towards the dialysate. The pore structure of the membrane serves to exclude certain molecules (permeability).

A flow diagram of the dialysis process is shown in figure 5. The beer to be dealcoholized is fed continuously into the dialysis module. There, the alcohol is separated from the retentate into the dialysate flowing in the opposite direction. The final alcohol concentration can be adjusted to the desired level by varying the flow rate of both fluids. The dialysate is routed through a rectification column, where the alcohol is removed. After the alcohol has been removed, the dialysate is returned to the loop to be recirculated.

Substances in beer of a molecular size and polarity similar to ethanol are separated by means of a membrane in the dialysis process. Although dialysis is essentially a process that occurs without pressure, a certain amount of pressure must be maintained on the beer side as well as on the dialysis side, otherwise the liberation of CO_2 would interfere with the diffusion process. Due to the small differences in concentration, the dialysis technique is less suited for the production of beers very low in alcohol (< 0.2 % by volume). Dialysis is primarily used in the production of dietetic beer (made especially for diabetics), low gravity beers or those produced using interrupted fermentation [21, 23, 24].

Although the loss of aroma compounds is considerable with dialysis, their reduction is still less per unit alcohol than with vacuum evaporation. Another advantage of dialysis is that the retentate does not become concentrated because the concentration of water in the dialysate is higher than that of the retentate. The retentate contains the majority of the substances truly dissolved in the extract (carbohydrates, soluble nitrogen, acids, minerals, bitter substances, etc.).

It is helpful to infuse the dialysate with a small amount of carbon dioxide in order to avoid losing the carbon dioxide dissolved in the retentate. This lowers the risk of the beer taking up oxygen during the process. If non-alcoholic beer is used instead of water in the dialysate loop, the differences in concentration of many of the non-volatile substances in beer can be compensated for at the beginning of the process. It is sometimes necessary to operate at high flow rates in order to prevent concentration along the interfaces, which in turn can heat up the beer and require a subsequent cooling step [24].

The quality of non-alcoholic beer produced using dialysis is generally quite good, but it is by no means comparable to that of the source beer. The aroma profile loses a significant amount of balance and overall harmony, resulting in a sour taste. The mouthfeel, in particular, leaves something to be desired. To improve this situation, some breweries have developed a special recipe for the source beer to be dealcoholized. For example, these recipes involve the production of wort with higher levels of non-fermentables. Also, this wort is often only fermented partially. These beers have a better mouthfeel after the dealcoholization process but are lacking in freshness and drinkability. Sometimes special malts (e.g. caramel or roasted malts) are utilized to enhance mouthfeel and produce more rounded beers. Higher fermentation temperatures and stronger original gravities can be used to increase the aroma in beer. Moreover, dialysis offers the option of adding the aroma collected from the rectification column back into the dealcoholized beer.

Figure 5: Schematic of dialysis with rectification

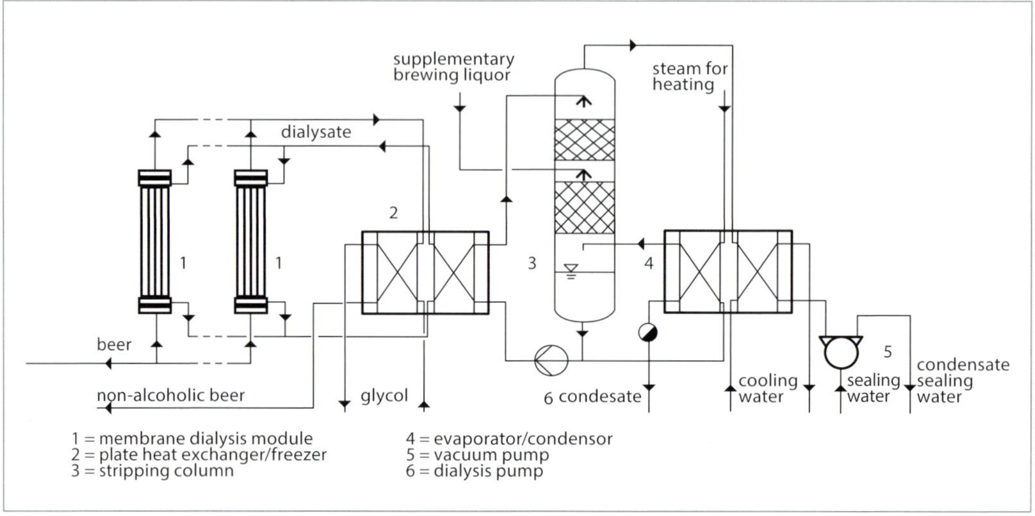

2.2 BIOLOGICAL METHOD

With the biological method, a low alcohol content is achieved by interrupting fermentation before the target alcohol concentration is reached. If the interrupted fermentation process is carried out at a particularly low temperature (e.g. 0 °C), this is referred to as the "cold contact" method. Production is usually performed in batches; however, continuous systems are employed in some situations. Recently, methods have been utilized in which special microorganisms selectively ferment a portion of the available fermentable extract.

The primary advantage of the biological method is that no additional energy must be expended to remove the alcohol. Due to the incomplete fermentation of the extract in the wort, the resultant beer is generally sweet and worty in character with more mouthfeel than conventional beer.

2.2.1 INTERRUPTED FERMENTATION

A number of special measures may be utilized with interrupted fermentation to produce non-alcoholic beers with characteristics more similar to conventional beer. One option is to carry out special mashing regimens. The "spring" mashing method is of particular importance here (cf. Mashing, section 2.2.1). This decoction method involves eliminating the maltose rest (60–65 °C) to produce wort with a low apparent attenuation (E_A below 70 %). As a result, the sweet, worty flavor is less pronounced. This is to the smaller difference between the final degree of attenuation and the original gravity compared to beer produced from conventionally brewed wort. Since the amount of available fermentables in this type of wort is still relatively high, it is recommended that the original gravity be reduced to 5–8 %. Thus, it is also possible to increase the degree of attenuation while achieving the same alcohol content, leading to significant improvement in the sensory characteristics of the beer [27, 28].

During the boil and subsequent handling of the wort, all measures designed to evaporate and eliminate the formation of undesirable wort aromas such as, DMS, Strecker aldehydes and other carbonyl compounds, have proven favorable. This can be achieved through a gentle boiling process, in which wort aroma compounds are actively evaporated while keeping thermal stress on the wort at a low level (cf. Wort Boiling and Wort Boiling Systems). Lowering the temperature of the wort in the whirlpool, in addition to downstream evaporation of the wort, also benefits the flavor. The high levels of residual extract present in these beers also can offset increased bitterness. Late additions of aroma hops can lend a pleasantly hoppy note to the beer [29].

Additional (biological) acidification of the wort is necessary due to the low fermentation activity of the yeast. The pH of the finished beer should be between 4.2 and 4.5. A pH reduction is necessary to improve the flavor but is also important for microbiological and chemical-physical stability (cf. Acidification with Natural Lactic Acid). Preferentially, *Sauergut* or wort naturally acidified through the addition of pure lactic acid bacteria cultures is added at the end of the boil, so that the desired chemical conversions (cleavage of the DMS precursor, isomerization of the hop bitter acids) are not hindered and excessive amounts of protein are not eliminated. If proper whirlpool operation is compromised by the low pH of the wort, the *Sauergut* can be added between the whirlpool and the wort chiller. If the wort is brewed to a higher original gravity, dilution to the desired gravity should take place prior to fermentation [29].

Fermentation is generally carried out in such a manner that contact of the yeast with wort is not only intensive but can occur over a longer period of time (cold contact method). During this time, particulates adsorb to the surface of the yeast cells, exerting a clarifying effect on the beer. In addition, the yeast also possesses reducing properties, causing the carbonyl compounds in the wort to be converted to their corresponding alcohols. A significant decline in Strecker aldehydes is observed after a short contact time (fig. 6). The lowest concentrations are achieved if fermentation is allowed to progress until the target alcohol concentration is reached.

Figure 6: Concentrations of aroma compounds in wort during interrupted fermentation (fermentation temperature: 0 °C, yeast cell count at pitching: 25 million cells/ml)

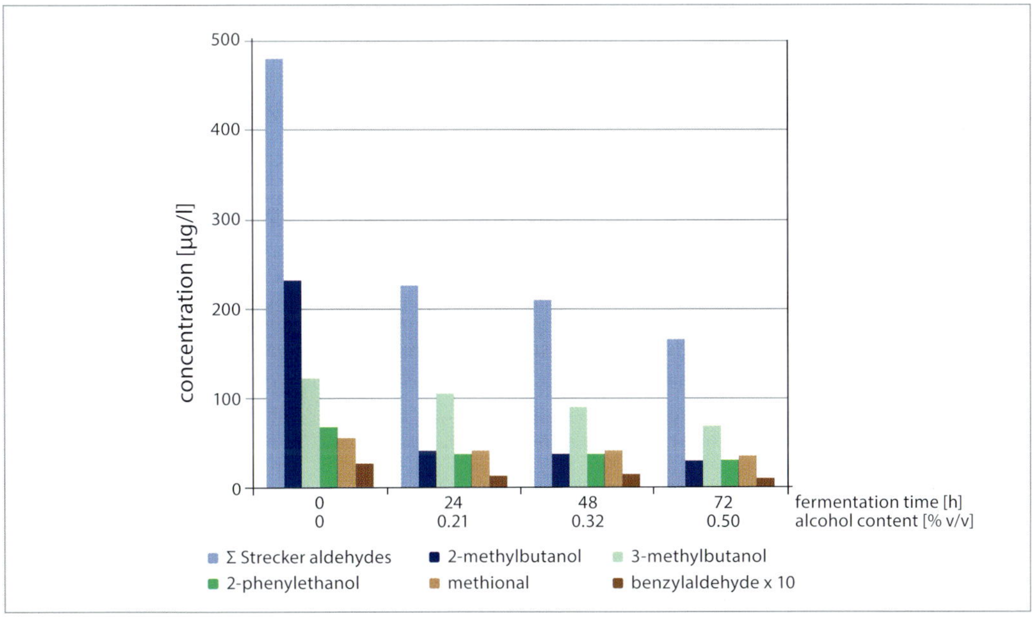

Contact time with the yeast can be extended significantly by maintaining low fermentation temperatures (fig. 7). A more pronounced decline in Strecker aldehydes is measured at the same alcohol concentration. Despite the lower concentration of higher alcohols, beers produced with cold fermentation are deemed to exhibit a more typical beer flavor and are more balanced than those subject to warmer fermentation temperatures.

Generally speaking, pitching a relatively large quantity of vital yeast is advantageous. This enables reductions in wort aeration and the suppression of diacetyl formation. Washing the yeast prior to pitching removes residual beer trapped in the yeast, thus lowering the amount of alcohol introduced to the beer and permitting longer contact times with the yeast. In this context, stripping the green beer with CO_2 also has a positive impact, as it intensifies the contact between beer and yeast and removes some of the wort and green beer aromas.

Figure 7: Concentrations of aroma compounds in wort during interrupted fermentation (yeast cell count at pitching: 25 million cells/ml)

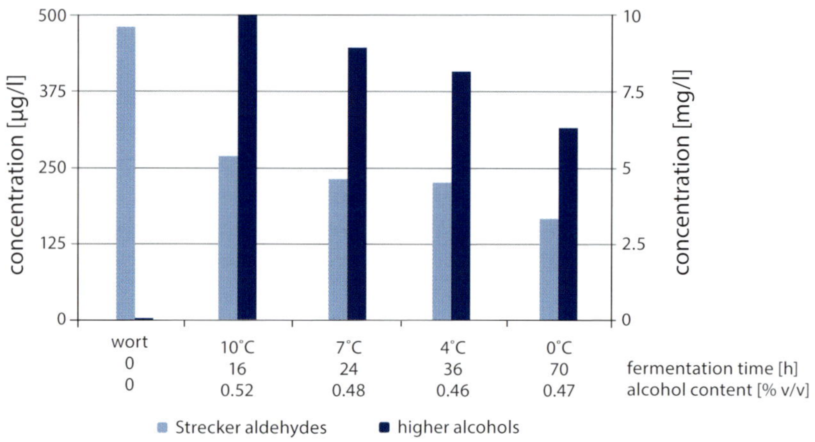

To halt fermentation, the yeast is removed before the target alcohol concentration is reached in the green beer. Centrifuges are generally used to separate the yeast from the beer. Typically, the yeast cell count is lowered to below 0.1 million cells/ml. Alternatively, this can be accomplished through filtration or flash pasteurization if the fermentation process has to be stopped for organizational reasons, e.g. on the weekend. If a flash pasteurizer is used, the yeast must subsequently be removed from the beer. After clarification, the green beer can be stored for an extended period, e.g. for one to three weeks at 0 °C. This is recommended due to the poor colloidal stability and the gushing tendency (insufficient precipitation of calcium oxalate). Stripping with CO_2 during the lagering period improves the sensory profile and the colloidal stability of the beer. Stabilization with silica gel and PVPP should be carried out during filtration and any uptake of calcium should be avoided [27].

Since beers produced with interrupted fermentation still contain a profusion of nutrients for potential beer spoilers, they require close microbiological attention. The microbiological safety of the packaged beer can only be guaranteed by pasteurization in excess of 30 PU or by cold sterilization.

Along with pure brewing yeast strains, special microorganisms can also be employed in the production of non-alcoholic beers. The first option and best suited to this task are yeast strains incapable of converting maltose or maltotriose to ethanol. Secondly, yeast mutants may be utilized, such as those exhibiting a defect in the citric acid cycle, which produce large amounts of organic acids. Thirdly, *Lactobacillus* cultures may also be used for this application [9].

2.2.2 BIOREACTORS

Bioreactors with immobilized yeast allow fermentation and/or maturation to occur continuously in the brewing process. The degree of attenuation can be controlled through the flow rate, temperature, substrate composition and loading of the carrier material. The immobilization of yeast or other microorganisms (cf. section 2.2.1, p. 260) creates the effect of a greater biomass to reactor volume. This enhances fermentation performance and leads to shorter residence times in the system, thus lowering the investment and operating costs [30, 31, 32, 33].

The most common carrier materials employed in the immobilization of yeast include DEAE cellulose, calcium alginate, calcium pectate or sintered glass. Fixed bed reactors are predominately used in conjunction with immobilized yeast. In these reactors, the substrate enters the reactors from below and flows through the carrier material in the bed at a relatively slow rate. Unfortunately pockets of carbon dioxide gas generated during fermentation may collect in the bed. Further, disadvantages include a lack of mixing and the tendency of the bed to become blocked. Reactors designed to create a forced flow (fluidized bed reactors, gas lift reactors, modular loop reactors) achieve more uniform products and higher material conversion efficiency [33, 34]. In order to preserve the vitality of the immobilized yeast cells, revitalization can be carried out through recirculation under special conditions (oxygen injection).

The following points must be considered with regard to managing the processes occurring in a bioreactor [35]:

- controlled oxygen supply to the yeast
- selection of the most homogenous reactor system possible
- flexibility in reacting to any changes that may occur

2.3 COMBINED TECHNOLOGIES

The combination of different technologies offers attractive options for improving the flavor of non-alcoholic beers. The negative aspects of non-alcoholic beer produced by interrupted fermentation (typical wort character) and dealcoholized beer (empty, sour flavor) can be avoided by blending the two. A further technological option is to produce wort with a low original gravity and to ferment it with a pure brewing yeast culture. The alcohol is subsequently removed from the fermented beer in a vacuum process and blended with a standard-strength beer to the target alcohol concentration.

Food additives such as (reclaimed) beer aroma, hop extracts, colorants, caramel color or sugar can also provide assistance in adjusting the aroma and flavor characteristics of the beer. These may be used in regions not subject to the *Reinheitsgebot*. The disadvantages of combined processes include more effort and higher equipment costs.

2.4 ECONOMIC ASPECTS

The most economical option for the production of non-alcoholic beer is the interrupted fermentation method. It requires the least equipment and the lowest quantity of malt. The prerequisite for microbiological stability is pasteurization of the packaged beer. Continuous methods also possess further economic advantages but usually incur higher operating costs and require more in-depth quality control due to the length of time the reactor is in operation.

After interrupted fermentation, vacuum evaporation is the most economical method for removing alcohol from beer. However, vacuum evaporation is not equally appropriate in all situations. For throughputs below 5 hl/h, membrane separation is generally the more economical solution. In cases where extensive dealcoholization (below 0.05 % by volume) of beer is required, membrane systems are not suitable [9, 24].

3 SUMMARY

Beers with an alcohol concentration of less than 0.5 % by volume can be produced with any of the methods described here. The quality of the resultant beers conforms to the expectations normally placed on non-alcoholic beer. Both production methods, interrupted fermentation and dealcoholization of fermented beer, are distinguished by significant sensory and analytical differences. Non-alcoholic beers produced by means of interrupted fermentation are somewhat sweeter and worty in character, possessing only a slight hint of the aroma typical of beer. Also, a thin layer of sediment tends to form in bottles of this beer. While continuous processes do offer a number of potential advantages over batch processes, they are often plagued by constant fluctuations in quality [31, 32, 35]. Non-alcoholic beers produced by the subsequent removal of alcohol from fermented beers exhibit an empty flavor compared to conventional beers. The absence of several important beer aroma compounds causes less desirable notes reminiscent of bread and wort or a more pronounced sourness to come to the fore. The suitability of membrane separation methods is limited to a maximum reduction in the alcohol content down to 0.5 % by volume [10, 20, 24, 29].

In the end, it is the consumer who determines which of the principal methods for producing non-alcoholic beers is preferable. The less desirable flavors exhibited by the beer produced using the methods described above seem to be accepted by consumers. Nevertheless, it would be desirable to find ways to improve the flavor of non-alcoholic beer. With this in mind, a number of technological measures are detailed below which can enhance the quality of non-alcoholic beer. The disadvantages of individual methods can be partially compensated for by employing a combination of these methods. The decision to implement a specific method should be made on the basis of a corresponding cost-benefit analysis. Although non-alcoholic beers rarely achieve the premium sensory quality of regular standard-strength beers produced through conventional fermentation, with the methods available today it is possible to produce distinctive, high quality products, which will continue hold a place in the beverage market.

4 OVERVIEW

Interrupted fermentation

Attribute/ Problem	Cause	Techniques/Measures
sweet flavor	high residual extract content (sugar)	utilize the "spring" mash method (cf section 2.2.1), lower the original gravity, use specialty malt, adjust hop bitterness
worty aroma	high concentration of wort aroma compounds	increase the intensity of the boil, evaporation through the release of pressure, lower the temperatures in the whirlpool, pitch more yeast with a high vitality, strip with CO_2, avoid oxygen uptake
atypical beer aroma	lacking in fermentation by-products	allow fermentation to progress until the precise target alcohol concentration is reached, use specialty malt, avoid excessive pasteurization, increase hop aroma
lacking in freshness, harsh flavor	pH is too high, CO_2 content is too low, residual extract content is too high	natural acidification with *Sauergut* to a pH < 4.5, check yeast vitality, ensure adequate carbonation, lower the original gravity, reduce thermal stress, minimize oxygen uptake
presence of sediment	inadequate clarification due to low fermentation performance	boil wort longer, wort acidification should be carried out early and in sufficient amounts, extend cold storage period, strip with CO_2, protein and polyphenol stabilization, employ filter aids low in calcium and iron, employ gentle filling and pasteurization procedures

Vacuum evaporation

Attribute/ Problem	Cause	Techniques/Measures
deficits in aroma	lacking in beer aroma compounds (esters, higher alcohols)	adjust alcohol precisely to the target alcohol concentration by blending with the source beer, use beer rich in aroma or kräusen, reintroduce the beer aroma collected during the dealcoholization process
lacking in mouthfeel and palate fullness	low extract content, too few aroma and flavor compounds	raise the original gravity, lower the degree of attenuation, use specialty malt, natural acidification with *Sauergut*, increase contact with yeast, add wort or sugar
bread-like, worty or caramel flavors	lacking in volatile beer aroma compounds, high concentration of Strecker aldehydes and aging compounds	ensure that the source beer is of high quality with good flavor stability, produce wort containing fewer carbonyl compounds, lower the evaporation temperature (vacuum), exclude oxygen, strip with CO_2, increase contact with yeast and dealcoholized beer
atypical fatty acid flavor	high concentration of medium chain fatty acids and aging compounds	ensure that the source beer is of high quality, pitch yeast with a high vitality, use specialty malt, avoid thermal stress, minimize oxygen uptake, strip the dealcoholized beer with CO_2
turbidity	subsequent protein precipitation due to thermal effects	employ measures to stabilize the source beer (wort boiling time, wort acidification, cold storage, stripping with CO_2, treatment with stabilizing agents, avoid calcium and iron uptake from filter aids), increase contact with yeast and dealcoholized beer, reduce thermal stress, minimize oxygen uptake

Membrane separation

Attribute/ Problem	Cause	Techniques/Measures
deficits in aroma	lacking in beer aroma compounds (esters, higher alcohols)	remove alcohol precisely to the target concentration, selectivity of the membrane, membrane operating time, source beer should be rich in aroma prior to dealcoholization, low temperatures
lacking in mouthfeel	low extract content, lacking in aroma and flavor compounds	raise the original gravity, raise the fermentation temperature, lower the degree of attenuation, use specialty malt, natural acidification with *Sauergut*, increase contact with yeast
bread-like or wort-like flavor	lacking in volatile beer aroma compounds, high content of aging compounds	ensure that source beer is of high quality, avoid exposure to oxygen, increase contact with yeast and strip the dealcoholized beer with CO_2
sour flavor, fatty acid flavor	lacking in beer aroma compounds, high concentration of fatty acids and Strecker aldehydes	produce wort containing fewer carbonyl compounds, use specialty malt, pitch yeast with a high vitality, strip the dealcoholized beer with CO_2, avoid oxygen uptake, add sugar
turbidity	poor stabilization of the source beer before dealcoholization	employ measures to stabilize the source beer (wort boiling time, wort acidification, cold storage, treatment with stabilizing agents, minimize calcium and iron uptake from filter aids), increase contact with yeast, filter the dealcoholized beer, minimize oxygen uptake

5 REFERENCES

[1] Deutscher Brauerbund e.V., Berlin.

[2] Bund für Lebensmittelrecht und Lebensmittelkunde e.V.: Leitfaden zur Nährwertkennzeichnungsverordnung, 1996.

[3] Biersteuergesetz. BGBl I 1992, 2150, 2158 (1993, 169).

[4] Piendl, A.: Langzeitstudien Alkohol und Gesundheit. Berichte aus der Ernährungswissenschaft, Fachverlag Hans Carl, 2001.

[5] Piendl, A.: Zielgruppen für alkoholfreies Bier: Alkoholfreies Bier für Sportler. Brauwelt, 133 (1991), no. 43, pp. 2198–2206.

[6] Tretzel, J.: Beer meets function. Wellness und Fitness – Ein Zukunftsmarkt für Brauereien, Teil 1. Brauindustrie 88 (2003),no. 6, pp. 34–36; Teil 2, Brauindustrie 88 (2003), no. 7, pp. 26–29.

[7] Porta, S.: Bier verändert das Stoffwechsel-Regenerationsmuster nach sportlicher Belastung. Handbuch zum Symposium "Bier – ein gesunder Genuss", München, 5. Mai 2004.

[8] Egle, G., Zarnkow, M., Kreisz, S., Back, W., Krahl, M.: Isotonie bei Getränken – Beispiel eines isotonischen Sportgetränks auf Malzbasis, Getränkeindustrie 61 (2007), no. 7 pp. 30–31.

[9] Leibhard, M.: Optimierung verschiedener biologischer Verfahren zur Herstellung alkoholfreier Biere. Dissertation, Technical University of Munich (Weihenstephan), 1991.

[10] Zufall, C.: Verfahrenstechnische Parameter bei der Entalkoholisierung mittels Fallstromverdampfung und ihr Einfluß auf die Bierqualität. Monatsschrift für Brauwissenschaft 53 (2000), no. 7/8, pp. 124–137.

[11] BGBl 1 1998, 384, §§ 4–54 Brennereiordnung (Anlage zur Branntweinmonopolverordnung).

[12] Berger, H. F.: Theoretische Grundlagen der Dünnschichtverdampfung zur Alkoholreduktion von Bier am Beispiel Centri-Therm. Monatsschrift für Brauerei 35 (1982), no. 3, pp. 98–101.

[13] Oliver-Daumen, B.: Praktische Erfahrungen mit dem Centri-Therm zur Alkoholreduktion von Bier, Monatsschrift für Brauerei 35 (1982), no. 3, pp. 101–106.

[14] Oliver-Daumen, B.: Optimierung der Verfahrensparapmeter zur Alkoholreduktion mittels Centri-Therm. Monatsschrift für Brauwissenschaft 39 (1986), no. 4, pp. 132–142.

[15] Hochberg, U.: Bierentalkoholisierung durch Verdampfung. Brauindustrie 70, 1985. pp. 480–485.

[16] Kessler, H. G.: Lebensmittel- und Bioverfahrenstechnik, 3th ed., Verlag A. Kessler, München 1988.

[17] Piper, H. J., Bruchmann, E., Kolb, E.: Technologie der Obstbrennerei, Ulmer, Stuttgart 1993.

[18] Fischer, U.: Massenbilanz der Aromastoffe bei der Entalkoholisierung von Wein: Korrelation chemischer und sensorischer Datensätze. Dissertation, Universität Hannover, 1995.

[19] API Schmidt-Bretten, SIGMATEC Entalkoholisierungsanlagen, Firmenschrift no. 372, 10/2004.

[20] Eckert, M.: Entalkoholisierung von Bier mittels Verdampfung bzw. Hochdruckextraktion unter besonderer Berücksichtigung der Aromastoffe. Dissertation, Universität Hohenheim, 1990.

[21] Jäger, P.: Technologische Maßnahmen im Hinblick auf die Bildung einiger Aroma- und Geschmacksstoffe zur Herstellung von alkoholreduzierten Bieren bzw. alkoholfreien Braugetränken, Mitt. der Versuchsstation f. d. Gärungsgewerbe Wien, 37 (1983), no. 9/10, pp. 100–106, no. 11/12, pp. 130–133; (1984), no. 1/2, 43–46.

[22] Medina, I., Martínez, J.-L.: Declaration of Cider by supercritical extraction with carbon dioxide, J. Chem. tech. Biotechnol. (68) 1997, pp. 14–18.

[23] Schwinghammer, G.: Herstellung von alkoholfreien und alkoholreduzierten Bieren aus ausstoßreifem Vollbier. Brauindustrie, 73 (1988), no. 11, pp. 1222–1224.

[24] Zufall, C.: Die Entalkoholisierung von Bier mittels Dialyse – Verfahrenstechnische Beeinflussung der Bierqualität. Monatsschrift für Brauwissenschaft 53 (2000), no. 9/10, pp. 164–179.

[25] Adler, K. W.: Ein neues Verfahren zur Herstellung von alkoholvermindertem Bier. Brauwelt 114 (1974), no. 22, pp. 443–446.

[26] Wucherpfennig, K., Neubert, S.: Zur teilweisen Entalkoholisierung von Bier mittels Umkehrosmose. Brauwelt 116 (1976), no. 41, pp. 1326–1328; no. 43, pp. 1419–1423; no. 47, pp. 1573–1579.

[27] Narziß, L., Miedaner, H., Kern, E., Düll, F.: Schankbiere, alkoholarme und alkoholfreie Biere – Technologie und Erscheinungsbild. Part 1, Brauwelt 123 (1988), no. 20, pp. 866–870; part 2, no. 21/22, pp. 916–926.

[28] Offenlegungsschrift DE 3213056 A1, Deutsches Patentamt, 1983.

[29] Narziß, L., Back, W., Gastl, M., Zarnkow, M.: Abriss der Bierbrauerei, 8th ed. Wiley-VCH, Weinheim 2017.

[30] Pittner, H., Back, W., Swinkels, W., Meersman, E., van Dieren, B., Lommi, H.: Continous production of acidified wort for alcohol-free beer using immobilised lactic acid bacteria. Proceedings of the 25th European Brewery Convention Congress, Brussels, 1995. pp. 301–308.

[31] Pajunen, E.: Immobilized yeast lager maturation: DEAE-Cellulose at Sinebrychoff. European Brewery Convention, Symposium Immobilized Yeast Applications in the Brewing Industry, Espoo, 1995. Monograph XXIV, pp. 41–49.

[32] Back, W., Krottenthaler, M., Braun, T.: Untersuchung zur kontinuierlichen Bierreifung. Brauwelt 138 (1998), no. 3/4, pp. 70–73.

[33] Fitzner, M.: Der Einsatz immobilisierter Hefen in der Brauerei unter besonderer Berücksichtigung der Angärphase und ausgewählter Reaktortypen und Trägermaterialien. Dissertation, TU Berlin 1998.

[34] Andries, M., van Beveren, P. C., Goffin, O., Masschelein, C. A.: Design of a multipurpose Immobilized Yeast Bioreactor System for application in the brewing process. European Brewery Convention, Symposium Immobilized Yeast Applications in the Brewing Industry, Espoo, 1995. Monograph XXIV, pp. 134–143.

[35] Ludwig, A.: Zur Verbesserung der Langzeitstabilität von Verfahren mit immobilisierter Hefe bei der Hauptgärung. Dissertation, TU Berlin 2003.

BEER RECOVERY

1 INTRODUCTION

Breweries have the opportunity to minimize internal beer losses by utilizing beer streams that would otherwise be lost during the production process (beer recovered from yeast, tank bottoms and returned beer). Furthermore, beer recovery helps conserve valuable resources and lowers wastewater costs. By volume, the most significant beer streams include beer recovered from yeast slurry or tank bottoms as well as from preliminary or post-filtration runs and changeovers. This amounts to > 5 % of total beer production annually. However, the key requirement for reintroducing recovered beer into the brewing process is to ensure that the quality of the finished beer is not negatlvely impacted.

2 UTILIZING BEER RECOVERED FROM THE PRODUCTION PROCESS

2.1 THE SIGNIFICANCE OF BEER LOSS

In order to gain an overview of the actual losses of beer and wort throughout the entire production process, they need to be determined volumetrically, beginning with the hot cast-out wort and ending with the packaged beer. This key metric is not only relevant for monitoring the economic aspects of the brewing process, but more importantly, it forms the basis for calculating the taxable volume of beer produced. For example, in Germany, legislation governing the taxation of beer (§10 Abs. 2 BierStG, dated January 1st 1993) stipulates that tax may not be levied on beer that has been documented as spoiled. Therefore, in this case, this volume of spoiled beer is considered as beer loss [1].

Even if the loss in volume has been determined, the actual loss in terms of real extract cannot be calculated. This calculation involves comparing the extract content of the hot cast-out wort or the grain bill with the finished, packaged beer. The equations used to calculate beer loss are given in table 1. Beer loss is usually determined at the end of the month and is separated into groups according to beer style.

Table 1: Calculating the loss in extract beginning with the cast-out wort and malt in the grain bill, respectively [2]

$E_I = V_{cast\text{-}out\ wort,\ hot}$ (hl) x 0.96 x ρ (% m/v)	equation 1: extract content in the cast-out wort volume
$E_{II} = V_{beer,\ total}$ x ρ (%, m/v)	equation 2: extract in the total volume of beer
$E_{III} = grist_{total}$ (kg) x BY (%)	equation 3: extract in the mass of malt in the grain bill
$Loss_{starting\ with\ cast\text{-}out\ wort} = \frac{E_1 - E_{11}}{E_{11}} \times 100$	equation 4: extract loss starting with cast-out wort
$Loss_{starting\ with\ malt\ in\ grain\ bill} = \frac{E_1 - E_{11}}{E_{11}} \times 100$	equation 5: extract loss starting with malt in grain bill

V = volume, BY = brewhouse yield, ρ = density

2.2 RECOVERING BEER FROM SURPLUS YEAST

A yeast suspension can, for the most part, be separated into its two components – yeast and beer – by means of two mechanical process techniques: centrifugation and filtration. Today, separation is primarily performed with disc centrifuges and decanter centrifuges. Among the various filtration techniques available, microfiltration with a vibrating membrane filter has recently been employed along with the more common technique of cross-flow microfiltration. Filter presses are no longer used to recover beer for qualitative and microbiological reasons. The operating principle behind each piece of equipment is described in the literature [3, 4, 5, 6, 7, 8].

By and large, the negative impact on the quality of the beer as a result of the type of system employed in beer recovery has been minimized through constant improvement. Today, the focus has shifted to the quality of the yeast suspension subject to separation. Physical stress factors such as time and temperature as well as some chemical factors degrade the physiological condition of the yeast cells. Along with ethanol, chemical factors include other toxic metabolites (e.g. acetaldehyde, acetic acid) [9] and medium chain fatty acids [10]. Cell death leads to autolysis and subsequent release of the yeast cell contents (e.g. polysaccharides, nucleic acids and lipids) into the medium [11]. The activity of proteinase A (PrA) is of special significance, since this enzyme has a negative impact on beer foam. Lipid transfer protein 1 (LPT1) has been identified as the primary substrate of PrA in beer [12] (cf. Beer Foam). PrA is typically found in cell vacuoles and is released from the cell as an inactive precursor in response to stress. This precursor undergoes automatic catalysis, creating the active form of the enzyme which then degrades foam-positive proteins [13].

Figure 1 provides a schematic overview of equipment and the process involved in beer recovery.

Figure 1: Process flow diagram depicting a system for beer recovery

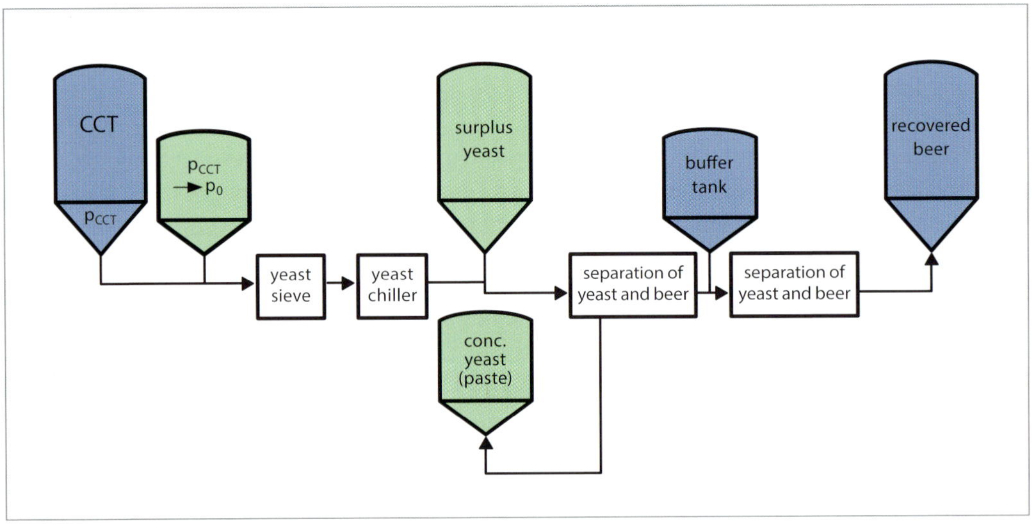

The individual pieces of equipment necessary for beer recovery are described in detail below.

2.2.1 EXPANSION TANK

After the yeast has been cropped, the pressure must be released gradually to liberate the CO_2 formed by the yeast cells.

2.2.2 YEAST SIEVE OR FILTER

Installing a yeast sieve or filter has proven beneficial, as it enables the removal of undesirable solids. This passive aeration also allows for the further liberation of CO_2. However, active aeration of the yeast should definitely be avoided because a lack of available sugar will cause the yeast to use glycogen stored in the cells as an energy source, thereby weakening them (cf. Yeast Technology and Fermentation).

2.2.3 YEAST CHILLER

As particularly high temperatures lead to a deterioration in the physiological condition of the yeast suspension and concurrently the quality of the beer to be recovered, it is imperative that the cropped yeast must be cooled to a temperature of < 4 °C by means of a yeast chiller. This especially applies to surplus top-fermenting yeast. Yeast chillers should be designed as tubular chillers or free-flow chillers with adequately sized gaps between the plates to ensure the gentle handling of cropped yeast.

2.2.4 YEAST STORAGE TANK (SURPLUS YEAST)

Yeast storage tanks should be cylindroconical in design. The tanks must be cooled in some way and they should be outfitted with a homogenization device or agitator to ensure that the contents remain well mixed. Homogenization can be achieved with an agitator or better yet, a recirculation loop. With a recirculation loop, the return line is of the utmost importance; it needs to be mounted at an angle of approximately 23° relative to the radius in order to achieve the desired level of mixing. An adequate level of homogenization is imperative; otherwise, the correct function of the centrifuges used to recover beer might be impaired.

2.2.5 BEER RECOVERY SYSTEMS

Recommendations regarding beer recovery systems will not be made here, since the decision to install a certain type of system depends on many factors specific to a particular brewery (cost, microbiology, processability of yeast suspensions and brewery philosophy). However, it can generally be stated that all of the newer beer recovery systems currently on the market do not have a negative impact on the quality of the recovered beer. Nevertheless, one should confirm that the system does not damage the surplus yeast stored for subsequent sale. The physiological condition of the yeast cells may be of importance to the companies purchasing surplus yeast for further processing.

2.2.6 FLASH PASTEURIZATION

The recovered beer must be flash-pasteurized to ensure its microbiological safety. Furthermore, this eliminates the risk of any proteinase A activity and its negative effects on beer foam. In order to be on the safe side with regard to microbiology as well as enzymatic activity, the beer should be pasteurized at approximately 40–100 pasteurization units (cf. Microbiology). It may be necessary to install a small buffer tank upstream from the flash pasteurizer. The buffer tank allows the flash pasteurizer and the beer recovery system to be operated independently, eliminating the necessity to maintain similar conditions between the systems such as pressure and throughput.

2.2.7 STORAGE TANKS FOR RECOVERED BEER

Storage tanks containing recovered beer must possess cooling jackets, though the geometry of the tanks plays a less critical role.

2.2.8 SURPLUS YEAST TANK

Ideally, the surplus yeast collected during the brewing process is stored in a cylindroconical tank until it is sold. The tank should be equipped with a device or system to keep the contents homogenized, e.g. an agitator or external recirculation system. Mixing keeps the yeast cells in suspension and prevents disruptions in the operation of any pumps in the system.

2.2.9 QUALITATIVE ASPECTS OF RECOVERED BEER

In particular, longer storage periods and higher temperatures negatively impact the physiological condition of the cells in a yeast suspension. This is accompanied by a significant decrease in pleasant ester aromas (hexanoic acid ethyl ester, 3-methylbutyl acetate), elevated concentrations of undesirable short and medium chain fatty acids (butyric acid, 3-methylbutyric acid, decanoic acid, lauric acid) and a significant increase in the activity of the foam-negative enzyme proteinase A. Therefore, it is recommended that yeast suspensions be stored at temperatures ranging from 0 to 3 °C and for the shortest time possible. Recovered beer should be flash pasteurized in order to deactivate proteinase A and subsequently added to the pitching wort.

2.2.10 ECONOMIC ASPECTS OF BEER RECOVERY

In a brewery, two liters of yeast are produced for each hl of beer sold. This means, for example, that a brewery with a total annual production of 500,000 hl will produce 10,000 hl of yeast. Depending on the recovery system, beer yields of up to 75 % can be achieved. An overview of potential yields is provided in table 2.

Table 2: Values for efficiency and yield for various beer recovery systems

Parameter	Decanter centrifuge	Disc centrifuge	Natural sedimentation	Cross-flow microfiltration	Vibration microfiltration
efficiency [%]	~ 100	~ 100	–	–	–
yield [%]	~ 75	~ 75	~ 60	~ 60	~ 60

The efficiency E or the yield Y of the beer recovery process can be calculated with the following equations:

$$E = 100 \times \left(1 \frac{C_{v,out}}{C_{v,in}} \right)$$

$C_{v,\,out}$ = volume of solids discharged [% v/v]
$C_{v,\,in}$ = volume of solids entering the system [% v/v]

Equation 6: Efficiency

Determining the degree of efficiency is only possible with centrifugation techniques, since no solids exit the system when using membrane filters.

$$Y = \left(\frac{DM_{YP} - DM_{in}}{DM_{YP} - E_R} \right)$$

DM_{YP} = dry matter in concentrated yeast (paste) [% w/w]
DM_{in} = dry matter entering the system [% w/w]
E_R = real extract in beer [% w/w]

Equation 7: Yield

2.3 BEER LOSSES DURING FILTRATION

Losses during filtration amount to approximately 6 % of the volume of the beer being filtered. This volume may be higher if several types of beer are being filtered, due to the more frequent changeovers. Beer recovered from filtration operations is problematic because diluted beer (with an extract content of approx. 4 % w/w) possesses fewer attributes which inhibit the growth of microbes (hop bitter compounds, pH, alcohol, etc.) [14]. As a result, obligate and potential beer spoilers take the opportunity to multiply in these beers and possibly spread to other production areas. The time the filtration system is in operation is also a decisive factor in determining the volume of beer losses during filtration. For this reason, the time the filter is in operation should be extended by means of the proper selection of raw materials and the appropriate production techniques. To further reduce the burden on the beer filter, a centrifuge can also be installed upstream to lower the yeast concentration in the beer prior to filtration [15] (see section 2.5, p. 274).

Adding recovered beer immediately back into the ongoing filtration process is recommended whenever possible. This method requires brewing to a slightly higher original gravity in the brewhouse, so that the desired gravity can still be reached when adding diluted beer directly back into the filtration process.

2.3.1 REQUIRED EQUIPMENT

2.3.1.1 Pre-run and post-run collection tank

It is essential that these tanks can be cooled, or that they are placed in a cooled space. Tank geometry is of lesser importance.

2.3.1.2 ADDING RECOVERED BEER DURING PRODUCTION

Recovered beer can be added based on volume (< 10 %) or through the adjustment of the gravity (extract content). For larger breweries, it is advisable to install a unit for adding deaerated water, which enables a precise adjustment of the extract content. At the same time, such a unit guarantees the correct dilution of the beer in the event that the beer obtained from filtration operations cannot be added back into the production stream.

2.4 OTHER RECOVERED BEER STREAMS

All other recovered beer streams should be added on the hot side of the brewing process to ensure microbiological safety. However, beer recovered from the brewing process when producing low volume products of a different color (e.g. dark beers) should be disposed of. Additions during wort boiling are not recommended, primarily due to excessive evaporative losses (aroma compounds, ethanol). Aside from reducing the efficiency of the wort boiling process, the flavor is also compromised. It is better to add the recovered beer in the whirlpool, although a loss in the ethanol content of approximately 30 % has been observed as a result of subsequent evaporation or through retention in hot break material. A further consideration is that the addition of recovered beer causes a drop in the temperature of the hot wort. It is therefore more practical to introduce recovered beer into the production stream between the whirlpool and the wort chiller. The metering pump will normally turn off automatically once the temperature drops below 85–87 °C, though this should not occur at a dosage rate of about 1–2 % relative to the total volume. In order to ensure sufficient microbiological safety, the required number of pasteurization units must be reached by means of a sufficiently long holding time at the target temperature. This method is advantageous since none of the volatile compounds are lost if the recovered beer is added at a point in the process under pressure (fig. 2).

The reintroduction of recovered beer on the hot side of the brewing process clearly has a negative impact. This can be seen most distinctly in the sharp decline in the flavor stability of the finished beers as they age.

Figure 2: Schematic view of recovered beer dosage into the hot wort line

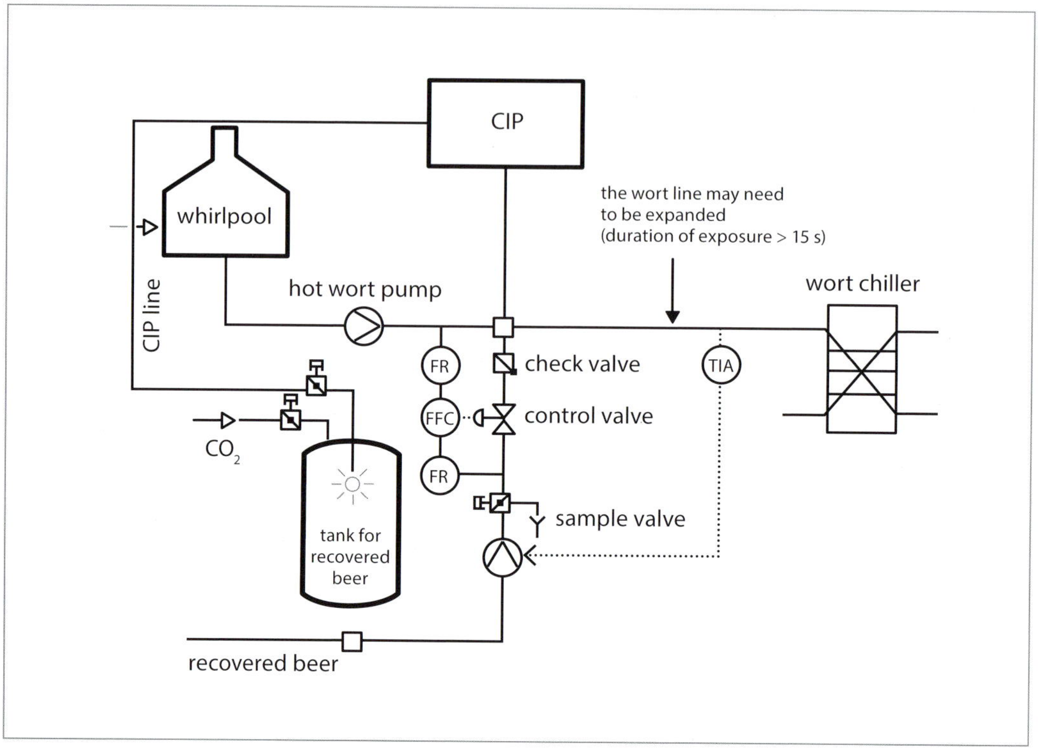

2.5 AVOIDING BEER LOSS AND SUBSEQUENT BEER RECOVERY

"Prevention is better than recovery". This approach is certainly one of the greatest concerns associated with beer recovery. In general, it applies to the processes of fermentation, storage and filtration, as these stages of production are where the largest volumes of beer are recovered.

Green beer centrifuges can reduce the volume of yeast accumulating in the lager cellar to such a degree that there is a sufficient number of yeast cells available for secondary fermentation, and yet the amount of yeast sediment is reduced significantly. Moreover, it is also possible to recover beer from freshly harvested yeast during fermentation. The yeast is collected in a storage tank and sent to the green beer centrifuge for beer recovery. During this process, it is important to maintain a constant dosage of yeast slurry (a homogeneous yeast suspension in the collection tank is essential!), so that a consistent volume of suspended solids is moving through the centrifuge during operation. Using a green beer centrifuge may also offer technological advantages. Firstly, a lower amount of solids relieves the burden on the beer filter, which in turn, positively impacts filter operations. The ratio of beer from preliminary and post-filtration runs is improved relative to the volume of filtered beer. Secondly, yeast autolysis products (medium chain fatty acids, proteinase A, alkaline amino acids) which may end up in the beer during maturation and secondary fermentation, may also be minimized by reducing the volume of the yeast present.

Ideally, the beer recovered during filtration should be added directly back into ongoing filtration operations (see section 2.3, p. 273). Planning and filtering batches in the proper sequence can minimize the volume of beer losses recovered from changeovers during filtration.

When designing new filter cellars, kieselguhr-free filtration systems based on a membrane filter unit should be considered. With membrane filtration, the volume of beer loss during filtration is negligible. Membrane systems meet almost all of the criteria for microbiological safety and economic performance in practice. In existing plants, pre-filtration is generally carried out with a centrifuge [16].

3 SUMMARY

Utilizing recovered beer enables a brewery to reduce the internal volume of beer losses. In the beer recovery process, it is important to keep a watchful eye on quality criteria and microbiological safety. In particular, flavor and foam stability can be negatively impacted. On the other hand, the economic efficiency of recovering and utilizing this beer must be carefully weighed. Table 3 provides an overview of the recommended points where additions of the respective streams of recovered beer can be made.

Table 3: Utilization and handling of various recovered beer streams in a brewery

Recovered beer streams	Point of addition	Reason behind addition at this point in production
beer recovered from surplus yeast or tank bottoms	pitching wort	reduction of short and medium chain fatty acids as well as FAN
beer collected during the filtration process (pre-runs, changeovers, and post-runs)	between the whirlpool and the wort chiller	process technology, microbiology
all other beer streams from production	cast-out wort line (alternative: whirlpool)	microbiology
beer returned to the brewery (non-spoiled)	cast-out wort line (alternative: whirlpool)	microbiology
beer returned to the brewery (spoiled)	correct disposal away from production areas	microbiology

4 OVERVIEW

Problems associated with recovered beer	**Cause**	**Suggested corrective measures/ reference values (RV)**
recovered beer is low in quality	yeast population possesses an insufficient level of overall vitality; conditions were unsuitable for storing the yeast suspension prior to preparation	improve yeast vitality; optimize storage conditions for the yeast (shorter storage time; lower temperatures (0–3 °C])
proteinase A activity in recovered beer is too high (> 50 units x 10^{-5}/ml)	yeast population exhibits poor physiological condition; flash pasteurizer is not functioning correctly	improve the physiological condition of the yeast; check effectiveness of flash pasteurization process (RV: PE > 40)
uncharacteristically high values measured for original gravity	intracellular glycogen is metabolized to alcohol during storage of the yeast suspension	shorten the storage period for yeast
turbidity levels are too high in recovered beer (centrifuges)	carryover of particulates due to fluctuations in the solids content at the centrifuge inlet (yeast cells)	maintain a constant concentration of solids entering the centrifuge
negative impact on foam stability in the finished beer	proteinase A activity is too high in the recovered beer	flash pasteurize recovered beer soon after recovery (> 40 PE)
negative flavor and aroma profile of the blended beer (harsh, lingering bitterness, cheesy aroma)	the recovered beer is low in quality	improve the condition of the yeast population
the yeast suspension in the bulk storage tank (surplus yeast to be sold) cannot be pumped	the dry matter content is too high in the surplus yeast storage tank	dilute contents to a pumpable consistency with water; store surplus yeast for sale in tanks with suitable geometry (conical!)
negative impact on the flavor and aroma profile of the blended beer	the quality of the recovered beer is poor; the proportion of recovered beer used in blending is too high, microbiological contamination, oxidation	monitor the quality of the recovered beer; lower the proportion of recovered beer for use in blending; if beer shows evidence of a microbial contamination, then dispose of the contaminated beer outside of the production area; strip with CO_2 gas, treat with activated charcoal (50 g/hl)
negative impact on the aging behavior of the blended beer	add recovered beer on the hot side of production	blend recovered beer with pitching wort
negative impact on the foam stability of the blended beer	proteinase A activity in recovered beer, blending proportion is too high (> 10 %), possibly too many medium chain fatty acids present	flashpasteurize recovered beer lower the proportion used in blending (< 5 %)
disintegration of the hot break (trub) cone in the whirlpool	the percentage of recovered beer added to the whirlpool is too high (excessive cooling of the hot wort)	either reduce the proportion of recovered beer or add it directly into the hot wort stream upstream from the wort chiller

5 REFERENCES

[1] BierStG i. d. F. vom 01.01.1993.

[2] Schneeberger, M.: Verwertung von Prozessbieren in der Brauindustrie. Dissertation, Technical University of Munich (Weihenstephan), 2006.

[3] Hansen, N. L.: From Filter Presses over Centrifuges to Cross Flow and Vibrating Membrane Filtration. MBAA Technical Quarterly 38 (2001), no. 2, pp. 115–121.

[4] Colesan, F., Paterson, S.: Bierrückgewinnung aus Überschusshefe mit dem Flottweg-Sedicanter. Brauwelt 139 (1999), no. 8, pp. 300–302.

[5] Hansen, N. L.: Recovery of Beer from Surplus Yeast by Membrane Filtration. MBAA Technical Quarterly 26 (1989), pp. 8–13.

[6] Potgieter, D., Parsotam, R., Perry, M.: Extract recovery from surplus yeast. MBAA Technical Quarterly 32 (1996), no. 2, pp. 110–114.

[7] Schneider, J., Weisser, H.: New technology for beer recovery from spent yeast. MBAA Technical Quarterly 37 (2000), no. 4, pp. 477–481.

[8] Methner, F.-J., Peters, U., Stettner, G., Lotz, M., Ziehl, J.: Untersuchungen zur Bierrückgewinnung aus Überschusshefe. Brauwelt 144 (2004), no. 17, pp. 470–474.

[9] Walker, G. M.: Yeast Physiology and Biotechnology, Wiley-VCH, Weinheim 1998.

[10] Stratford, M., Anslow, P. A.: Comparison of the inhibitory action on *Saccharomyces cerevisiae* of weak-acid preservatives, uncouplers, and medium-chain fatty acids. Federation of European Microbiological Societies Microbiology Letters 142 (1996), pp. 53–58.

[11] Menzel, H.: Einfluß von Druck und Temperatur bei der Reifung und Lagerung auf die Exkretion von Hefeinhaltstoffen. Thesis, Technical University of Munich (Weihenstephan), 1979.

[12] Leisegang, R.: Proteinase A und Schaumstabilität – biochemische Untersuchungen der Wechselwirkungen von Proteinase A aus *Saccharomyces cerevisiae* und schaumaktivem Lipidtransferprotein. Dissertation, TU Berlin, 2003.

[13] Kondo, H., Yomo, H., Furukubo, S., Kawasaki,Y., Nakatani, K.: Advanced method for measuring Proteinase A in beer. Proceedings of 25th Convention of the Institute of Brewing – Asia Pacific Section, 1998, pp. 119–124.

[14] N. N.: Restbier-(Restwürze)-Verwertung. Brauwelt 133, Nr. 9 (1993). pp. 375–377.

[15] Narziß, L.: Filtration and stabilization of beer. European Brewery Convention Symposium, Waste Reduction in Brewery Operations, Rheinfelden 1992. Monograph XIX, pp. 101–112.

[16] Bohak, I., Gaub, R., Back, W.: Microbiological performance of DE-free filter systems. Proceedings of the 30th European Brewery Convention Congress, Prague 2005.

BEER AND HEALTH

1 INTRODUCTION

Beer is a pure, wholesome, natural food product, which has a positive effect on one's sense of well-being and physical health, if it is consumed in moderation. The raw materials – malt, hops, yeast and water as well as the natural brewing process – are responsible for the health benefits that beer offers. In beer production, malting, mashing, lautering, boiling, fermentation and filtration all serve as cleansing processes, ultimately removing undesired substances such as pesticide residues, mycotoxins and pathogens.

2 THE EFFECTS OF BEER ON HUMAN HEALTH

The raw materials used to produce beer – malt, hops, yeast and water – provide a number of important nutritional substances, which are generally present in sufficiently relevant concentrations. Among these substances, numerous polyphenols are primarily transferred to the beer via malt, though some are derived from the hops. A number of the vitamins in beer also come from the malt while others are produced by the yeast. Malt and water are responsible for supplying the beer with minerals. Malt is the sole source of dietary fiber in beer. Maillard products are also contributed by the malt in addition to being formed during the boiling process. A fascinating array of hop oils and valuable bitter compounds stem from hops.

2.1 NUTRITIONALLY RELEVANT SUBSTANCES IN BEER

Beer contains a multitude of nutritionally relevant substances, a number of which are described in this chapter. Using modern brewing technology, renewed efforts are being devoted to promoting the retention of these valuable substances derived from the raw materials in beer, in order to produce high-quality, nutritive products.

2.1.1 POLYPHENOLS

Polyphenols are classified as secondary plant extracts and include phenolic acids – for example ferulic acid and gallic acid – as well as flavonoids like quercetin and xanthohumol (XN). The composition of these substances differs greatly, as do the effects they have on the body. Their antioxidative qualities are of special interest, since they offer some protection against free radicals and thus also against cardiovascular diseases, among other conditions [1, 2, 3]. Furthermore, they have anti-inflammatory, antithrombotic and antimicrobial effects [4, 5]. Even antiviral effects have been identified [6, 7]. Several polyphenols are considered to have anticarcinogenic qualities and this activity has been documented during various stages of carcinogenesis [8, 9]. Lastly, polyphenols also comprise estrogens (e.g. 8-prenylnaringenin) as well as substances that have an anti-estrogenic effect (e.g. xanthohumol) [10, 11, 12].

With regard to the composition of polyphenols, the total quantity is not the deciding factor, but rather the wide variety of effects afforded by specific polyphenols [13]. Table 1 provides an overview of several polyphenols found in beer that are particularly relevant to health, along with their effects on the human body.

Table 1: Selected polyphenols and their effects on health [3, 9, 14]

Polyphenol	Effect
quercetin	antimicrobial, antioxidative anti-osteoporotic anticarcinogenic during the promotion and progression of carcinogenesis
xanthohumol	antimicrobial anti-inflammatory antiviral anticarcinogenic, antioxidative in all phases of carcinogenesis anti-osteoporotic helps prevent diabetes estrogen blocker
isoxanthohumol	antimicrobial antioxidative estrogen-promoting
gallic acid	antimicrobial, antioxidative
ferulic acid	antioxidative anti-inflammatory

2.1.2 VITAMINS

Vitamins are essential for maintaining vital functions in the body. Certain vitamins are not produced at all by the body or in insufficient quantities. Therefore, it is necessary to ingest them through nutritional intake. Beer primarily contains water-soluble vitamins in the B group [14]. These have a positive effect on the cardiovascular system, stimulate metabolism, support the generation of red blood cells and help improve mental concentration [15]. Several vitamins are undersupplied in the average diet, such as folic acid. A folic acid deficiency compromises performance and can lead to digestive problems as well as anemia [15, 16, 17].

Table 2: Selected vitamins in beer and their function in the body [14, 15, 16, 17]

Vitamin	Function
niacin (B_3)	important for carbohydrate, fatty acid and protein metabolism essential for the growth and maintenance of cells a constituent of coenzymes can be synthesized in the body from tryptophan
pyridoxine (B_6)	involved in the formation, conversion and degradation of amino acids and neurotransmitters supports nerve tissue, important for the formation of hemoglobin, supports the immune system against infection
cobalamin (B_{12})	active in protein and fatty acid metabolism and as a coenzyme in the cell aids in the formation of blood cells B_{12} is essential for the conversion of folic acid into a physiologically active form involved in cell division, necessary for DNA synthesis

Vitamin	Function
pantothenic acid (B_5)	plays a central role in energy metabolism in the cell component of coenzyme A protein and fat synthesis essential in the production of cholesterol and several hormones
folic acid (B_9)	protein metabolism, synthesis of structural and functional proteins, conversion of homocysteine involved in cell growth and the formation of new blood cells involved in the generation of new blood cells, supports nerve tissue
riboflavin (B_2)	classified as a coenzyme carries out important functions in the metabolism of proteins, carbohydrates and fats produces energy from carbohydrates and fats required for cell growth and nourishment
biotin (H)	formation and degradation of fatty acids synthesis of glucose formation of amino acids involved in DNA synthesis acts as a coenzyme
vitamin D (calciferol)	important for healthy teeth and bones increases calcium uptake and storage in the body regulates phosphate metabolism strengthens the immune system
thiamine (B_1)	important for energy and carbohydrate metabolism nerve tissue, important for the maintenance of multiple neurotransmitters, supports heart and muscle function

2.1.3 ALCOHOL

Alcohol is the only nutritionally questionable substance in beer. In the prevention of alcohol abuse, responsible drinking also plays a significant role in determining the health effects of beer. From a medical point of view, moderate alcohol consumption is often cited as being beneficial. Accordingly, it aids in the prevention of neurological diseases and anxiety by creating a liberated, relaxed disposition in the drinker [18]. Alcohol affects the metabolism of fats and can help prevent diabetes to a certain degree [23, 24]. In addition, it has a positive effect on kidney function, enabling the increased excretion of potentially undesirable substances in the urine [25, 26]. Due to the effect alcohol has on coagulation factors (antithrombotic effect), among other things, alcohol consumption helps reduce the risk of heart attack or cardiovascular disease by 25–55 % (L-shaped curve, fig. 1). Simultaneously, the risk of cancer increases once the quantity of 30 mg of alcohol per liter has been exceeded (J-shaped curve, fig. 1). It follows that a clear correlation can be identified in support of moderate alcohol consumption (vertex of the U-shaped curve, fig. 1) [19, 20, 21]. This measure varies depending on gender and constitution. In a study by Keil [22], it was determined that a consumption of 30 g alcohol/day for males and 25 g alcohol/day for females is beneficial. The German Nutrition Society (Deutsche Gesellschaft für Ernährung – DGE) recommends slightly lower amounts, namely 10 g alcohol/day for females and 20 g alcohol/day for males.

Figure 1: Relationship between alcohol consumption and risk of mortality [20, 22]

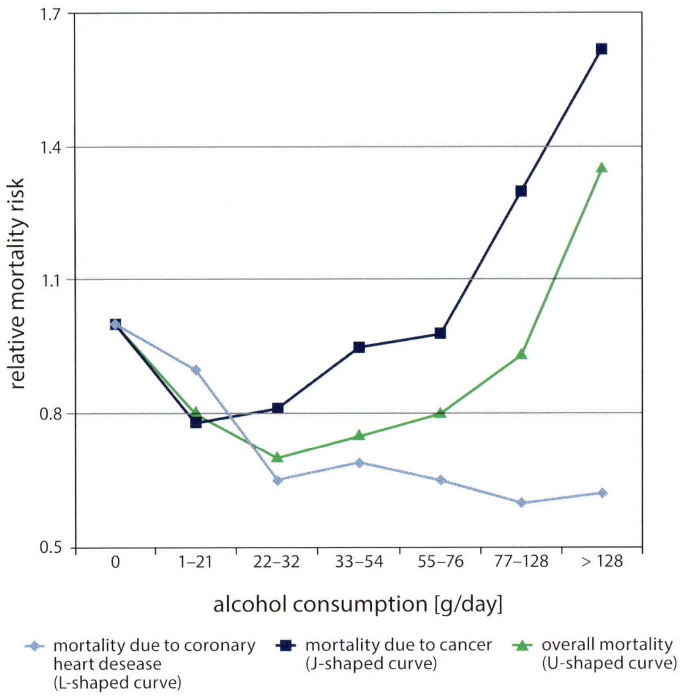

2.1.4 CARBON DIOXIDE

When enjoying carbonated beverages, the release of carbon dioxide in the mouth stimulates blood circulation and imparts a refreshing sensation. The longer the mucus membranes are in contact with the CO_2, the greater the stimulating effect on the digestive system [24]. Moreover, aromas are released more rapidly in carbonated beverages.

2.1.5 WATER

In order to stay hydrated, the DGE recommends drinking 2.5 l of water per day or more, depending on the ambient temperature and physical exertion. At 93 % of the total volume, water is the main constituent of beer. The minerals contained in the beverage – the favorable ratio of potassium to sodium and the low levels of residual sugar – facilitate rapid resorption of the liquid [27]. Some non-alcoholic beers are even isotonic.

2.1.6 MINERALS

Minerals are vital for metabolism and for maintaining good health. However, they are not fully effective on their own and function only in combination with vitamins and other nutrients.

Silicon improves the elasticity of bones, slows their deterioration and is instrumental in bone growth and development. Beer is considered one of the best sources for bioavailable silicon. This is because silicon is transformed to silicic acid through its prolonged contact with water, which is very easily resorbed. In contrast, the silicon found in solid foods (wheat, barley, oats, bananas, etc.) is much less bioavailable, despite the fact that it occurs in higher concentrations in these foods [4, 17]. This is, in part, because

liquid foods pass through the stomach faster than solid foods.
Sodium is a key factor in maintaining the acid-base equilibrium and for numerous other bodily functions (table 3). Sodium is, however, consumed in excessive amounts in the average diet, which consists largely of industrially processed foods, such as meat, cheese, soups, etc. [15]. For individuals who are sensitive to sodium, this can lead to high blood pressure or to water retention in otherwise healthy individuals [15, 17]. The vast majority of beers are low in sodium (< 40 mg Na/l).

A selection of nutritionally relevant minerals and trace elements also found in beer and their health effects are described in table 3.

Table 3: Selected minerals and their function in the body [14, 15, 17]

Mineral	Function
phosphorus	aids in the development of bones and teeth regulates muscle contraction promotes energy generation in the cell blood clotting nerve impulses brain activity physical endurance
potassium	regulates the water balance in the body (acts as an antagonist to sodium) lowers the risk of developing high blood pressure involved in the acid/base equilibrium in the body stimulates nerve impulses and muscle movement involved in the production of proteins
magnesium	physical fitness and mental performance energy metabolism oxygen transport circulation of blood in the body formation of bones and teeth
chloride	involved in the acid/base equilibrium of the body a component of bones aids in the formation of stomach acid promotes hormone transport
fluoride	reduces dental caries maintains skeletal structure in adults
sodium	maintains the acid-base equilibrium (primary factor) regulates the osmotic pressure in the body important for nerve and muscle function, serves as a central factor for the propagation and transmission of nerve impulses and for muscle relaxation required for the resorption of glucose and the transport of other nutrients
chromium	involved in carbohydrate, protein and lipid metabolism cell division, especially important for the formation of RNA

2.1.7 MAILLARD PRODUCTS

Maillard products, among them melanoidins, are a result of the malting process and wort boiling. There are differing opinions regarding their nutritional relevance. Many of these substances have an antioxidative effect, can neutralize free radicals and reduce lipid peroxidation [28]. Pronyl-lysine is a valuable substance among the Maillard products and also has an antioxidative effect [29]. This substance can be found primarily in bread crusts and beer.

2.1.8 DIETARY FIBER

Generally speaking, dietary fiber is the indigestible component of plant-based food, which regulates intestinal function. Furthermore, it can help reduce the risk of a heart attack by absorbing and thus facilitating the elimination of substances such as cholesterol from the body. Compounds found in the cell walls of barley, including β-glucan and arabinoxylan, represent a significant portion of the soluble dietary fiber found in beer.

Research has shown that dietary fiber helps regulate insulin levels and lower blood pressure. Animal trials have also proven that it helps boost immunity against infections [30, 31].

2.1.9 BITTER ACIDS IN HOPS AND HOP OILS

The bitter compounds in hops appear to promote digestion, possess antiseptic qualities and thus have a beneficial impact on digestion and the intestinal tract. The bitter acid humulone also has an antimicrobial and antioxidative effect, while delaying the onset of osteoporosis and preventing the development of kidney stones [32, 33]. It has been proven that isohumulone has an antiviral effect and helps prevent diabetes by positively influencing blood sugar levels and the breakdown of fatty acids. Hops have also long been considered to act as a sedative. This calming and stress-reducing quality is attributed, in part, to the hop oils [34, 35].

2.2 BEER – A VALUABLE CONTRIBUTION TO THE HUMAN DIET

Although not all of the previously mentioned substances found in beer are present in significant concentrations, there are still numerous positive health benefits. These can be explained, for example, through the synergistic effects the individual substances have on each other. For instance, trials with extreme athletes showed that the regular consumption of non-alcoholic wheat beer containing a high concentration of polyphenols helped reduce the risk of inflammation in general, as well as inflammation of the upper respiratory tract in particular [36]. Other antioxidants and the alcohol in conventional beer are also potentially responsible for this effect [37].

Silicon is a substance whose full effects are expressed in a beer matrix. The silicon extracted from the malt during the brewing process takes the form of silicic acid in beer, a form which is highly bio-available. This was proven in human trials through its resorption and detection in blood plasma [4, 38]. Also, the ratio of the minerals present in beer plays a role in the body. Beer is normally considered to be a very low-sodium food (< 40 mg Na/l) and therefore possesses a positive sodium-potassium ratio. It follows that the consumption of beer can aid in the prevention of high blood pressure and renal diseases [14]. Figure 2 provides an overview of several proven risk factors (*in vivo*), which were reduced as a result of moderate beer consumption.

Figure 2: Potential reduction of health risk factors through consumption of beer [2, 4, 21, 22, 31]

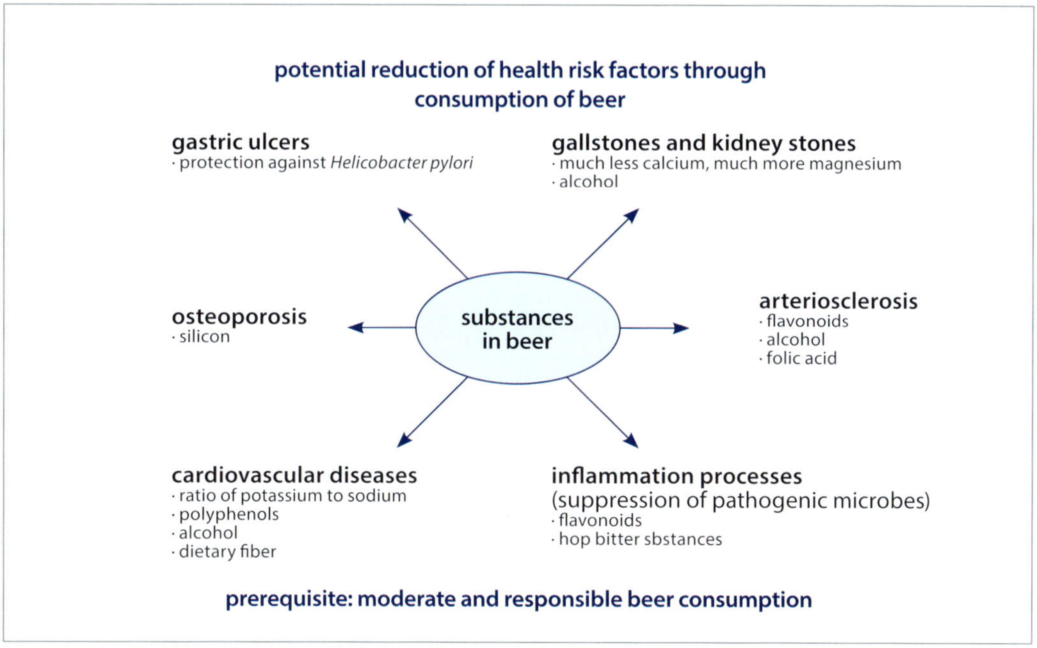

Specialty beers may also demonstrate a range of effects in particular dietary situations. As an example, enjoying beers containing yeast or non-alcoholic beers (produced using interrupted fermentation) is not recommended for individuals with high levels of uric acid in their blood due to the elevated concentrations of purines present in these products. However, *Kristallweissbier* or filtered wheat beer contains comparatively low concentrations of purines, while light beers brewed with rice, sorghum and rye are even more beneficial [39]. On the other hand, non-alcoholic beers produced by means of interrupted fermentation are usually well suited as isotonic or sports drinks.

Beer consumption is commonly associated with a high caloric intake. However, 250 ml of beer contains a mere 108 kcal, which has the lowest caloric content by volume when compared with other common beverages, such as apple juice (118 kcal), milk (165 kcal) or red wine (192 kcal) [14]. It has been medically proven that the so-called "beer belly" is not the result of drinking beer but rather the hearty food that often accompanies it [40]. This is in part due to the fact that beer stimulates the appetite and digestive system.

Table 4: Daily allowances for selected dietary substances beneficial to health and the percentage found in beer [14, 15]

	Ø Recommended daily allowance[1]	Incidence in 1 liter of pilsner beer	Demand met by 1 liter of pilsner beer (%)
calories	2400 kcal	434 kcal	18.1
water from beverages	2600 ml	919.6 g	35.4
fat	80 g	trace	unknown
protein	56 g	5.0 g	8.9
carbohydrates	300 g	28.0 g	9.3
dietary fiber	30 g	1.529 g	5.1
phosphorous	700 mg	319 mg	45.6
potassium	2000 mg	554 mg	27.7
magnesium	350 mg	96 mg	27.4
chloride	830 mg	174 mg	21.0
fluoride	3.8 mg	0.49 mg	12.9
sodium	550 mg	44 mg	8.0
chromium	100 µg	6.8 µg	6.8
manganese	5 mg	0.16 mg	3.2
zinc	10 mg	0.06 mg	0.6
niacin (B_3)	17 mg	7.733 mg	45.4
pyridoxine (B_6)	1.5 mg	0.619 mg	41.3
cobalamin (B_{12})	3.0 µg	0.82 µg	27.3
pantothenic acid (B_5)	6 mg	1.49 mg	24.8
folic acid (B_9)	400 µg	86 µg	21.5
riboflavin (B_2)	1.5 mg	0.335 mg	20.9
biotin (H)	60 µg	12 µg	20
vitamin D	5 µg	0.9 µg	18.0
thiamine (B_1)	1.3 mg	0.029 mg	2.3

[1]for men, 19-65 years of age, body weight 70 kg

Of course, when consumed in excess, even beer has a negative impact on one's health (e.g. increasing the risk of cancer [41]). Table 4 shows several substances that can promote health and the average recommended daily allowance for males between 19 and 65 years of age, compiled by the German Nutrition Society (Deutsche Gesellschaft für Ernährung – DGE) [15]. With regard to alcohol, the DGE recommends 10 g/day for females and 20 g/day for males. Yet according to other sources for medically approved recommended allowances, up to 20 g/day for females and 40 g/day for males are considered acceptable [42]. One liter of pilsner contains approximately 40 g of alcohol. Thus, moderate beer consumption is a realistic option. Or, one can simply resort to drinking non-alcoholic beer. It should be noted that the term "non-alcoholic" or "alcohol-free" is not applied in a uniform manner around the world (cf. Non-alcoholic Beer).

Since almost all of the qualities of beer known to afford health benefits are retained in non-alcoholic beers, such beers can also contribute to a balanced diet. These substances are present in their natural form in beer. The vitamins, antioxidants and easily digestible carbohydrates found in beer are all convincing arguments for making it part of a health-conscious diet. The prolactin from the malt and the calming effect of the hop oils also makes it an attractive option for nursing mothers. As mentioned previously, non-alcoholic beers are ideal sports drinks, thanks to their isotonic and slightly hypotonic qualities (fig. 3).

Tonicity refers to the osmotic pressure of liquids, (expressed in terms of osmolality in mmol/kg), as compared to the osmotic pressure of blood. Liquids and the dissolved substances they contain are absorbed most rapidly into the blood through the intestinal tract if balanced osmotic conditions prevail, in other words, under isotonic conditions.

Drinking isotonic liquids is beneficial during periods of physical exertion, as it quickly provides the body with energy – especially in the form of carbohydrates. Additionally, the water and minerals depleted as a result of sweating can be replenished more rapidly. The average osmolality of blood is 290 mmol/kg. Liquids with an osmolality between 245 and 335 mmol/kg are considered to be isotonic. Liquids with a lower value are considered hypotonic, while those with a higher value are classified as hypertonic. In order to absorb electrolytes from hypertonic liquids, water is extracted from the blood to thin the liquid in the intestinal tract in order to establish an isotonic equilibrium. Since water needs to be added to the blood during periods of physical exertion, rather than extracted, drinking a hypertonic beverage would be rather counterproductive.

As shown in figure 3, the tonicity of alcohol-free beers varies according to the dealcoholization process (cf. Non-alcoholic Beer). Thermally dealcoholized beers are classified as hypotonic, whereas beers produced by means of interrupted fermentation are frequently isotonic. Combined processes may help to regulate deficits in tonicity, in addition to improving the sensory profile of the beer. Furthermore, the osmolality of conventional and light beers is 550 mmol/kg or more, depending on the original gravity and the concentration of alcohol and extract.

Figure 3: Osmolality of commercially available non-alcoholic beers

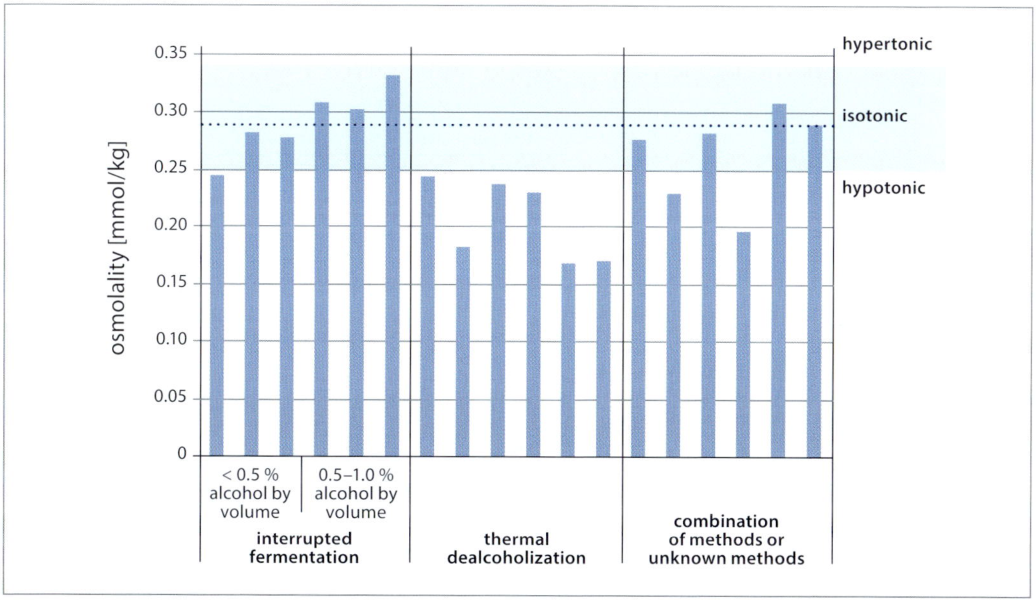

2.3 ENRICHING OR EXCLUDING PARTICULAR SUBSTANCES IN BEER

Certain substances that provide health benefits can also be enriched in beer – as proven by the "XAN" product (beer enriched with xanthohumol) and the techniques for increasing the folic acid content in beer. Similarly, it is also possible to reduce or even eliminate undesirable substances (e.g. gluten). Techniques and strategies for enriching or preventing the occurrence of particular substances are described in more detail below.

2.3.1 INCREASING XANTHOHUMOL CONTENT

Normally, there are only very low levels (< 0.2 mg/l) of xanthohumol (XN) in beer, as it is mostly removed over the course of the standard brewing process. In accordance with the *Reinheitsgebot*, it is only possible to enrich beer with XN through hop additions rich in XN and by altering the brewing process. XN is a nonpolar substance, which only dissolves to a limited degree in polar media and is isomerized to isoxanthohumol during boiling.

When applying "XAN" technology, the resultant losses, which are mostly due to brewing to a higher original gravity, can be limited by adding more hops in later additions, or by adding cold brewing liquor (fig. 4). One deciding factor is to cool the wort down quickly at the end of the boil to 80 °C, in order to minimize the isomerization of XN. Adding hops for the purpose of enriching the XN content in beer can be achieved with commercially available ethanol extracts or products fortified with XN, depending on the XN content desired.

Another important step is the use of roasted malts. As the proportion of roasted malt rises, the concentration of XN clearly increases, too. The roasted substances are responsible for this, as they keep the XN stable so that it remains in solution throughout the rest of the brewing process. When producing pale beers, roasted malt can only be utilized in miniscule amounts. XN content can be retained in the finished beer by reusing yeast or refraining from using stabilizers. In keeping with the *Reinheitsgebot*, it is still possible to achieve XN levels of 1 to 3 mg/l in unfiltered beers without using roasted products. When combining the use of roasted malt and "XAN" technology, levels of over 10 mg XN/l can be achieved in filtered beer [43].

Figure 4: "XAN" technology – brewing techniques for increasing the XN content in beer

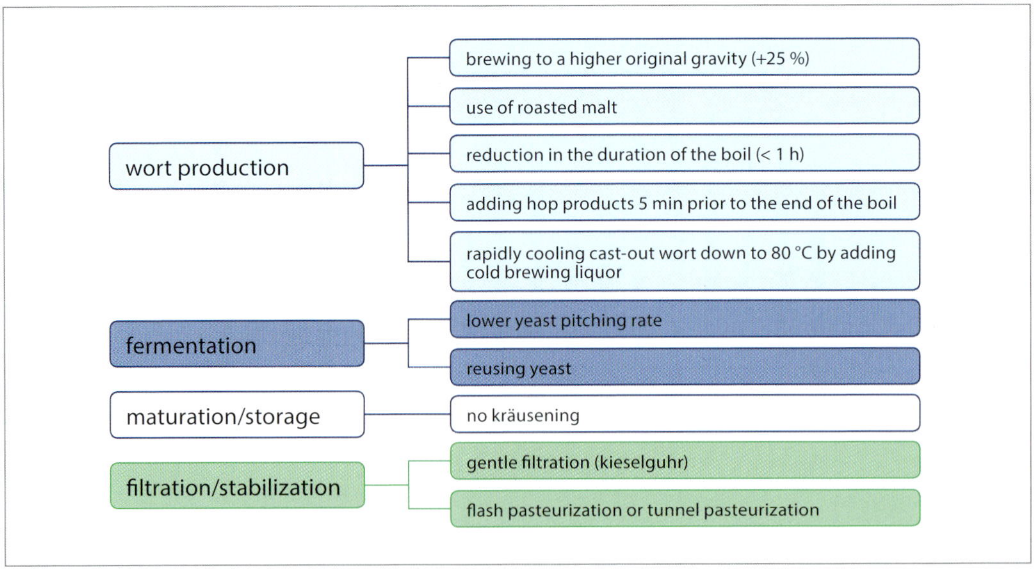

2.3.2 INCREASING FOLIC ACID CONTENT

The folic acid (vitamin B_9) content of beer is determined primarily by how much folic acid is found in the malt. However, the mashing process and the yeast strain also play a significant role. Malt contains a maximum of 4.0 mg folic acid/kg. Excellent results can be achieved by diluting the mash (e.g. a grist to brewing liquor ratio of 1 : 5) and mashing in at low temperatures. This promotes the activity of the proteinases and the associated release of vitamin B_9.

A further increase in folic acid concentrations can be expected through the utilization of a suitable yeast strain and by avoiding exposure to air during maturation and lagering. The yeast itself does not take up any folic acid during fermentation and lagering. Instead, it releases the folic acid that it has produced into the surrounding medium. However, the fermentation, maturation and storage temperatures common for bottom-fermenting beers are not high enough to promote the release of folic acid by the yeast. In pilsner beers, concentrations of approximately 80 to 120 µg of folic acid/l of beer can be reached (the usual concentrations of B_9 in pilsner beer are 60 to rarely 100 µg/l). By contrast, it is possible to achieve folic acid levels of approximately 180 to 220 µg/l in top-fermented beers (commercial southern German-style wheat beers contain a maximum concentration of 100–140 µg of folic acid/l). This significant difference can be attributed to higher maturation temperatures and a lengthier maturation, both of which lead to an increase in the folic acid content in such top-fermenting beers. No difference could be determined between secondary fermentation carried out in tanks versus bottles.

The technological factors that influence the folic acid concentration in beer have been summarized in figure 5 [44]. Since folic acid is a vitamin which is sensitive to temperature, these beers should not be pasteurized.

Figure 5: Brewing techniques affecting folic acid content [44]

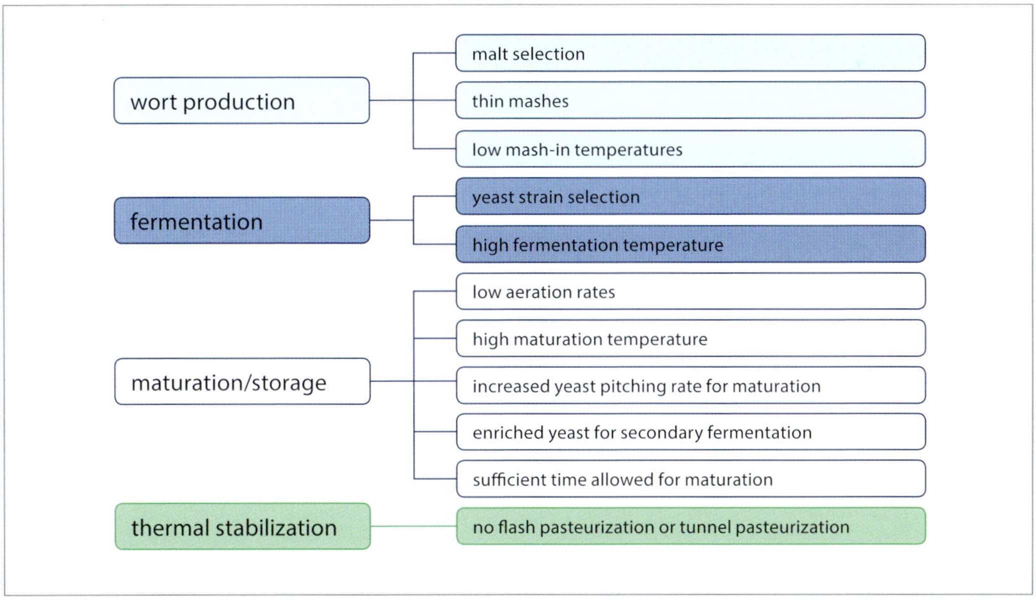

2.3.3 AVOIDING GLUTEN

Nowadays, many technologies and processes are available for producing gluten-free beers. Certain individuals suffer from ailments causing sensitivity to specific substances present in foods. These illnesses include celiac disease or the skin condition known as *dermatitis herpetiformis* (Duhring's disease). Individuals with these conditions are able to enjoy gluten-free beer. Celiac disease is a food-induced immunological illness, which stems from an intolerance of particular amino acid sequences of the prolamin fraction of wheat gluten (gliadin) and the corresponding fractions of related grains (barley, rye, triticale, etc.). Conventionally produced beers can therefore compromise the health of those who suffer from such a condition. Several different strategies can be employed in the production of gluten-free beers (table 5). According to the Codex Alimentarius, the gluten content must be below 20 ppm in order for the product to be designated gluten-free [45].

Figure 6: Methods for producing gluten-free beer [45, 47]

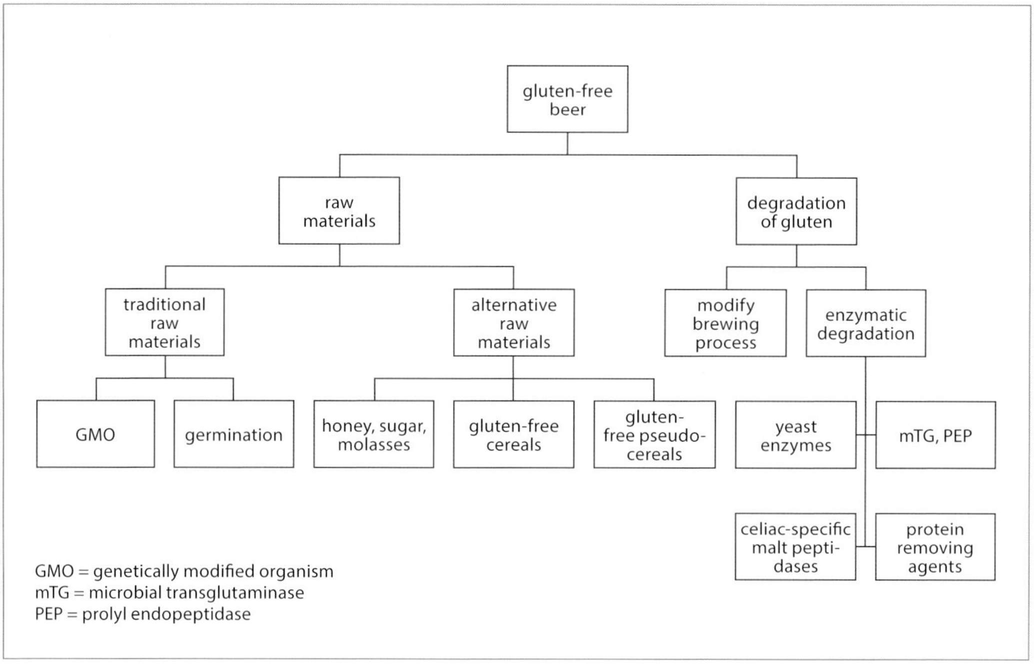

Table 5: Strategies for the production of gluten-free beer and evidence from the beer market [46]

Strategy	Remarks
utilization of raw materials modified through breeding (gluten-free)	Only one raw material relevant to this application has been identified to date.
removal of gluten present in conventional beers using enzymes (transglutaminases, prolyl endo-peptidases, among others) or using clarifying/stabilizing agents (e.g. silica gel)	Several products are on the market and new concepts are in development. Most of the typical beer characteristics have been retained.
fermentation of raw material containing sugar from gluten-free sources (e.g. honey, sugar, syrup, etc.)	The resulting product is not similar to beer. In Germany, alcoholic beverages made from honey are regulated by the laws governing wine.
utilization of gluten-free grains, rich in carbohydrates	unmalted grains, e.g. amaranth, buckwheat, millet, corn, rice, quinoa [see above] addition of enzymes, alginates, etc.
utilization of malt made from gluten-free grains, rich in carbohydrates	amaranth (e.g. *Amaranthus hypochondriacus*), buckwheat (*Fagopyrum esculentum*), millet (e.g. *Panicum miliaceum*, *Setaria italica*), sorghum (e.g. *Sorghum bicolor*), corn (*Zea mays*), rice (*Oryza sativa*) and quinoa (*Chenopodium quinoa*) all possess distinctive flavor notes; optimization of the brewing process is necessary

3 OVERVIEW

Beer possesses a great number of nutritionally relevant substances that aid in the prevention of cardiovascular diseases as well as in counteracting the formation of gallstones and kidney stones. It should be noted that the individual substances present in beer are not limited to a single effect, but also bring about synergistic effects in combination with other substances. Different beer styles each have their own unique benefits (e.g. the isotonic benefits of certain non-alcoholic beers). Techniques for enriching or avoiding certain substances in beer – like xanthohumol, folic acid and gluten – have already been developed and are the focus of ongoing research. Even though numerous beneficial substances can be found in beer, it is advisable to consume beer in moderation due to its alcohol content, amounting to approximately 5 % by volume.

4 REFERENCES

[1] Gorinstein, S., Caspi, A., Zemser, M., Trakhtenberg, S.: Comparative contents of some phenolics in beer, red and white wines. Nutrition Research 20 (2000), no. 1, pp. 131–139.

[2] Vinson, J. A., Jang, J., Yang, J., Dabbagh,Y., Liang, X., Serrg, M., Proch, J., Cais, S.: Vitamins and especially flavonoids in common beverages are powerful in vitro antioxidants which enrich lower density lipoproteins and increase their oxidative resistance after ex vivo spiking in human plasma. Journal of Agricultural and Food Chemistry 47 (1999), pp. 2502–2504.

[3] Rice-Evans, C. A., Miller, N. J., Pagana, G.: Structure-antioxidant activity relationship of flavonoids and phenolic acids. Free Radical Biology and Medicine 20 (1996), no. 7, pp. 933–956.

[4] Walker, C. J., Baxter, E. D.: Health-promoting ingredients in beer. Technical Quarterly – Master Brewers Association of the Americas 37 (2000), no. 2, pp. 301–305.

[5] Ghiselli, A., Natella, F., Guidi, A., Montanari, L., Fautozz, P., Scaccini, L.: Beer increases plasma antioxidant capacity in humans. Journal of Nutritional Biochemistry 11 (2000), no. 2, pp. 76–80.

[6] Wang, Q., Ding, Z.-H., Liu, J.-K., Zheng, Y.-T.: Xanthohumol, a novel anti-HIV-1 agent purified from Hops Humulus lupulus. In: Antiviral research 64 (2004), no. 3, pp. 189–194.

[7] Buckwold, V. E., Wilson, R. J. H., Nalca, A., Beer, B. B., Voss, T. G., Turpin, J. A., Buckheit, R.W., Wei, J., Wenzel-Mathers, M., Walton, E. M., Smith, R. J., Pallonsch, M., Ward, P., Wells, J., Churala, L., Sloane, S., Paulman, R., Russell, J., Hartmann, T., Ptak, R.: Antiviral activity of hop constituents against a series of DNA and RNA viruses. Antiviral research 61 (2004), no. 1, pp. 57–62.

[8] Horvathova, K., Vachalkova A., Novotny, L.: Flavonoids as chemoprotective agents in civilization diseases. Neoplasma 48, 2001, pp. 435–441.

[9] Gerhauser, C., Alt, A., Heiss, E., Gamal-Eldeen, A., Klimo, K., Knauf, J., Neumann, I., Scherf, H. R., Frank, N., Bratsch, H., Becker, H.: Cancer chemopreventive activity of xanthohumol, a natural product derived from hop. Molecular Cancer Therapeutics 1 (2002), no. 11, pp. 959–969.

[10] Stevens, J. F., Page J. E.: Xanthohumol and related prenylflavonoids from hops and beer: to your good health! Phytochemistry 65 (2004), no. 10, pp. 1317–1330.

[11] Milligan, S. R., Kalita, J. C., Pocock, V., Van De Kauter, V., Stevens J. F., Deinzer, M. L., Rong, H., De Keukeleire, D.: The endocrine activities of 8-prenylnaringenin and related hop (Humulus lupulus L.) flavonoids. Journal of Clinical Endocrinology and Metabolism 85 (2000), no. 12, pp. 4912–4915.

[12] De Keukeleire, D., Milligan, S. R., du Cooman, L., Heyerick, A.: The oestrogenic activity of hops (Humulus lupulus L) revisited. Pharmaceutical and Pharmacological Letters 7 (1997), no. 2/3, pp. 83–86.

[13] Walker, C. J., Bolshaw, L., Chandra, S.: Healthy Drinks? Beer and Cider Antioxidants. Proceedings of the 28th European Brewery Convention Congress, Budapest, Hungary 2001. pp. 92–101.

[14] Piendl, A.: Physiologische Bedeutung der Eigenschaften des Bieres. Fachverlag Hans Carl, Nürnberg 2000.

[15] Deutsche Gesellschaft für Ernährung (DGE): Die Nährstoffe: Bausteine für Ihre Gesundheit. Wahrlich Druck, Meckenheim 2004.

[16] Bässler, K. H.: Vitamine. Dr. Dietrich Steinkopff Verlag, Darmstadt 1989.

[17] Burgerstein, L.: Handbuch Nährstoffe. Karl F. Haug Verlag, Stuttgart 2002.

[18] Cupples, L. A., Weinberg, J., Beiser, A., Auerbach, S. H., Volicer, L., Cipolloni, P. B., Wells, J., D'Agostino, R. B., Growdon, J. H., Wolf, P. A., Farrer, L. A.: Effects of smoking, alcohol and APOE genotype on Alzheimer disease: the MIRAGE study. Alzheimers Reports 3 (2000), pp. 105–114.

[19] Fagrell, B., de Faire, U., Bondy, S., Criqui, M., Gronback, M., Jackson, R., Klatsky, A., Salonen, J., Shaper, A. G.: The effects of light to moderate drinking on cardiovascular diseases. Journal of Internal Medicine 246 (1999), pp. 331–340.

[20] Rimm, E. B., Williams, P., Fosher, K., Cirqui, M., Stampf, M. J.: Moderate alcohol intake and lower risk of coronary heart disease: meta-analysis of effects on lipids and haemostatic factors. British Medical Journal 319 (1999), pp. 1523–1528.

[21] Stuttaford, T.: To Your Good Health! The Wise Drinker's Guide. Farber & Farber, London 1997.

[22] Keil, U., Chambless, L. E., Döring, A., Filipiak, B., Stieber, J.: The relation of alcohol intake to coronary heart disease and all-cause mortality in a beer drinking population. Epidemiology 8 (1997), no. 2, pp. 150–156.

[23] Muntwyler, J., Hennekes, C. H., Buring, J. E., Gaziano, J. M.: Mortality and light alcohol consumption after myocardial infarction. Lancet 352 (1998), pp. 1882–1885.

[24] Perry, I. J., Refsum, H., Morris, R. W.: Prospective study of serum total homocysteine concentration and risk of stroke in middle-aged British men. Lancet 346 (1995), pp. 1395–1398.

[25] Buday, A. Z., and Denis, G.: The diuretic effect of beer. Brewers Digest 49 (1974), no. 6, pp. 56–58.

[26] Piendl, A., Wagner, I.: Beer drinking and diuresis. Brauindustrie 70 (1985), no. 11, pp. 1082–1084.

[27] Piendl, A.: Beer as an athlete's beverage. Brauwelt 130 (11), 1990. pp. 370–372.

[28] Goupy, P., Hugues, M., Boivin, P., Amiot, M. J.: Antioxidant composition and activity of barley (Hordeum vulgare) and malt extracts and of isolated phenolic compounds. Journal of the Science of Food and Agriculture 79 (1999), no. 12, pp. 1625–1634.

[29] Lindenmeier, M., Faist, V., Hofmann, T.: Structural and Functional Characterization of Pronyl-lysine, a Novel Protein Modification in Bread Crust Melanoidins Showing in Vitro Antioxidative and Phase I/II Enzyme Modulating Activity. Journal of Agricultural and Food Chemistry 50 (2002), no. 24, pp. 6997–7006.

[30] Lasztity R.: Oat grain: a wonderful reservoir of natural nutrients and biologically active substances. Food Reviews International 14 (1998), pp. 99–119.

[31] Bamforth, C. W.: Beer: Health and Nutrition. Blackwell Science Ltd, Oxford 2004.

[32] Holbrook, T. L., Barrett-Connor, E.: A prospective study of alcohol consumption and bone mineral density. British Medical Journal 306 (1993), pp. 1506–1509.

[33] Chen, W.-J., Lin, J.-K.: Mechanisms of Cancer Chemoprevention by Hop Bitter Acids (Beer Aroma) through Induction of Apoptosis Mediated by Fas and Caspase Cascades. Journal of Agricultural and Food Chemistry 52 (2004), pp. 55–64.

[34] Kubisch, U., Ullrich, N., Müller, A.: Therapie von Schlafstörungen mit einem Hopfen-Baldrian-Extrakt. Wirksame Alternative zu Benzodiazepinen. Zeitschrift für Phytotherapie 24 (2003), no. 2, pp. 63–69.

[35] Buchbauer, G., Jirowetz, L., Jager, W., Dietrich, H., Plauk, C.: Aromatherapy: evidence for sedative effects of the essential oil of lavender after inhalation. Zeitschrift für Naturforschung C 46 (1991), no. 11–12, pp. 1067–1072.

[36] Scherr J, Nieman DC, Schuster T, Habermann J, Rank M, Braun S, Pressler A, Wolfarth B, Halle M.: Nonalcoholic beer reduces inflammation and incidence of respiratory tract illness. Med Sci Sports Exerc. 44 (2012), no. 1, pp. 18–26

[37] Frank, N.: Ist XN im Bier krebspräventiv? In: Handbuch zum 37. Technologischen Seminar, Freising-Weihenstephan 2005.

[38] Rico, H., Gallego-Lago, J. L., Hernández, E. R., Villa, L. F., Sanchez-Atrio, A., Seco, L., Gervas, J. J.: Effect of silicon supplement on osteopenia induced by ovariectomy in rats. Calcified Tissue International 66 (2000), no. 1, pp. 53–55.

[39] Zarnkow, M.: Influence on the Purine Content of Beers. ASBC Meeting, Fort Myers 2011.

[40] Janssens, J., Shapira, N., Debeuf, P., Michiels, L., Putman, R., Bruckers, L., Renard, O., Molenberghs, G.: Effects of soft drink and table beer consumption on insulin response in teenagers. European Journal of Cancer Prevention 8 (1999), pp. 289–295.

[41] Renaud, S., de Lorgeril, M.: Platelets and dietary prevention of coronary heart diseases (CHD). Acta Cardiologica 48 (1993), no. 5, pp. 446–448.

[42] Walzl, M., Hlatky, M.: Jungbrunnen Bier. Verlagshaus der Ärzte, Wien 2004.

[43] Wunderlich, S., Zürcher, A., Back, W.: Enrichment of xanthohumol in the brewing process. Molecular Nutrition & Food Research 49 (2005), no. 9, pp. 874–881.

[44] Walker, C., Herrmann, M., Faulkner, A., and Back, W.: Investigations on the folate (vitamin B_9) content of Bavarian wheat beers. Proceedings of the 29th European Brewery Convention Congress, Dublin 2003, p. 135.

[45] Zarnkow, M.: Glutenfreies Bier – Stand der Technik. Vortrag am 38. Technologischen Seminar. Freising-Weihenstephan 2005.

[46] Bamforth, C.W.: In Arendt, E.K., Dal Bello, F. (Eds.), The Science of Gluten-Free Foods and Beverages. AACC 2009, pp. 113–117.

[47] Zarnkow, M., Kerpes, R., Becker, T., Knorr, V. Köhler, P. Eintrag vermeiden – Strategien zur Herstellung von glutenfreien Bieren. Brauindustrie 98 (2013), no. 4, pp. 20–23.

MICROBIOLOGY

1 INTRODUCTION

Although numerous microorganisms can be found in a brewery, only a few are capable of multiplying in beer and causing damage. This is primarily due to the selective properties of beer [1, 2, 3]:

- a low pH value, around 4.5
- CO_2 content (anaerobic environment)
- alcohol content, around 5 % by volume
- bitter compounds from hops
- scarcity of easily utilizable sugars and amino acids
- low temperatures in the production area

In expressing the microbiological selectivity of beer as a percentage, the bitter compounds derived from the hops make the most significant contribution (30 %), followed by the pH (25 %), the anaerobic environment (20 %) and the alcohol (10 %), while the scarcity of nutrients and other compounds necessary for nutrition and growth account for 15 %. These selective properties prevent pathogens and heat-resistant microbes from growing in beer. This allows pasteurization to be carried out at moderate temperatures (62–72 °C) (fig. 17, p. 313) to achieve the necessary degree of microbiological stability [3, 4].

The microorganisms listed in figure 1 are primarily pure cultures (bottom-fermenting and top-fermenting yeast strains as well as cultures used in the production of natural lactic acid) along with the relevant spoilage microorganisms. As the latter must be detected in trace amounts, the capability to identify special indicator microorganisms is very important in microbiological quality control.

Figure 1: Microflora in the brewery

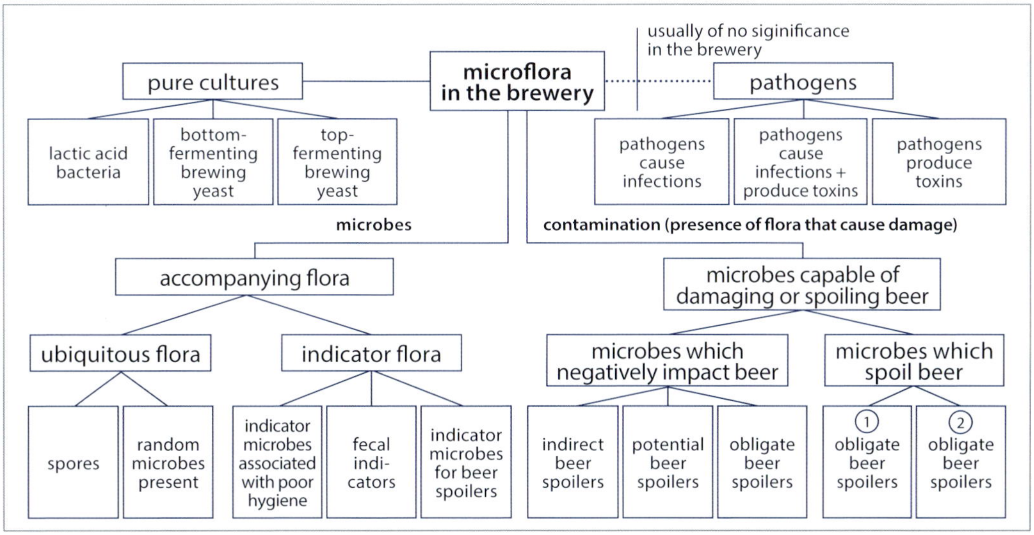

[1] causes nausea
[2] renders beer und rinkable

Here, *Gluconobacter frateurii* and *Acetobacter pasteurianus*, both acetic acid bacteria, play a key role as indicator microorganisms, since they encapsulate their cells in a layer of slime and prefer to grow in an environment where beer is present. Therefore, they grow where beer spoilers can also be found. Given the fact that acetic acid bacteria require oxygen, they are incapable of growing in beer as it is almost devoid of oxygen.

These indicator microbes, however, settle into various niches in production areas where they come into contact with beer. They produce slime, forming biofilms, which then become colonized by typical beer spoilers. These, in turn, grow and multiply in the safe haven of the biofilm, eventually adapting to their environment and developing the ability to utilize beer as a substrate (fig. 2).

Figure 2: Development of beer-spoiling microorganisms

Presence of microbes, niches of colonization, swab sample collection areas	Microorganisms	Analysis results for swab samples in NBB-AM broth (color change from red to yellow), 20–28 °C	Analysis results for swab samples in lactose broth (LMC), (color change from violet to yellow), 36–37 °C
normal conditions 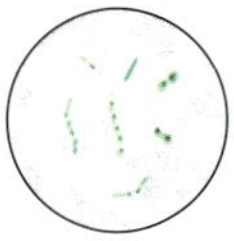	**acetic acid bacteria** *Gluconobacter* spp. *Acetobacter* spp. along with other ubiquitous slime-producing microbes	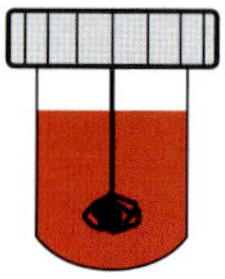 ⊖ after 3 days	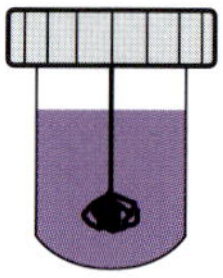 ⊖ after 2 days
stage 1 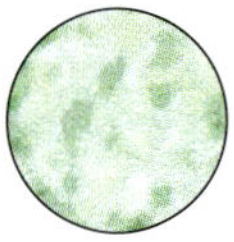	**slime deposits** (acetan) often in association with other slime producers (mannan, dextran)	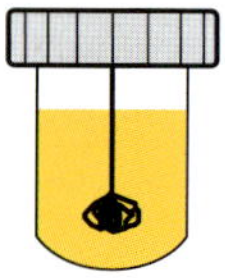 ⊕ after 3 days	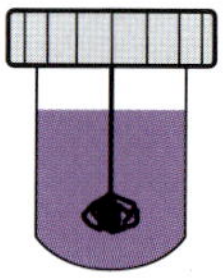 ⊖ after 2 days
stage 2 	**biofilms** persistent layer slime deposits, encapsulation	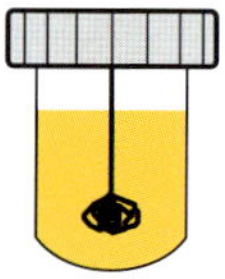 ⊕ after 2 days	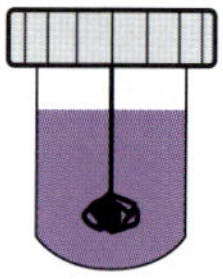 ⊖ after 2 days
stage 3 	**biofilms** persistent layer ⊕ yeasts ⊕ lactic acid bacteria ⊕ coliform bacteria	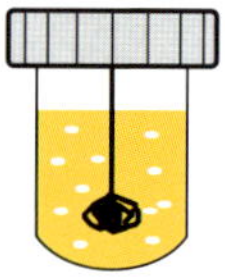 ⊕ after 1–2 days + gas	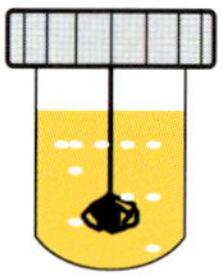 ⊕ after 2 days + gas, coliforms
	adaptation!		
stage 4 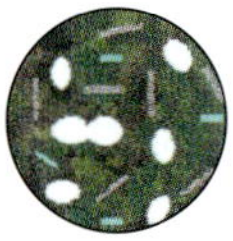	**biofilms** persistent layer ⊕ yeasts, lactic acid bacteria ⊕ *Pectinatus, Megasphaera* ⊕ coliform bacteria as well as "potential pathogens"	 ⊕ after 1 day + gas	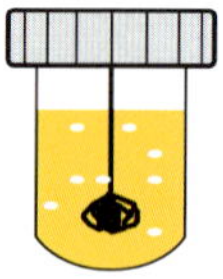 ⊕ after 1–2 days + gas, coli-forms

Biofilms can also arise on moist surfaces with no contact to beer. However, other kinds of slime-forming microorganisms are more commonly present in these areas, e.g. bacteria strains typically found in water (e.g. *Pseudomonas, Enterobacter*), ubiquitous yeasts (e.g. *Rhodotorula, Cryptococcus*) and molds (e.g. *Dematium, Aureobasidium*).

Beer spoilers are generally divided into three categories based on their level of risk: obligate, potential and indirect beer spoilers. Of these, the most dangerous are the obligate beer spoilers. They negatively impact beer quality, since they can tolerate the selective properties specific to beer and can usually grow in beer without any prior adaptation.

Potential beer spoilers can also cause problems, usually in beer which lacks selective properties. In the brewery, if these microbes remain in contact with beer over a longer period, they can adapt to it as a substrate. They would then be viewed as obligate beer spoilers.

Indirect beer spoilers are not able to multiply in packaged beer. They are, however, capable of growing in production areas, causing some initial damage to the quality of the wort, yeast or green beer. This compromises the finished beer and is carried over to the packaged product. Ultimately, consumers may find the damage done by these indirect beer spoilers unpleasant, thereby prompting them to submit complaints.

In addition to classifying microbes based on their level of beverage spoilage, categorizing them as primary or secondary contaminants can be quite helpful for purposes of microbiological quality control. Primary contaminants present in production areas include *Pediococcus damnosus* and *Lactobacillus lindneri*, among others. In certain situations, these microorganisms can enter the beer stream during the filtration process and end up in the packaged beer. As a rule, the number of microbes declines dramatically as a result of filtration. Hence, analysis results from microbiological tests often appear negative initially. However, microbial contamination becomes evident during incubation, because the turbidity increases over an extended period of time (e.g. four weeks). Locating the source of primary contamination is often very difficult, since by the time complaints about a product are received, the microbes have often spread throughout the entire production area.

Secondary contaminations generally occur in the filling area [5], especially near the filler or capper. Biofilms that harbor beer spoilers tend to form in these areas due to the constant contact with beer.

Figure 3: Development of *Pectinatus* and *Megasphaera* contaminations in the brewery

Phase 1: Microbes are brought into the brewery, e.g. through air currents or on contaminated returnable bottles and crates.

Phase 2: Microbes colonize corners and crevices near the filling/capping area, on conveyor belts or in damaged flooring where physical and chemical cleaning has restricted access.

Phase 3: Such niches are difficult to clean and always harbor a mixture of microorganisms (accompanying flora such as non-beer spoiling aerobic microbes, often with lactic acid bacteria).

↓

The aerobic microorganisms consume all of the oxygen (e.g. acetic acid bacteria), creating anaerobic ecological niches, accumulation of nutrients (e.g. lactic acid) and slime layers (biofilms).

↓

Optimal growth conditions for *Pectinatus* and *Megasphaera* include an anaerobic atmosphere, the availability of lactic acid as the preferred carbon source and an increasing pH as a result of the autolysis of dying microbes.

Phase 4: Microbes and biofilms can contaminate individual bottles during filling operations through the rotation of moving parts, spraying and creation of aerosols: "scattered contamination".

Phase 5: Complaints and possible legal ramifications

Individual microbes or entire biofilms can be sprayed into the air through the action of rotating machinery. The wet environment then facilitates their growth. Consequently, contamination can easily spread to open bottles. This results in a level of contamination of varying intensity scattered across individual bottles. Microbes such as *Lactobacillus brevis* and *L. casei* as well as *Pectinatus* sp. and *Megasphaera* sp. (fig. 3) are usually present as secondary contaminants. Contamination can be combatted through targeted cleaning regimens to achieve the required level of hygiene, e.g. by spraying down equipment in the filling area with hot water (75–85 °C) at regular intervals.

2 MICROORGANISMS

Pure top-fermenting and bottom-fermenting brewing yeast cultures are crucial to producing untainted beer. The golden rule of quality control is to maintain a high level of yeast purity and vitality since the general condition of these cultures is pivotal to the quality of the beer. Pitching yeast with a high vitality guarantees a rapid fermentation, a swift and extensive degradation of diacetyl and a pleasant flavor profile along with a suitable level of flavor and foam stability. Vital brewing yeast with a strong fermentative capacity is able to successfully compete with beer spoilers (cf. Yeast Technology and Fermentation).

The morphological, biochemical and physiological properties as well as the taxonomic criteria of bottom-fermenting and top-fermenting brewer's yeast (*Saccharomyces pastorianus* var. *carlsbergensis*, *S. cerevisiae*) are described in extensive detail in the book "Colour Atlas and Handbook of Beverage Biology" [2, 3]. A closer examination of the metabolic properties of the yeast strains along with the relevant fermentation techniques can be found in "Mikrobiologie der Lebensmittel, Getränke" [6].

The cultures used to produce natural lactic acid (primarily *Lactobacillus amylolyticus*) [7] have gained in importance over the past few years – not only for adjusting the pH of the mash and wort, but also because of the numerous other advantages attributed to them (cf. Acidification with Natural Lactic Acid).

In order to prevent disruptions in the fermentation process and problems with their beer, every brewery should constantly be monitoring its production process for contamination with beer spoilers. Several strains of lactic acid bacteria are classified as primary beer spoilers along with various foreign and wild species of yeasts and certain Gram-negative bacteria, particularly *Pectinatus* sp. as well as *Megasphaera* sp.

Generally speaking, molds [8] in a brewery are viewed as harmless contaminants. So-called cellar molds (*Aureobasidium*, *Cladosporium*, *Moniliella*, among others) are considered to be a nuisance because their unsightly growth on wet walls and vessels necessitates special cleaning measures from time to time. Indirect beer spoilers include *Fusarium* species, particularly *Fusarium graminearum* and *F. culmorum* (other *Fusarium* species may be present depending on the field crop from the previous season, the grain's provenance and the year it was harvested). These microorganisms are classified as indirect beer spoilers, because they produce toxins and other undesirable substances capable of entering the brewing process, some of which may also be transferred to the beer. *Fusarium* species play a particularly important role in gushing. Recent research has revealed that more diversity exists in the most prevalent species of *Fusarium* than previously thought, and that currently *F. avenaceum* and *F. tricinctum* are the most common *Fusarium* species present in Central Europe. These findings have led to a reevaluation of the methods of visual assessment for barley, since throughout the process of the beer production, *F. avenaceum* and *F. tricinctum* can produce a range of secondary metabolites (e.g. enniatins) that are relevant to food safety. However, up to now, this has not been widely explored [9, 10]. The importance of *Fusarium* spp. as indirect beer spoilers is perhaps attributable to their ability to tolerate low-oxygen environments. This allows them to grow inside cereal grains, where they bring about overmodification of the endosperm through their cytolytic, proteolytic and amylolytic activity. In doing so, they also produce undesirable substances (most notably, mycotoxins).

2.1 PURE BREWING YEAST STRAINS

Brewing yeast strains encompass certain proprietary strains from individual breweries and breeding forms of *Saccharomyces cerevisiae* or *S. pastorianus* var. *carlsbergensis*, which have been identified as especially suitable for beer production based on their fermentation characteristics. After pitching, bottom-fermenting yeast strains consume the sugars and other compounds in the wort at low temperatures (5–15 °C). Toward the end of fermentation, bottom-fermenting yeast cells sediment out to varying degrees and accumulate on the bottom of the tank. Based on their sedimentation behavior (flocculation traits), yeast strains are grouped into one of two categories: powdery or non-flocculent strains and flocculent strains (this is not entirely clear-cut). The sedimentation of flocculent yeast during fermentation forms a fairly compact layer at the bottom. This simplifies yeast cropping – and beer filtration once maturation is complete. Powdery yeast strains, conversely, tend to remain in suspension. Although fermentation yields are frequently higher with powdery yeast, under certain circumstances difficulties may arise when harvesting the yeast or filtering the beer.

Normally, top-fermenting yeast strains ferment at higher temperatures, ranging from 15 to 25 °C. Fermentation is brisk and the yeast rises to form a thick layer over the surface of the fermenting wort. Different strains of top-fermenting yeast are employed to produce many different beer styles. For example, there are special top-fermenting strains for the production of weissbier and *Altbier*. In contrast to bottom-fermenting strains, branching chains of linked cells are characteristic of most top-fermenting strains.

Both bottom-fermenting and top-fermenting brewing yeasts form rounded to oval cells approximately 5 to 10 by 5 to 12 µm in size. In some instances, elliptical or cylindrical cells 3.5 to 9.5 by 5.0 to 20.0 µm in size have been observed. In rare cases, the cells may be elongated (up to 40 µm) or even tubular. The average cell size can vary significantly from strain to strain. Lipid droplets are frequently visible inside the cells, with obvious vacuoles in older cells. Other unusual or taxonomically relevant cell structures are not present. The branching cells of top-fermenting yeasts are linked together in multilateral chains.

Bottom-fermenting yeasts occur primarily as single cells or as pairs of cells. Large chains of cells may be present individually, but agglomerates (clumps consisting of random cells) are much more common.

Both top-fermenting and bottom-fermenting strains of brewing yeast do not grow on agar containing crystal violet, lysine or copper sulfate and cannot assimilate nitrogen. *S. pastorianus* var. *carlsbergensis* can be distinguished from *S. cerevisiae* through its ability to fully utilize melibiose and raffinose. *S. cerevisiae* ferments one third of the raffinose molecule, leaving a melibiose residue. *S. pastorianus* var. *carlsbergensis* exhibits no growth at 37 °C and an absence or rare occurrence of ascospores.

2.2 FOREIGN YEAST STRAINS

In some cases, certain strains of foreign yeast can proliferate in packaged beer, creating undesirable aromas and off-flavors, prompting their classification as beer spoilers. These usually include numerous wild strains of *S. cerevisiae* and others, especially those belonging to the genera *Saccharomyces* and *Brettanomyces* [2, 3].

In figure 4, yeasts are grouped into categories according to their applications in brewing technology. Foreign yeast strains exhibiting strong fermentative activity are of great significance, since they are often present as obligate beer spoilers and negatively impact the product.

Figure 4: Classification of yeast strains according to their application in brewing technology

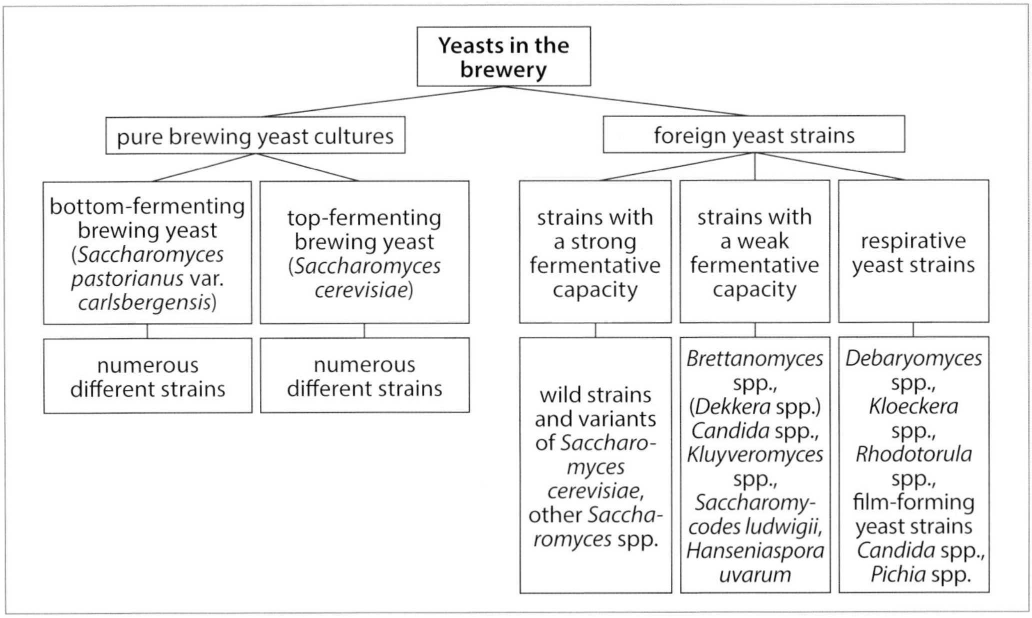

The morphology of foreign yeast strains often differs from that of brewing yeast. These yeast cells are smaller and elongated and are usually present as branching chains. For example, *S. c.* var. *diastaticus occurs* primarily as smaller oval, ellipsoidal or cylindroidal cells, and they may also be elongated. The branched chains are more linear and often consist of only a few cells. The fermentation spectrum is largely similar to top-fermenting yeast strains; however, foreign yeasts can also ferment dextrins and starch with the help of amylases and amyloglucosidases. Consequently, the contamination of bottled beer by foreign yeast equipped with such enzymes results in overattenuation, which is characterized by the development of haze after bottling, sediment, off-flavors and atypical aromas. The additional conversion of residual extract creates an empty beer with a harsh bitter note that cannot be masked, since the yeast has also consumed the dextrins. These foreign yeasts are predominantly found in beer upstream from the filter, but their small cell size also allows them to pass through the filter quite easily. Therefore, in some instances, they may be found as secondary contaminants in the filling area.

Saccharomyces bayanus, *S. uvarum* and various wild strains of *S. pastorianus* number among the most feared wild yeast contaminants. Their cells are mostly cylindroidal and can assume irregular club shapes or allantoidal (sausage-like) forms. Branched chains of cells are seldom. Instead, the cells tend to occur predominantly in pairs, with the cells frequently exhibiting a distinctive signpost pattern. These yeasts usually possess a bottom-fermenting character; however, like top-fermenting yeast strains, they cannot utilize melibiose and therefore only one-third of the raffinose molecule. Often, these yeast strains can utilize maltotriose and maltotetraose. Beer with a low original gravity going into fermentation is at risk of contamination. Signs of contamination include the formation of haze after bottling and harsh, bitter off-flavors. These yeasts are mostly found in unfiltered beer and as contaminants in pure brewing yeast cultures.

Non-*Saccharomyces* yeasts appear less frequently in the role of beer spoilers; however, the presence of *Brettanomyces/Dekkera* strains (e.g. *B. anomalus*) can cause problems in beer. These microorganisms are slow fermenters and are commonly found in Belgian gueuze and lambic beers, in ales and porters as well as in *Berliner Weisse*, where they are responsible for creating the desirable ester aromas characteristic of these beer styles. In other beers stored over longer periods under warm conditions, these microbes can cause massive off-flavors and unpleasant aromas through the formation of acetic acid and acetic acid esters. The cells can be oval, ellipsoidal, cylindroidal or elongated, frequently with ogival budding (budding on the shoulder of the broad side of the mother cell). Furthermore, a pseudomycelium is often present, which may contain blastospores occurring as singles or as multiples arranged in a whorl.

Methods for distinguishing the yeasts found in the brewery are shown in figure 5. Although only five test media are used here, this is a meaningful tool for brewery microbiologists in their efforts to identify yeast strains based on their growth habits. Numerous physiological and biochemical tests are available if a precise identification of the strain is required [2, 3, 4, 11, 12].

Figure 5: General differentiation of yeast strains by membrane filtration and cultivation

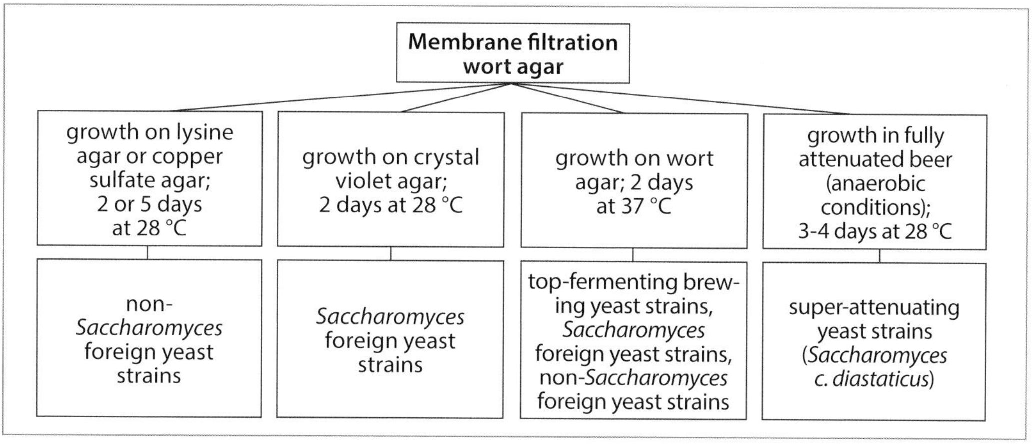

2.3 BEER-SPOILING BACTERIA

Of the beer-spoiling bacteria, lactobacilli and pediococci are considered to be the main culprits [15], though Gram-negative strains belonging to the genera *Pectinatus* and *Megasphaera* also pose a threat (fig. 6).

The most important obligate beer spoilers include the *Lactobacillus* strains *L. brevis* (along with *L. brevis* var. *brevisimilis*), *L. backii* [13], *L. lindneri* and *L. (para)casei*. Less prevalent are *L. (para)collinoides* and the slime-forming *L. rossiae* and *L. brevis* ssp. *frigidus*. More detailed information about the occurrence, importance and frequency of contamination of various beer-spoiling strains can be found in the relevant scientific literature [16, 17, 18, 19].

Figure 6: Classification of beer-spoiling bacteria [1, 2, 3, 16, 17, 18, 19]

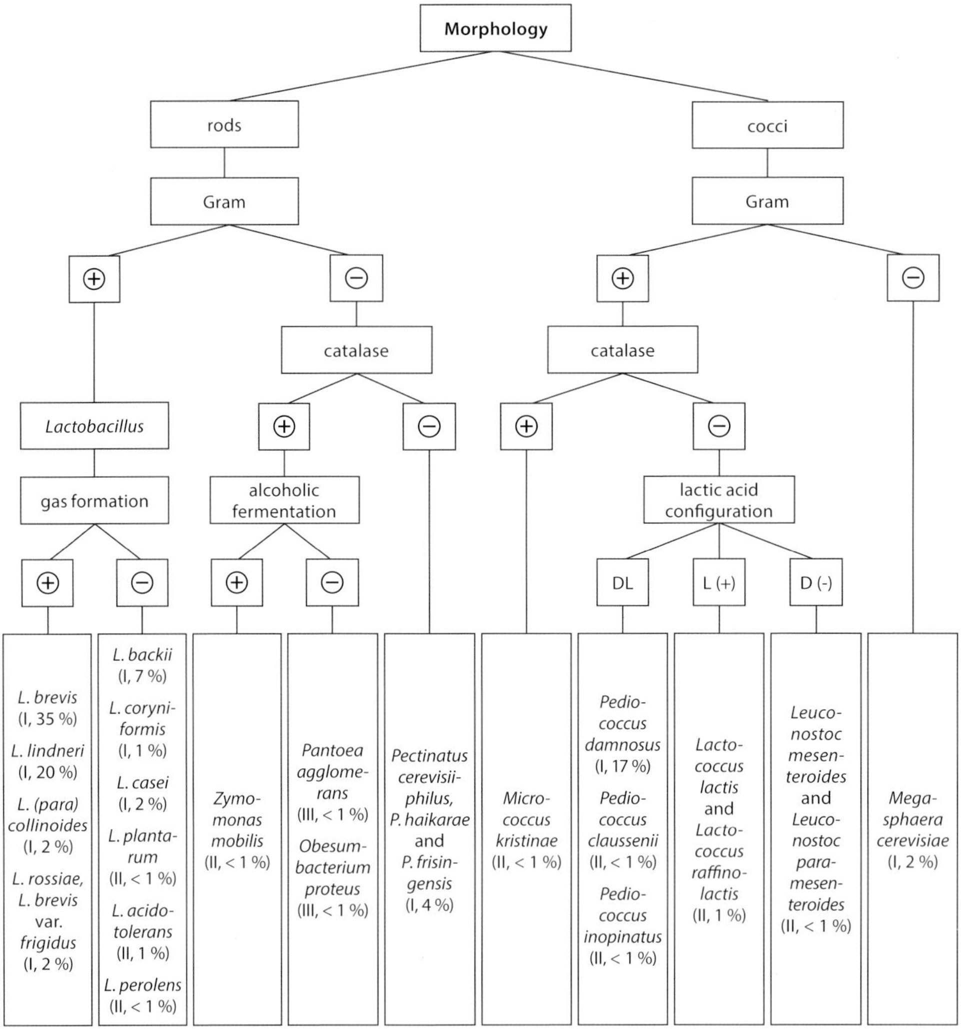

I = obligate beer spoiler; II = potential beer spoiler; III = indirect beer spoiler
The percentages provided represent the frequency for each species identified in conjunction with complaints regarding spoiled beer (documented between 1980-2017).

L. brevis usually forms longer, parallel-walled, rod-shaped cells with rounded ends (0.7 x 4.0 μm), which are present as single cells or pairs. If they occur as a pair, the cells are frequently found in a zigzag or signpost orientation. Chains are atypical for this species. In some instances, extremely long rods (up to 50 μm) have been observed in beer. Metabolic characteristics of *L. brevis* include the formation of gas (heterofermentative), utilization of pentoses and melibiose as well as the cleavage of arginine. *L. brevis* is commonly encountered as a spoilage microorganism in beer. It causes haze and sediment while lowering the pH value significantly, thus imparting a sour flavor to the beer. However, it does not produce diacetyl. Generally, these microorganisms are classified as secondary contaminants.

With *L. lindneri*, short somewhat irregular or coccoid cells are usually present in longer chains. Occasionally, long rods may also appear, particularly in beer samples. This heterofermentative species ferments mainly glucose and maltose and does not have the ability to cleave arginine. Although small amounts of sediment and haze form in beer contaminated with *L. lindneri*, this is not usually accompanied by any off-flavors. It is a common primary contaminant typically associated with yeast propagation, fermentation and lager cellars. The cells are often very tiny which allows them to slip through the filter and remain in the beer.

Recently, complaints have arisen regarding a contamination with *L. rossiae* in wheat beers. Like *L. brevis* ssp. *frigidus*, this heterofermentative obligate beer spoiler also produces slime. The recently characterized potential beer spoiler *L. cerevisiae* is rarely found in breweries (< 1 %) [20]. Beer spoilers *L. acetotolerans* [21] and *L. perolens* [22] are most prevalent in shandies.

The obligate beer spoiler *L. backii* has been observed with noticeable frequency over the past few years. It is a facultative heterofermentative microbe (producing gas from gluconic acid, but not from glucose and other sugars), present mostly as short rods and short chains, creating haze and sediment in beer, but no diacetyl.

L. (para)casei, *L. coryniformis*, *L. perolens/harbinensis*, *L. plantarum* and *L. perolens* [22] are likewise facultative heterofermentive strains. They generally form short rods that are often connected in long chains. These strains are not as prevalent, appearing more in weakly hopped beers (wheat beers), and cause serious off-flavors in beer through the production of diacetyl. For the most part, these microbes are secondary contaminants. At this point, the homofermentative strain *L. acetotolerans* [21] should be mentioned as well (test results for gas production are always negative). It has been detected on a number of occasions in Chinese beers with a low hop rate and an extremely low pH (< 4.0) [23]. It causes haze, sediment and an astringent sour flavor in beer.

Cells arranged in tetrads are characteristic of *Pediococcus damnosus*, a typical primary contaminant frequently appearing in brewing yeast cultures and in unfiltered beer as well as in wheat beer. The cells can also be carried through the filter into the packaged beer. In beer, this microorganism generates diacetyl (butter aroma), lowers the pH and creates haze and obvious sediment. Two more potential beer-spoiling *Pediococcus* strains, *P. inopinatus* and *P. claussenii*, demonstrate similar effects. The latter also produces slime in beer. Both strains, however, are rarely seen [24, 25].

The Gram-negative, catalase-negative, strictly anaerobic strain *Megasphaera cerevisiae* is characterized by large oval or spherical cells (1.2–1.6 μm), which appear as diplococci or as short chains. Fructose, pyruvic acid and lactic acid are the primary substrates fermented by this microorganism. The main metabolic products include butyric acid, hexanoic acid, acetic acid, propionic acid and valeric acid as well as CO_2 and H_2. *Megasphaera cerevisiae* also produces H_2S. Although only a slight haze is formed in contaminated beer, the metabolic products listed above contribute to severe off-flavors and unpleasant odors (fecal). This strain shows some sensitivity to alcohol (< 5 % by volume) and prefers conditions with higher pH levels (> 4.4). It is a typical secondary contaminant and tends to be found in the proximity of filling and capping equipment.

Figure 7: Sample collection regimen for microbiological quality control

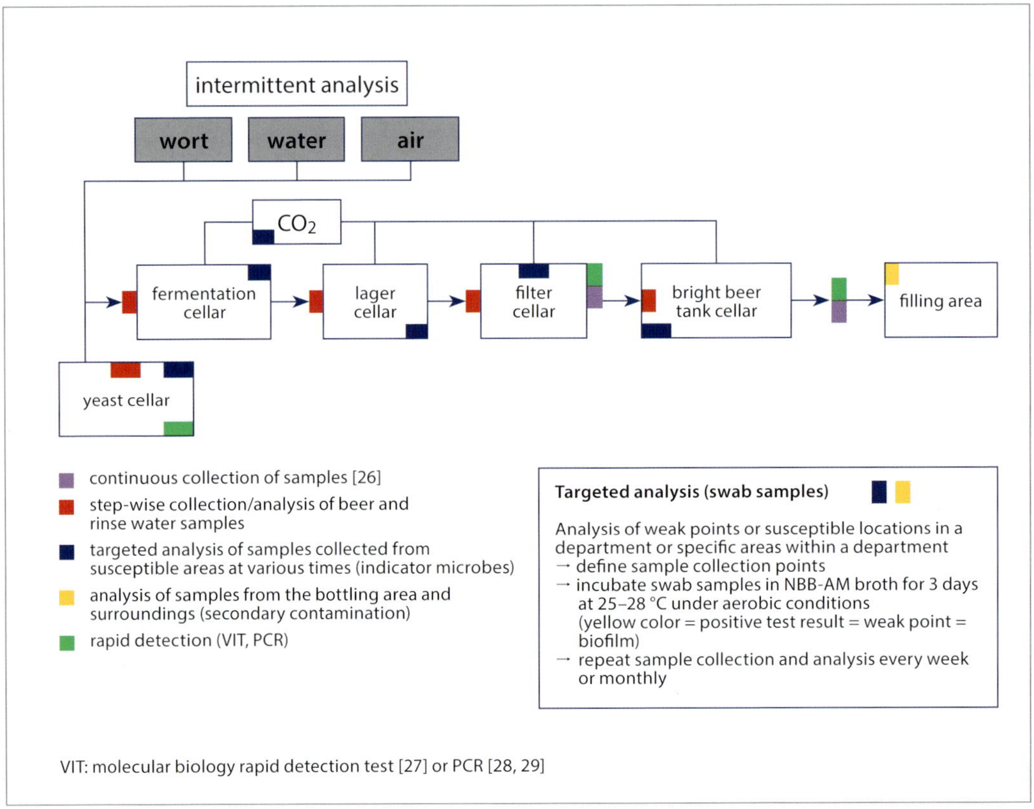

Pectinatus cerevisiiphilus, *Pect. frisingensis* and *Pect. haikarae* are similar to the aforementioned microorganism in that they are also strictly anaerobic, Gram-negative and catalase-negative and can cause similar adverse effects in beer. The cells are slender (0.8 x 4.0 µm) with parallel walls and pointed ends. The cells may be slightly curved or have snake-like or helical shapes with unilateral flagella. Like *M. cerevisiae*, these microorganisms grow at temperatures ranging from 15 to 40 °C (optimum: 28–32 °C) and can ferment various sugars, sugar alcohols and organic acids (especially pyruvate and lactate). Primary metabolic products include propionic acid, acetic acid, succinic acid, acetoin and CO_2. Not only are larger amounts of haze and sediment or even clumps evident in contaminated beer (pH > 4.3, alcohol concentration < 5.5 % by volume), they impart strong unpleasant odors (fecal) and off-flavors to the beer. Like *M. cerevisiae*, these *Pectinatus* strains are typically secondary contaminants, primarily occurring in filling areas.

2.4 DETECTION OF BEER SPOILERS

The prerequisite for a high level of microbiological safety is the implementation of a quality control regimen tailored to a specific production facility (fig. 7). Samples should be collected regularly (at least once a week) at defined points located throughout the entire production process. Since beer spoilers are generally present in very small numbers in the samples, analyses allowing for the detection of trace contamination must be performed. Therefore, the largest sample volume possible must be tested. Samples must be collected in a systematic manner and analyzed.

2.4.1 DETECTION OF BEER-SPOILING BACTERIA

Detection of beer-spoiling bacteria is typically carried out using NBB media [1, 3], which have been evaluated internationally. Experts in the field recommend these media for their excellent reliability and rapidity in detecting microbes and their discriminating specificity in beer [15, p. 1216]. Simple processing of the samples and straightforward, clear interpretation of the results are added advantages. The NBB detection media promote the rapid proliferation of beer spoilers while simultaneously suppressing the growth of brewing yeast strains and other harmless flora in the sample.

Beer-specific selectivity in NBB is targeted to facilitate only cultivation of the microbes specified in the analysis, i.e. those important for microbiological quality control in the brewery.

In order to successfully detect trace contamination (with the largest possible sample volume), various types of samples are tested using detection methods developed precisely for each specific type of sample. For example, clear beer samples and rinse water samples are filtered through a membrane and incubated on NBB agar in an anaerobic jar at 27 °C for approximately 5 days.

Figures 8a and 8b: Detection of beer-spoiling bacteria in yeast samples

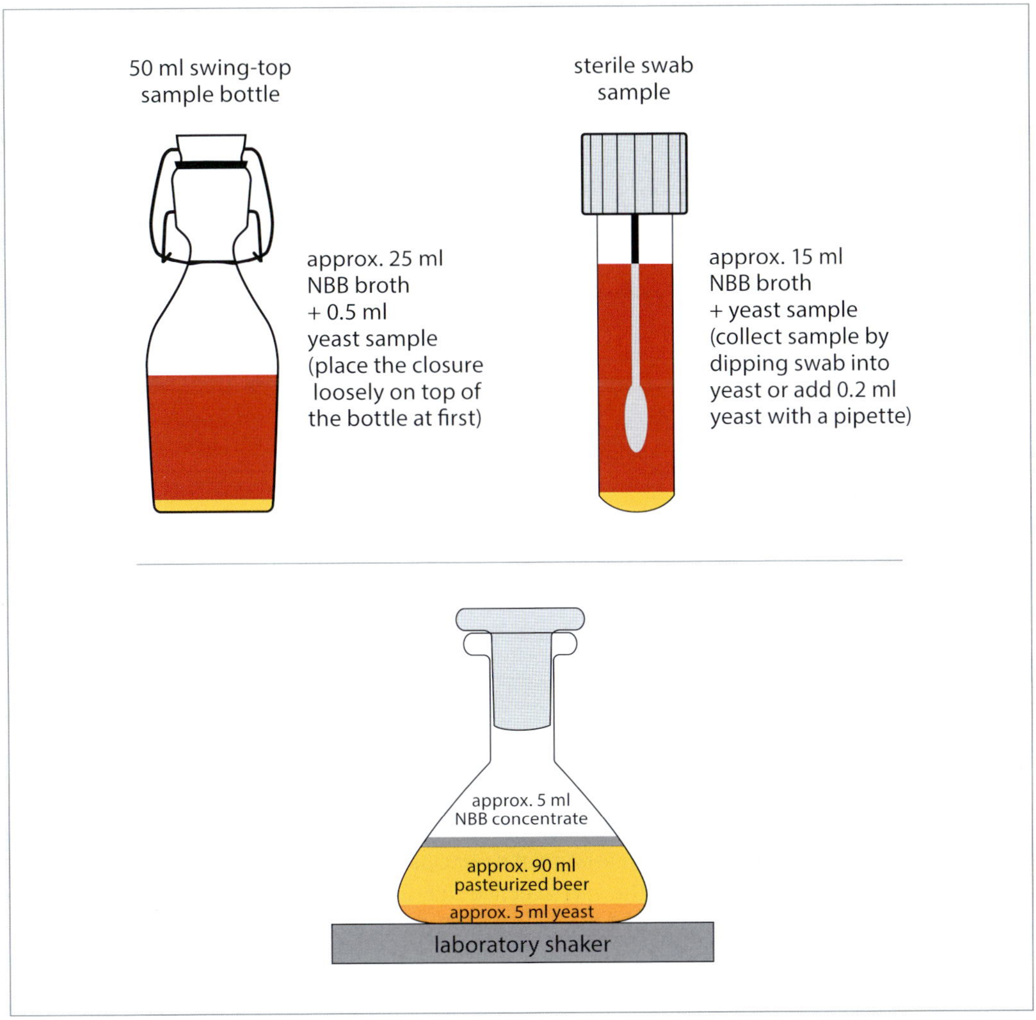

NBB broth is ordinarily utilized for the detection of beer spoilers in yeast samples. To prepare a sample, approximately 0.1 ml of yeast is added to 10 ml of broth and incubated as described above, the only exception being that it is not necessary to provide anaerobic conditions (fig. 8a). Generally after 2–5 days of incubation, a yellow color change occurs if beer spoilers are present. At high yeast cell counts, the competitive pressure may slow the growth of any beer spoilers or a color change may not occur despite some growth. Sure detection can still be carried out with a second enrichment culture (shake the first culture to mix and transfer 1–2 drops to a fresh test tube containing NBB broth). When analyzing yeast samples, care should be taken that the sample volumes are not too large.

NBB concentrate is normally employed for samples of unfiltered beer or beer containing yeast. A beer sample is collected and 5 % NBB concentrate is added (e.g. fill a 180 ml swing-top sample bottle almost to the rim with the beer sample and add the NBB concentrate). The sample is then incubated for 7 days at 27 °C. The test is very selective for obligate beer spoilers. The samples should be evaluated with a microscope, since protein haze can occur on occasion and unlike NBB agar or NBB broth media, the NBB concentrate does not contain an indicator. Potential and indirect beer spoilers can be detected by adding water (10–50 %) to the sample (fig. 9).

Figure 9: Selective detection of beer spoilers according to risk category

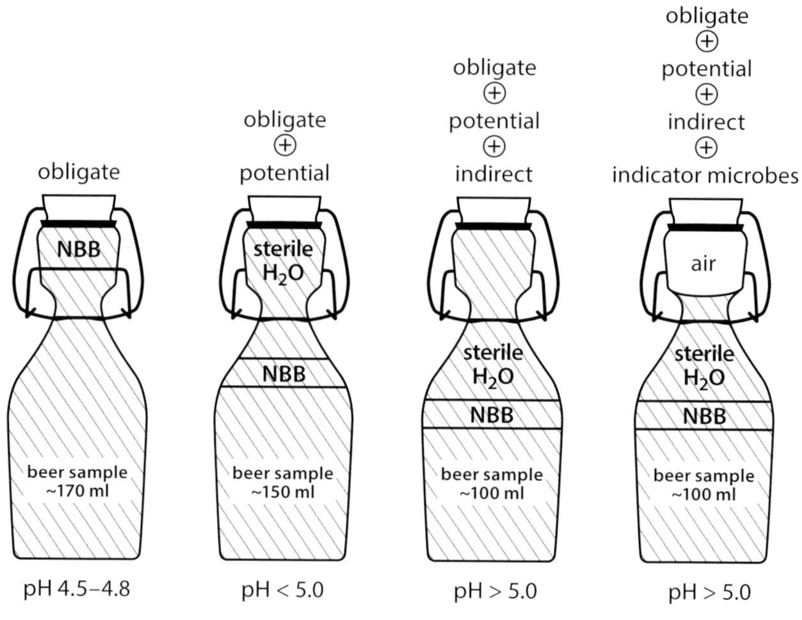

2.4.2 SWAB SAMPLES FROM BIOFILMS

Through the collection of samples with sterile swabs, weak points or areas prone to microbial contamination in production and filling areas are monitored. In addition to beer spoilers, indicator flora (especially acetic acid bacteria and wild yeasts) are the principal focus of detection in these areas, therefore aerobic incubation of samples in the less selective NBB-B-AM broth is recommended. The swab samples are incubated for a maximum of three days at 27 °C. Only tubes exhibiting a color change from red to yellow, indicating a positive test result, are evaluated. A positive test indicates the presence of a biofilm and thus a weak point, which could sooner or later be colonized by beer spoilers (fig. 2, p. 296).

Commonly encountered weak points in production areas include blind caps, dead-ends, portions of pipes in which water remains standing after drainage, CO_2 lines, pumps, by-passes and gaskets as well as carbonation and measurement devices (a swab test suitable for detecting the presence of biofilms found in such areas was developed by Back [3, p. 285, pp. 291–292, 5]).

The swabbing method for filling areas, as described above, is based on detection with NBB-B-AM broth. Over many years, this analysis method has become established around the world in the beverage industry. The method is used at breweries as well as at filling and packaging facilities for mineral water or other non-alcoholic beverages. Not only is the analysis uncomplicated in its implementation, the results also exhibit a high level of validity (fig. 10). Swab samples are taken from surfaces that come into direct and indirect contact with beer (e.g. starwheels, valves, lifting elements, threaded connections, vertical supports, empty spaces). It is recommended that approximately 20–30 weak points should be tested to provide a statistically meaningful sample set. Niches located on the filler and capper are particularly dangerous, since they are in constant contact with beer and provide ideal conditions for beer spoilers as well as for the formation of biofilms. Given the right conditions, persistent slime deposits can form even near the bottle washer (bottle outlet), bottle inspector and conveyor belts. These deposits, known as biofilms, subsequently become colonized with beer spoilers [5].

The practicality and expediency of this method are also attributable to the fact that through regular sample collection, weak points can be discovered by detecting indicator flora before beer spoilers have a chance to adapt and become established. For this reason, swab samples should be collected during the regular filling schedule (e.g. on Tuesday for evaluation on Friday) on a weekly basis if possible, so that the actual number and kinds of microbes present can be determined. The average number of positive samples (yellow tubes) over a period of several weeks should be below 30 % (safety limit), or preferably less than 20 %. From time to time, the effectiveness of the cleaning process should also be tested by collecting samples and evaluating them to determine the hygiene level of the equipment. After the filling area has been cleaned, the number of swab samples returning positive results (yellow) should be less than 10 %.

Inspection of the swab samples with a microscope is not absolutely necessary, since with NBB-B-AM broth, a color change to yellow within three days ("traffic light signal") can be interpreted as a relevant biofilm (fig. 11). The exact microbes involved in the formation of the deposit are of secondary importance at this point in the process. Hopefully, only indicator flora and no beer or beverage spoilers will be detected at locations that come into direct contact with beer.

The test results from the swab samples collected weekly can be expressed clearly as curves in a graph. In this form, it is immediately evident when the number of positive results is increasing, when the number is approaching the safety limit at 30 %, or when the limit has been surpassed. This is an indicator for persistent deposits and that a risk of secondary contamination exists. Individual outliers are not a problem if subsequent test results revert to below the 30 % mark (fig. 12, p. 309). A graphical depiction also brings with it the added benefit that the impact of negative events on hygiene conditions (e.g. construction work) – but also the effectiveness of cleaning measures – is immediately apparent. This affords insight into the level of hygiene in the filling area at all times. Furthermore, it allows the necessary countermeasures to be initiated in a timely manner.

Figure 10: Evaluation of swab samples

<u>Sample volume</u>
20–30 swab samples

<u>summer</u>
1–2 times per week

<u>winter</u>
once per week

<u>Sample collection points</u>
at weak points
(areas susceptible to contamination)

bottle washer – bottle outlet

bottle inspector

filler

capper

indirect contact:
area near filler/capper

keg filler

<u>Incubation:</u>
3 days in NBB-AM broth at 25–28 °C

<u>Evaluation of results:</u>
Samples collected after cleaning: max. 10 % of samples positive (yellow)

Samples collected during production: max. 20–30 % of samples positive (yellow)

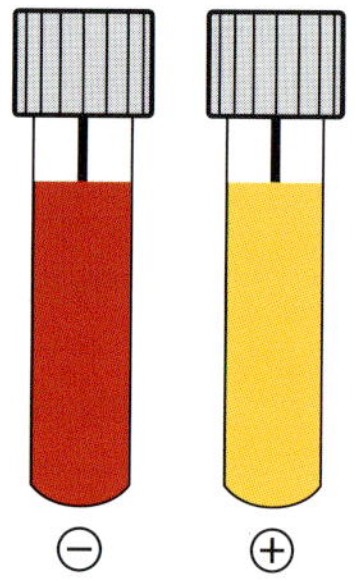

Figure 11: Levels of microbial contamination in the bottling area depicted on the basis of analysis results from swab samples

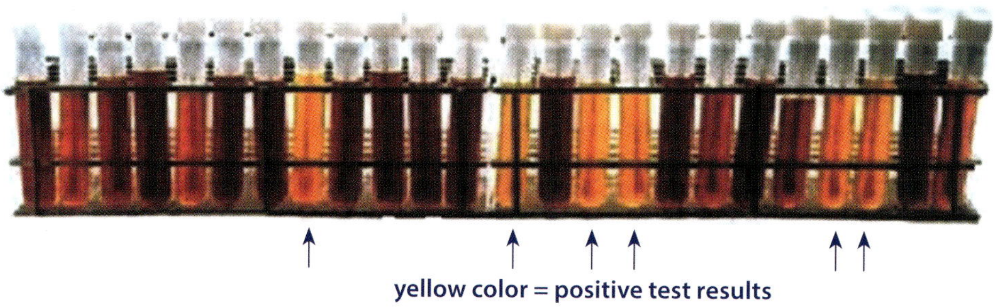

3 days of aerobic incubation in NBB-AM broth at 25–28 °C
(6 of 24 samples tested positive = 25 %)

Figure 12: Graphing positive analysis results for microbes detected in the bottling area

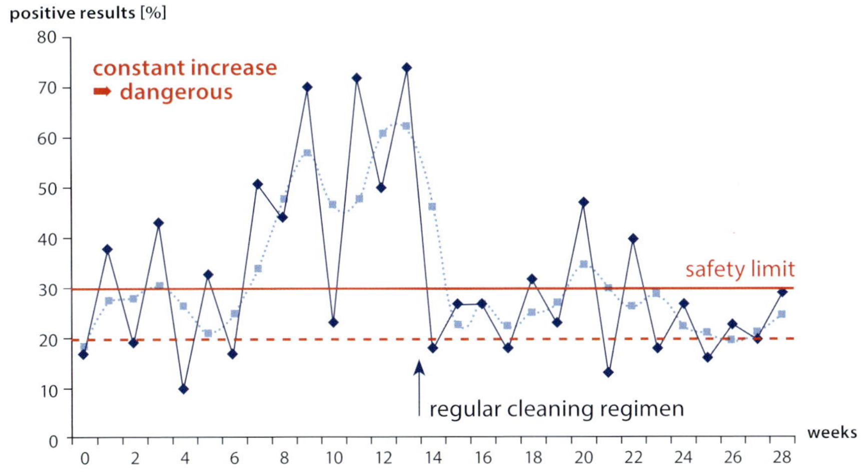

2.4.3 RAPID METHODS BASED ON MOLECULAR BIOLOGY

For certain samples, the rapid detection of contaminants and their identification is essential. A number of molecular biology analysis methods are available for this purpose, e.g. VIT® gene sensor technology with subsequent fluorescence microscopy [27] as well as PCR, Multiplex PCR and real-time PCR (Biotecon, Pall etc.).

The former have become well-established in laboratories to augment traditional cultivation methods (fig. 7). After the thermal digestion of the DNA sample, DNA sequences characteristic of the microorganism are localized by specific primers and amplified using polymerase. This process is repeated multiple times until a sufficient quantity of DNA material has been generated for analysis. The analysis is typically carried out by means of electrophoresis. In real-time PCR, the amplified DNA obtained during the PCR cycle is quantified through a fluorescens measurement in real time. The result can be evaluated by means of a melting curve.

Since it is trace contamination that must be detected in brewing microbiology, it is necessary to carry out a liquid enrichment of the sample over a two-day period at least. For trace contamination in yeast or in green beer samples, incubation of the sample for up to five days may be required. This liquid nutrient medium (e.g. NBB PCR broth) must be free of any DNA from beer-spoiling bacteria and should guarantee the most rapid growth possible of any microbial contaminants present in the sample.

Due to its quick and easy handling to identify the enriched beer spoilers (liquid culture or colonies on NBB), the Milenia PCR rapid test in lateral flow format (test strips) for beer-spoiling lactic acid bacteria has proven to be quite practical for use in brewery laboratories (supplier: DMD® Screen & Ident Kit, Döhler, 64295 Darmstadt, Germany). A positive hop-resistance strip test indicates bacteria with beer spoiling potential [28, 29].

Figure 13: DNA fingerprint of two *Lactobacillus brevis* strains

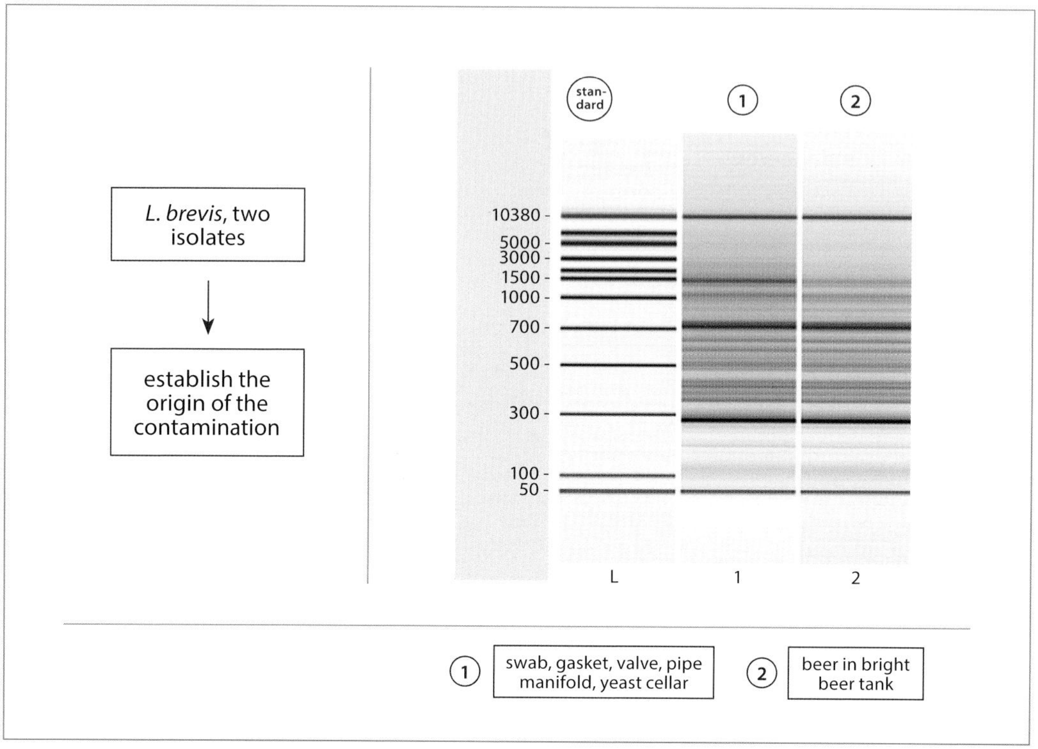

A contaminant can be identified on the basis of identical DNA fingerprint patterns for two *Lactobacillus brevis* strains found within the same production environment (*L. brevis* strain 2 from the bright beer tank is the same as *L. brevis* strain 1 from yeast propagation → the probable source of the contamination is the yeast cellar) [30].

2.5 DETECTION AND IDENTIFICATION OF YEAST STRAINS

The detection and cultivation of yeast is usually carried out by incubating samples on wort agar at temperatures ranging from 25 to 28 °C under aerobic conditions. To distinguish yeast strains capable of fermentation from respirative yeast strains, parallel tests are performed through the incubation of samples under aerobic and anaerobic conditions.

The detection of brewing yeast downstream from the filter is an important step when evaluating the effectiveness of the filtration process. Moreover, in some cases, it is necessary to test for the presence of foreign yeast in the pure brewing yeast culture (yeast from propagation, cropped yeast, yeast sediment in tanks). Here, different growth temperatures can be utilized to distinguish bottom-fermenting yeast strains (approx. 32 °C) from the majority of foreign yeasts (37–45 °C). This allows foreign yeasts incubated at a temperature of 37 °C to be selectively enriched from a sample of bottom-fermenting yeast. A multi-step enrichment process is recommended for greater certainty when testing for trace contaminations (fig. 14). After enrichment, a differentiation is made between the foreign strains of both *Saccharomyces* and non-*Saccharomyces* yeasts, top-fermenting brewing yeasts and super attenuating yeasts, by incubating the cultures for 1–2 days on crystal violet agar and lysine agar or on copper sulfate agar (fig. 15).

Foreign yeast detection in samples of top-fermenting brewing yeast (fig. 16) can be accomplished with pantothenate nutrient agar prepared without pantothenic acid. Since pantothenic acid is essential for top-fermenting yeast, a bottom-fermenting yeast contaminant can grow on this medium with no competition. The test media described above are described in detail in the book "Farbatlas und Handbuch der Getränkebiologie", Teil II [4, p. 183].

Figure 14: Detection of a trace contamination with foreign yeast in a sample of bottom-fermenting brewing yeast

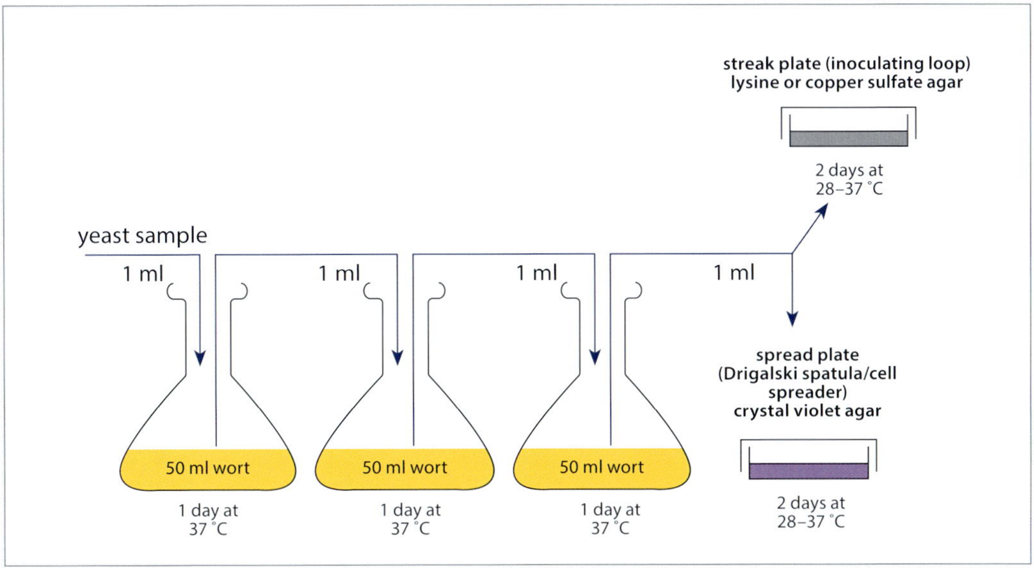

Figure 15: Detection and differentiation of a foreign yeast strain in bottom-fermenting yeast cultures

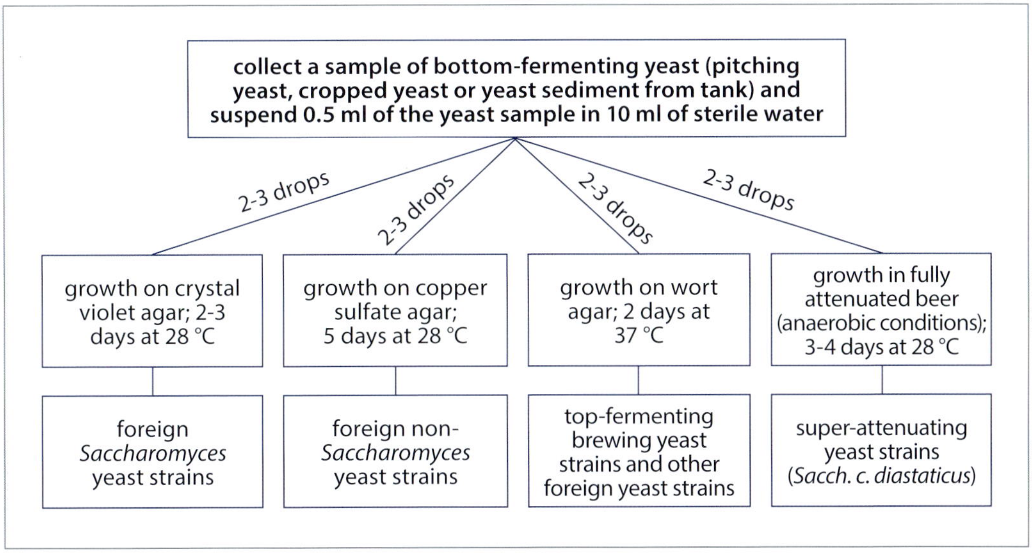

Figure 16: Detection and differentiation of a foreign yeast strain in a top-fermenting yeast culture

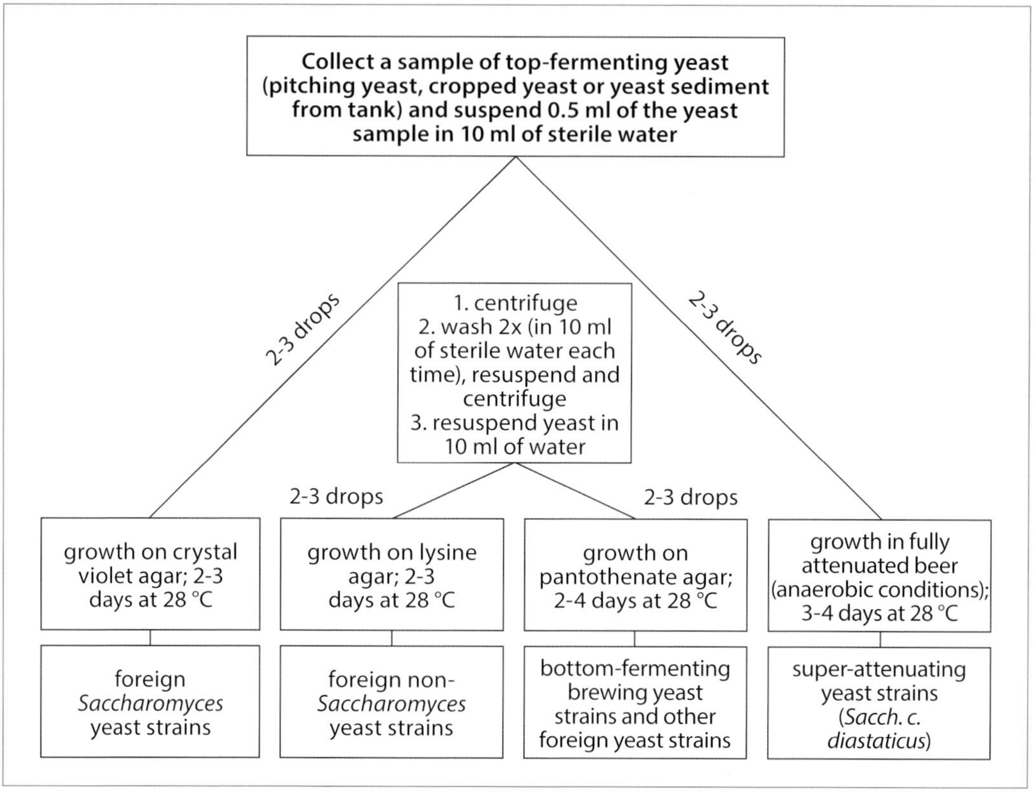

3 SUMMARY

In light of the highly selective conditions present in beer, only lactobacilli and pediococci as well as *Pectinatus* and *Megasphaera* species are of significance as spoilage organisms. Pathogens and heat-resistant spore-forming strains cannot multiply in beer. From a quality control perspective, classifying microorganisms as obligate, potential or indirect beer spoilers, as well as determining the type of contaminant, either primary or secondary, is a helpful tool. Additionally, slime-forming acetic acid bacteria are important due to their role as indicators for beer spoilers, since they frequently occur in communities with other slime-forming microbes. Together, they are primarily responsible for creating biofilms on surfaces that have direct and indirect contact with beer and thus play a key role in the identification and elimination of weak points in the production process.

The prerequisite for a successful microbiological quality control program is recognizing and understanding the relevant microorganisms in the brewery. Therefore, a detailed description of the key characteristics of pure brewing cultures, foreign yeasts and beer spoilers is provided above.

Finally, because the NBB methods are quite practical, they are recommended for the detection of beer-spoiling bacteria. Established methods and the most common differential media are also described for the detection of foreign yeast strains.

Figure 17: Pasteurization units required to kill beer-spoiling microorganisms [3]

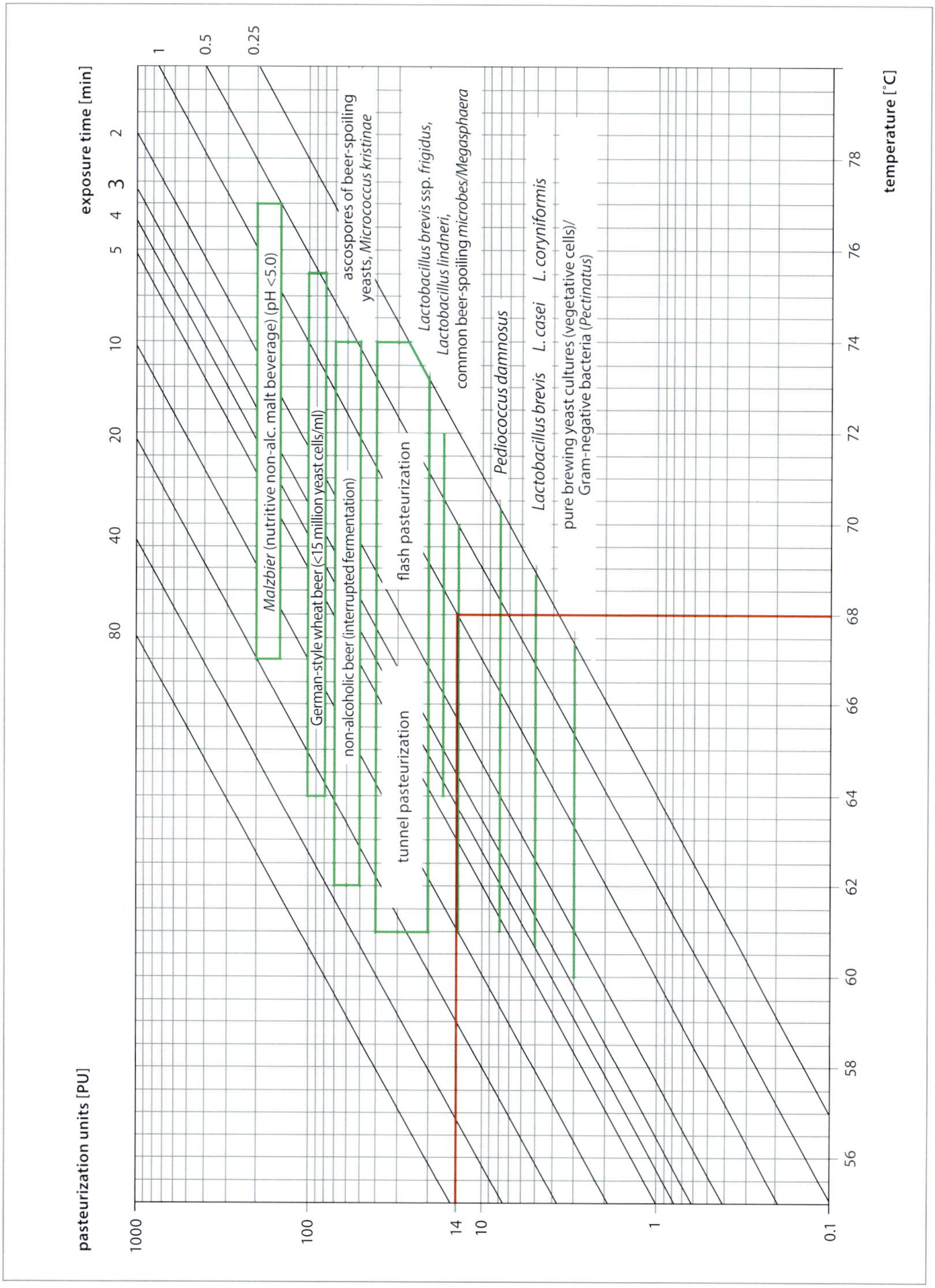

Example: hold at 68 °C for 1 minute = 14 PU (kills approx. 95 % of beer spoilers)
Flash pasteurization: minimum temperature of 66.4 °C ; minimum exposure time of 15 sec.
Tunnel pasteurization: minimum temperature of 61 °C; minimum exposure time of 4.5 min.

$PU = t \times 1.393^{(T-60)}$ [t = time in minutes; T = temperature in °C]

4 OVERVIEW

Attribute/ problem	Possible cause	Corrective measures
diacetyl in beer	poor yeast vitality; the yeast is too old; the yeast was used too many times; the yeast was stored at high temperatures or for too long	pitch only young, vital yeast suspensions
	contamination with *Pediococcus*; contamination carried over from the filter	replace yeast culture and carry out step-wise quality control and analysis in the unfiltrate area
	contamination with *Lactobacillus* species (e.g. *L. brevis*, *L. backii*, *L. casei*); secondary contamination at the filler or through the yeast added for bottle-conditioning (weissbier)	thoroughly clean the filler and capping equipment and surroundings; replace the yeast added for bottle conditioning
diacetyl in beer	Gram-negative members of the family Enterobacteriaceae are present in the brewing yeast; the yeast was stored at high temperatures or too long; the yeast was aerated as it was harvested	replace yeast culture; store yeast at low temperatures (yeast chiller); aerate prior to pitching
slow degradation of diacetyl	the yeast is too old; poor yeast vitality, the yeast was used too many times; the yeast was stored at high temperatures or for too long	use only young, vital yeast; store cropped yeast at low temperatures (0–3 °C)
DMS in beer	if no mistakes were made during brewing, then DMS is usually due to a wort infection: usually through contamination of the yeast with members of the family Enterobacteriaceae (more likely if the wort is exposed to air); the wort filter and flotation tank are weak points especially susceptible to contamination	check wort transfer lines, wort aeration equipment and other devices installed in the line; clean the dome of the tank and the sightglass for tank volume
harsh bitter aftertaste; empty-tasting beers	contamination with refermenting wild yeast strains (e.g. *S. c. diastaticus*)	eliminate weak points at the filler and also downstream from the filler if necessary
poor fermentation process in general	zinc deficiency	measure the zinc content of the wort (> 0.15 mg/l) and if necessary, take measures to increase the concentration (natural mash acidification)
the desired E_A is not achieved	the yeast is old or stressed; yeast vitality is lacking (ICP < 6; cf. Yeast Technology and Fermentation)	improve yeast management (methylene blue test: percentage of dead cells in the pitching yeast should be < 4 %)
fermentation progresses rapidly in the initial phases but later declines significantly	diauxie: a deficiency of zinc, amino acids, biotin and readily utilizable sugars (glucose/maltose ratio)	measure the concentration of zinc (reference value: > 0.15 mg/l in pitching wort) increase FAN (e.g. mash rest at 48 °C)

Attribute/ problem	Possible cause	Corrective measures
poor yeast harvest, frequently in conjunction with a lengthier fermentation	poor yeast vitality; bottom-fermenting yeast remains longer in suspension due to a slower fermentation and less CO_2 production	improve yeast management practices (cf. Yeast Technology and Fermentation)
poor filterability	low quality yeast; autolyzed cell; pump or centrifuge (shearing, cavitation) is unsuitable for yeast	check yeast vitality; examine the recirculation procedure used in yeast assimilation/propagation (aeration downstream from the pump); check pump size (too large or too small, incorrectly dimensioned)
low flavor stability, generally accompanied by poor foam stability; a harsh, lingering protein bitterness; perceptible sulfur flavor in fresh beer when it first enters the mouth	poor yeast quality; yeast cells are undergoing autolysis, releasing proteins and medium chain fatty acids as well as proteinases	improve yeast quality; the number of dead yeast cells in methylene blue test should be < 3 %
no detection of *Pectinatus* or *Megasphaera* despite contamination of filled bottles	difficult to detect because of trace contamination; microbes must be cultivated under strictly anaerobic conditions (with cultures obtained by means of membrane filtration, air was still present in the membrane); the sample was too old: *Pectinatus* and *Megasphaera* die rapidly after they start growing	detection by means of weak point analysis conducted in the area surrounding the filler (swab samples in NBB-AM broth); *Pectinatus* and *Megasphaera* also grow in the presence of acetic acid bacteria cultivated under aerobic conditions (the acetic acid bacteria make the medium anaerobic); flood the membrane filter with CO_2 or incubate in NBB broth under anaerobic conditions
beer-spoiling strains obtained from a culture collection exhibit no growth in beer-specific test media (NBB)	strains obtained from a culture collection frequently lose their resistance to hops (retention only through regular contact with beer)	only use original isolates from beer to validate the effectiveness of detection media; maintain the cultures in beer
certain lactic acid bacteria exhibit no growth in NBB	the strains in question are hop-sensitive, therefore they are not beer spoilers; occasionally found in beer, these latent bacteria do not grow in nutrient media specific to beer; some lactobacilli are somewhat hop-sensitive (e.g. *L. casei*, *L. coryniformis*, and *L. plantarum* can only tolerate a maximum of 15 EBC bittering units)	at most, these are potential beer spoilers, although they do not exhibit growth in beer; a relevant analysis of weak points is recommended

Attribute/ problem	Possible cause	Corrective measures
no detection in routine sample collection despite customer complaints; contamination spread throughout the production process	statistical problem due to the low ratio of sample to beer volume; trace contamination particularly after filtration; the number of cells is reduced per unit volume, but higher density of microbes after pressure surges; unfavorable fluid dynamics affecting flow; secondary contamination in the filling area	BM system [18] or sample collection downstream from a bend in a pipe (elbow) for reasons associated with fluid dynamics (turbulent flow), continuous sampling, weak point analysis using swab samples in NBB-AM broth
no detection of pediococci or lactobacilli in yeast samples	trace contamination; often the yeast sample is too large for incubation in NBB broth, resulting in inhibition of the yeast cells and selective pressure on isolated beer spoilers (inhibition of beer spoilers)	two-stage enrichment in NBB broth: after 2–4 days, 1–5 drops of sediment from the first enrichment culture are transferred to the fresh NBB broth and incubated for 3–6 days on a laboratory shaker (fig. 8b, p. 306)
certain lactic acid bacteria exhibit slow growth	trace contamination; cells in the lag phase; damaged cells (e.g. as a result of washing yeast with acid); at the most, these bacteria are only potential or indirect beer spoilers; *L. lindneri*, *L. brevis* var. *brevisimilis* and some pediococci generally grow more slowly than their counterparts	certain strains require special nutrients for growth; for some microbes, the pH of NBB is too high: mix the NBB with beer to lower the pH to below 5.5
	overheating nutrient media or autoclaving the media multiple times: the media turns brown and shows only a slight color change for positive results	heat nutrient media gently and for the shortest time possible (e.g. 5 min at 115 °C)
no color change in NBB broth despite microbial growth	the nutrient medium has been damaged by exposure to heat; growth of Gram-negative rods (mainly members of the family Enterobacteriaceae); the yeast sample is too large for use with NBB broth	heat the nutrient medium gently; no obligate beer spoilers, but possibly indirect beer spoilers. indication of weak points; improvement can be achieved with good yeast quality and thorough cleaning; inoculate each 10 ml of NBB broth with a maximum of 0.2 ml of yeast sample or with a sterile swab covered in yeast
no color change in yeast samples in NBB broth although pediococci were detected	the yeast sample is too large; trace contamination of the yeast sample with pediococci: the yeast begins to autolyze and the pH rises again	two-stage enrichment in NBB broth; microscopy results are usually conclusive due to the large numbers of cells present
sediment in NBB broth without microbial growth (especially at low pH values)	protein precipitation, because NBB is a natural nutrient substrate with a high concentration of protein	microscopic examination

Attribute/ problem	Possible cause	Corrective measures
turbidity observed in a sample in NBB concentrate after adding beer (usually after a short incubation time)	the beer added has been stabilized to a high level: protein precipitation occurs as a result of the addition of nutrient medium	microscopic examination; if necessary, lower the addition of NBB concentrate from 5 % to 2.5 %
moderate color change in NBB broth for samples containing yeast	vital yeast (at high cell counts) continues to lower the pH before inhibition begins	reduce the size of the yeast sample; examine with a microscope: if individual rods are present, make an enrichment culture in two stages (2–5 drops in fresh NBB broth)
slight color change in NBB broth/NBB-AM broth, often in conjunction with gas formation; unpleasant odor	growth of members of the family Enterobacteriaceae (DMS, celery flavor may be accompanied by gas formation), micrococci (orange color) and wild yeast	although these microbes are not obligate beer spoilers, if they appear multiple times, an analysis of weak points is recommended
yellow color in NBB broth/ NBB-AM broth, usually with a low level of turbidity; no gas generation	growth of aerobic acetic acid bacteria, often occuring in chains or as short rods, some of which are motile	no beer spoilers present; however, these are indicator microbes which point to a weak area of hygiene; incubate under anaerobic conditions
dead rod-shaped bacterial cells found in yeast and unfiltered samples (approx. 1–5 per visible field)	dead lactic acid bacteria were carried over with the addition of *Sauergut* or naturally produced lactic acid	no response is actually necessary; if uncertain, add 1 drop of the sample to fresh NBB broth and incubate for 2 days: if any living cells are present, then growth and a color change will be obvious within 2 days
atypical growth of colonies on NBB agar; yellow color change is variable	possible presence of potential beer spoilers or indicator microbes physiological tests: oxidase, catalase, Gram staining	if this occurs regularly, then perform a step-wise quality control analysis
formation of gas and a solvent-like flavor in the *Sauergut* added for natural acidification	*Candida kefyr* and *C. magnoliae* form CO_2 and ethyl acetate (adhesive odor)	increase the temperature of the *Sauergut* medium to 50 °C for multiple cycles; flood with CO_2 gas

5 REFERENCES

[1] Back, W.: Bierschädliche Bakterien. Nachweis und Kultivierung bierschädlicher Bakterien im Betriebslabor. Brauwelt, 120 (1980), pp. 1562–1569.

[2] Back, W.: Farbatlas und Handbuch der Getränkebiologie, Teil I. Fachverlag Hans Carl, Nürnberg 1994.

[3] Back, W.: Colour Atlas and Handbook of Beverage Biology, Fachverlag Hans Carl, Nürnberg 2005, pp. 285–293, 300–304.

[4] Back, W.: Farbatlas und Handbuch der Getränkebiologie, Teil II. Fachverlag Hans Carl, Nürnberg 2000, pp. 225–229.

[5] Back, W.: Sekundärkontaminationen im Abfüllbereich. Brauwelt, 134 (1994), no. 19, pp. 686–695. Secondary contaminations in the filling area. Brauwelt International, 12 (1994), no. 4, pp. 326–333.

[6] Back, W. (ed.): Mikrobiologie der Lebensmittel. Getränke. Behr's Verlag, Hamburg 2008. p. 113 ff.

[7] Bohak, I., Back, W., Richter, L., Ehrmann, M., Ludwig, W., Schleifer, K. H.: *Lactobacillus amylolyticus* sp. nov., Isolated from Beer Malt and Beer Wort. System. Appl. Microbiol. (1998), no. 21, pp. 360–364.

[8] Samson, R. A., Hoekstra, E. S., Frisvad, J. C., Filtenborg, O.: Introduction to food-borne fungi. Fourth edition. Central Bureau voor Schimmelkultures Baarn, 1995.

[9] Geissinger, C., Hofer, K., Habler K., Heß, M., Hückelhoven, R., Rychlik, M., Becker, T., Gastl, M.: *Fusarium* species on Barley Malt: Is Visual Assessment an Appropriate Tool for Detection? Cereal Chemistry Journal, 94 (2017), no. 4, pp. 659–669. 10.1094/CCHEM-08-16-0212-R.

[10] Habler K., Geissinger, C., Hofer K., Schüler, J., Moghari, S., Hess, M., Gastl M., Rychlik, M.: Fate of *Fusarium* Toxins during Brewing.J. Agric Food Chem., 65 (2017), no. 1, pp. 190-198. DOI 10.1021/acs.jafc.6b04182.

[11] Kreger van Rij, N. J. W.: The yeasts, a taxonomic study. Third revised and enlarged edition. Elsevier Science Publishers B.V., Amsterdam 1984.

[12] Barnett, J. A., Pankhurst, R.W., Yarrow, D.: Yeasts: Characteristics and Identification. Second edition. Cambridge University Press. Cambridge, New York, Port Chester, Melbourne, Sydney, 1990.

[13] Bohak, I., Thelen, K., Beimfohr, C.: Description of *Lactobacillus backii* sp. nov., an Obligate Beer-Spoiling Bacterium, Monatsschrift für Brauwissenschaft, 60 (2006), no. 3/4, pp. 78–82.

[14] Tohno, M., Kitahara, M., Irisawa, T., Masuda, T., Uegaki, R. Ohkuma, M., Tajima, K.: Description of *Lacobacillus iwatensis* sp. nov., isolated from orchardgrass (*Dactylis glomerata L.*) silage, and *Lactobacillus backii* sp. nov. Int. J. Syst. Evol. Microbiol., 63 (2013), pp. 3854–3860. DOI: 10.1099/ijs.0.051920-0.

[15] Sneath, P. H. A., Mair, N. S., Sharpe, M. E., Holt, J. G. (ed.): Bergey's Manual of Systematic Bacteriology. Volume 2. Williams & Wilkins Baltimore, Hong Kong, London, München, Philadelphia, San Francisco, Sydney, Tokyo, 1986.

[16] Hutzler, M., Koob, J., Grammer, M., Riedl, R., Jacob, F.: Statistische Auswertung der PCR-Analysen bierschädlicher Bakterien in den Jahren 2010 und 2011. Brauwelt (2012) no. 18/19, pp. 546-547.

[17] Hutzler, M., Müller-Auffermann, K, Koob, J., Riedl, R, Jacob, F.: Beer spoiling microorganisms – a current overview. Brauwelt International (2013), no. 1, pp. 23–25.

[18] Koob, J., Jacob, F., Grammer, M., Kleucker, A., Riedl, R., Hutzler, M.: PCR-Analysen bierschädlicher Bakterien 2012 und 2013. Brauwelt (2014), no. 10, pp. 288-290.

[19] Suzuki, K.: Grampositive spoilage bacteria in brewing. In: Hill, E.A. Brewing Microbiology (ed. Hill, E.A,) Woodhead Publishing, London 2015.

[20] Koob, J.; Jacob, F., Wenning, M. Hutzler, M.: *Lactobacillus cerevisiae* sp. nov., isolated from a spoiled brewery sample. Int. J. Syst. Evol. Microbiol. 67 (2017), pp. 3452–3457.

[21] Entani, E., Masai,H. Suzuki, K.: *Lactobacillus acetotolerans*, a New Species from Fermented Vinegar Broth. Int. J. Syst. Bacteriol., 36 (1986), no. 4, pp. 544–549

[22] Back, W., Bohak, I., Ehrmann, M., Ludwig, W., Pot, B., Kersters, K., Schleifer, K.H.: *Lactobacillus perolens* sp. nov., a Soft Drink Spoilage Bacteria. System. Appl. Microbiol, 22 (1999), no. 3, pp. 354–359.

[23] Yang Deng, Junyan Liu, Huiping Li, Lin Li, Jingxia Tu, Huijing Fang, Jiang Chen, Fei Qian: An improved plate culture procedure for the rapid detection of beer-spoilage lactic acid bacteria. J. Inst. Brew. 120 (2014); no. 2, pp. 127–132.

[24] Back, W., and Stackebrandt, E.: DNS/DNS-Homologiestudien innerhalb der Gattung *Pediococcus*. Arch. Microbiol., 118 (1978), no. 1, pp. 79–85.

[25] Dobson, C. M., Denner, H., Lee, S., Hemmingsen, S., Glaze, S., Ziola, B.: Phylogenetic analysis of the genus *Pediococcus*, including *Pediococcus claussenii* sp. nov., a novel lactic acid bacterium isolated from beer. International Journal of Systematic and Evolutionary Microbiology, 52 (2002), pp. 2003–2010. DOI: 10.1099/ijs.0.02191-0.

[26] Back, W., Pöschl, P.: Bypass-Membranfiltration (BM-System) – Verbesserung des Spuren- nachweises nach der Filtration. Brauwelt, 138 (1998), no. 46/47, pp. 2312–2315 and Brauwelt International, 17 (1999), no. 3, pp. 202–204.

[27] Thelen, K., Beimfohr, C., Bohak, I., Back, W.: Specific rapid detection test for beer-spoilage bacteria using fluorescently labelled gene probes. Brauwelt International, 20 (2002), no. 3, pp. 156–159.

[28] Breitbach, A., Jacob, F., Hutzler, M., Koob, J.: Einfacher Schnelltest zum Nachweis bierschädigender Organismen. Brauwelt, 155 (2015), no. 33, pp. 961-966.

[29] Koob, J., Breitbach, A., Hutzler, M., Jacob, F.: Molekularbiologischer Nachweis bierschädigender Organismen im Format eines Schwangerschaftstest. Der Weihenstephaner, 83 (2015), no. 4, pp. 155–159.

[30] Riedl, R., Dünzer, N., Michel, M., Jacob, F., Hutzler, M.: Beer enemy number one: genetic diversity, physiology and biofilm formation of *Lactobacillus brevis*. J. Inst: Brew. (2019), DOI 10.1002/jib. 553.

BREWING WITH MALTED AND UNMALTED CEREALS AND PSEUDOCEREALS

1 INTRODUCTION

For approximately 11,000 years, humans have cultivated circa 3 % of higher plant species (over 223,000). However, roughly only 30 species of these species provide 90 % of all plant-based food. Merely two species of plants are commonly utilized in malted form and three species in unmalted form in modern beer production around the world. The use of carbohydrate-rich grain serves as the primary source of extract, color and aroma in beer. Barley – principally malted barley – is of expceptional significance in this regard. No other malted grain is so well suited for brewing as *Hordeum vulgare*, especially in conjunction with hops and bottom-fermenting yeast. Thousands of years ago, various peoples came to understand this, as evidenced by archaeobotanical discoveries from a number of different countries. Early writings from Egypt and Mesopotamia also attest to beer production with *Hordeum vulgare*. They also indicate that it was specifically chosen for this purpose due to its merits as a brewing grain.

A prominent date in global brewing history is April 23rd 1516, when the original version of the Bavarian Purity Law or *Reinheitsgebot* was issued. The date is important insofar as it is a crucial step in the ascendancy of bottom-fermented lager beers. According to the *Reinheitsgebot*, only yeast, hops, water and germinated grain (malt) may be employed in beer production. Malt best suits this purpose, because it is through the malt's own enzymes that the compounds in the grain become soluble and, in turn, allow alcoholic fermentation to take place. Extract and thus alcohol can be generated with unmalted grain by means of thermal digestion but to a lesser extent than with malt [3].

On one hand, a consumer protection law governing food was issued, whose validity has remained unbroken since that time. On the other, everyone involved in beer production is faced with the task of striking the most favorable balance among these four ingredients – from plant breeders to farmers, maltsters and the brewers themselves to equipment manufacturers, scientists and those in trade and commerce. By doing so, these findings benefit all beer producers around the world.

The *Reinheitsgebot* serves as the basis for the modern *Vorläufiges Biergesetz* (provisional beer law). This modern law obliges all brewers producing bottom-fermented beer to exclusively employ barley malt in their grist. Top-fermented beer may be brewed with malt derived from a number of other cereal grains (§ 9 (4)) [4]. According to § 17 (4) of the provision for the implementation of the *Vorläufiges Biergesetz*, rice, maize (corn) and *Sorghum bicolor* (great millet) are not classified as cereals as defined in § 9 (3). In a commentary by Gerstenberg, sorghum is also referred to as milo [5]. Its common names also include durra, grain sorghum and sweet sorghum. *Sorghum bicolor* is grouped under the general term "sorghum" [6, 7].

The benefits of utilizing alternative starch/sugar sources, that is, those not permissible under the *Reinheitsgebot* (table 1), include their low price, their availability in countries where barley is not cultivated, particularly characteristics they may impart (such as flavor and coloring) when creating unique beer styles and which improve processing [8].

Table 1: Sources of extract permitted in selected countries for the purpose of brewing beer (as of 2002) [9]

Country	Adjunct	Maximum quantity permissible by law
Belgium Netherlands Luxembourg	adjuncts of wheat, maize (corn), rice and barley; sucrose, glucose, invert sugar, glucose syrup only permitted when used in combination with barley malt and/or wheat malt; sugar added as a sweetener after fermentation is not included in the quantity indicated	maximum 40 % by weight of the total weight of the raw materials (containing starch or sugar) being processed
Denmark	No regulations exist regarding the use of specific raw materials in beer production. The following are used: wheat, maize (corn), rice and barley adjuncts	
France	adjuncts of wheat, maize (corn), rice and barley; sucrose, glucose, invert sugar, glucose syrup only permitted when used in combination with barley and/or wheat malt; sugar added as a sweetener after fermentation is not included in the quantity indicated	maximum 40 % by weight of the total weight of the raw materials (containing starch or sugar) being processed
Great Britain	wheat, maize (corn), rice and barley adjuncts	no regulations regarding quantity exist
Italy	adjuncts of wheat, maize (corn), rice and barley; other raw materials containing starch as well as the addition of sugar and fruit juices are prohibited	maximum 25 % by weight of the total weight of the grain
Spain	adjuncts of wheat, maize (corn), rice and barley; sucrose, glucose, invert sugar, glucose syrup	maximum 30 % (by weight) of the total weight of the raw materials (containing starch or sugar) being processed
Portugal	none permitted (the following are explicitly prohibited: non-barley malts, glucose and invert sugar)	
Greece	none permitted	
Norway	none permitted	
Switzerland	adjuncts of wheat, maize (corn), rice and barley; sugar, invert sugar, dextrose, glucose syrup, starch	10 % 20 %
Austria	adjuncts of rice and maize (corn) or any products made with them	25 %
Japan	adjuncts of broken rice and maize (corn); starch, sugar, syrups	approx. 30 %
USA	unmalted cereals sugars and syrups	34 % 2.5 %

2 CARBOHYDRATE-RICH GRAINS

The following chapter provides insight into the great diversity present in cultivated, carbohydrate-rich caryopses (the dry, one-seeded fruit of cereal grasses), which are already or could reasonably be utilized in beer production – from a technological standpoint. These cereals are taxonomically described in more detail in figure 1. The subdivision into cereals (*Liliopsida* = monocotyledonous angiosperms) and pseudocereals (*Magnoliopsida* = dicotyledonous angiosperms) occurs at the level of the taxonomic class.

Figure 1: The taxonomy of carbohydrate-rich grains [10]

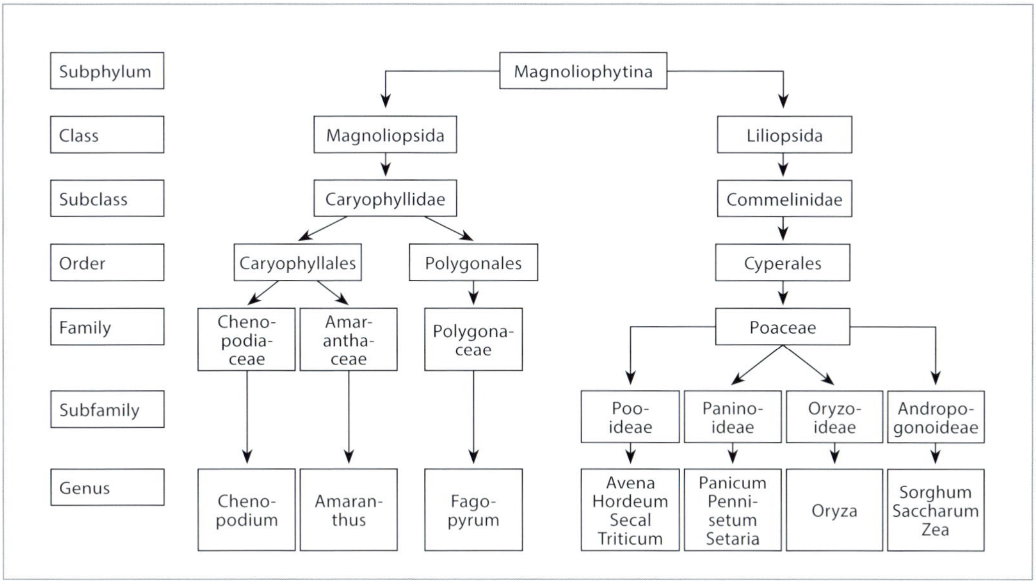

In principle, these alternative grains must meet the same requirements expected of barley and barley malt (cf. Malt). Despite the fact that barley has been selected and bred specifically as a brewing grain and these alternative grains have not, they must nevertheless be readily available at a reasonable price. Furthermore, they must be of a sufficient quality that they can, if necessary, be malted without involving extra time and effort spent in altering the requisite equipment to do so. The criteria for processing alternative grains as well as the key indicators for yield and quality must correspond to those in the malting and brewing processes.

One significant difference in the demands on raw materials used in brewing beer compared to those for the grains employed in bread and feed production is the raw protein content, which should fall between 9 and 12 % [11]. Ultimately, the quality of the beverages produced with these alternative raw materials is critical for consumer acceptance. These demands are largely satisfied by the grains described below, particularly as they also exhibit other unique attributes.

Table 2: **Analysis data for malt produced according to a standard process for cereals rich in carbohydrates. Values depicted in bold and in color are described in detail in the respective sections (analysis methods from the MEBAK Volume: Raw Materials – Barley, Adjuncts, Malt, Hops and Hop Products)**

		Attribute						
		barley	oats	proso millet	blue maize (corn)	black rice	rye	sorghum
Analysis method	**Unit**							
extract, moisture-free MEBAK 3.1.4.2.2	%	82.0	64.4	63.6	60.7	87.4	89.2	73.9
saccharification rate MEBAK 3.1.4.2.4	min	< 10	< 10	< 25	none	none	< 10	none
color MEBAK 3.1.4.2.8.2	EBC	2.9	3.7	2.1	3.9	7.6	7.7	7.5
nitrogen content, moisture-free MEBAK 3.1.4.5.1.1	%	10.5	12.6	13.6	14.6	8.4	10.4	7.8
viscosity MEBAK 3.1.4.4.1	mPa × s., adj. to 8.6 %	1.453	1.511	1.404	1.407	1.842	6.467	1.959
soluble nitrogen MEBAK 3.1.4.5.2.1	mg/ 100 g malt, d.m.	682	681	604	481	283	964	426
Kolbach index MEBAK 3.1.4.5.3	%	40.7	33.8	27.8	20.6	21.1	57.9	34.1
free amino nitrogen MEBAK 3.1.4.5.5.1	mg/ 100 g malt, d.m.	140	145	75	119	58	121	119
diastatic power, moisture-free MEBAK 3.1.4.6	WK units	311	269	78	72	82	177	83
α-amylase, moisture-free MEBAK 3.1.4.7	ASBC	55	24	11	7	6	18	8
limit of attenuation, apparent MEBAK 3.1.4.10.1.2	%	82.1	77.3	80.9	57.5	51.4	68.4	79.7

spelt	einkorn	emmer	kamut	triticale	tritor-deum	wheat	ama-ranth	buck-wheat	quinoa
84.3	85.1	88.9	74.8	88.0	81.4	86.2	79.7	52.9	83.2
< 10	< 10	10–15	10–15	10–15	10–15	10–15	none	none	none
3.1	3.5	2.8	2.0	6.1	4.4	3.2	5.6	2.5	5.0
13.6	14.3	12.3	15.9	10.6	15.1	12.1	15.2	15.4	13.7
2.034	2.920	1.704	1.732	2.219	1.587	1.756	1.969	3.507	1.520
654	756	843	797	840	898	725	1022	713	888
32.7	33.0	42.3	31.3	49.5	37.2	37.4	42.0	29.7	40.5
97	119	114	58	123	143	110	187	111	206
367	304	346	461	430	466	405	88	77	81
18	27	24	21	21	50	20	1	6	2
79.3	71.3	69.1	69.7	75.9	79.2	80.2	22.4	46.4	63.5

The malts in table 2 were produced according to a standard malting program (steeping: 5 h wet/14.5 °C, 19 h dry, 4 h wet/14.5 °C, 20 h dry, a moisture content of 45 % is reached through spraying; germination: 6 d/15 °C; kilning: 16 h/50 °C, 1 h/60 °C, 1 h/70 °C, 5 h/80 °C). For purposes of the analysis, the process was tailored to barley [9]. Wort produced from these malts was pitched with top-fermenting yeast (TUM 68 at 10 million cells per ml) and fermented at 20 °C until final attenuation was reached after circa one week. The yeast sediment was then removed, and 10 % *Speise* (wort fermenting with yeast in the log phase) was added to the beer for bottle conditioning. The beer was neither filtered nor stabilized in any way. Other information concerning the malting and brewing technology is described in the sections pertaining to the individual raw materials.

2.1 CEREALS

2.1.1 BARLEY (*Hordeum vulgare* L.)

Along with einkorn and emmer, barley has played an important role in the cultural history of Europe and Asia as one of the earliest cultivated cereals by humans. The oldest evidence for barley in prehistory has been found in the Peloponnesus and in the Euphrates River Valley in archaeological strata dating back to the Prepottery Neolithic (10,000 and 7000 BCE) [12].

Barley is one of the most cultivated cereals in the world. Basic differences distinguish the various types of barley: the number of rows (two or more), sowing and harvesting seasons as well as how they are utilized. In Europe, two-rowed barley is primarily utilized to produce malt. Barley thrives in a moderate climate on fertile, deep loamy soil with a good water supply. Barley exhibits a high degree of adaptability and is capable of being cultivated over a wide range of conditions, ranging from cold, wet areas with long daylight conditions (photoperiodism) to hot and dry steppe landscapes.

In beer brewing, unmalted barley serves as a source of extract in many countries outside of Germany. Due to the hardness of the kernels and the necessity for optimal extract yield, the barley must be finely pulverized using either a hammer mill or a dispersion mill. Adjunct mashes (common grist to water ratio: 1 : 4.5) are normally boiled in order to expedite solubilization [13]. Experimental trials conducted by the authors and new references in the literature [8] have shown that it is both economical (less energy usage and a shorter mash schedule) and practical from a quality standpoint to determine the gelatinization temperature prior to brewing [14]. Any type of available adjunct with a low gelatinization temperature is preferred, since significantly fewer enzymes from the malt or an exogenous source are needed. If the percentage of adjuncts in the grist amounts to less than 50 %, extract solubilization and protein degradation can be achieved solely by using the enzymes from the malt [15].

A typical mash schedule is depicted in figure 2. If the percentage of adjuncts in the grist is above 50 %, then exogenous amylolytic enzymes and a separate mashing procedure are obligatory. By increasing the quantity of unmalted barley in the grist, the β-glucan content rises linearly, while the viscosity climbs exponentially [16]. This necessitates further enzyme supplements (β-glucanases), otherwise complications arise with lautering and filtration processes.

Figure 2: A mash schedule for 50 % unmalted barley in the grist [8]

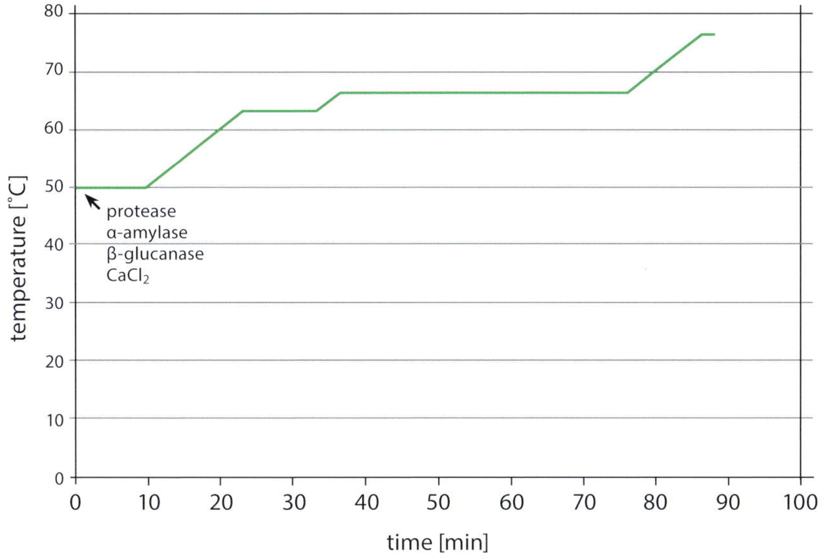

Unmalted barley can also be mixed into the grist in pre-gelatinized form. Pre-gelatinized adjuncts are exposed to infrared or microwave radiation and subsequently milled or flaked prior to being combined with the grist. Another type of treatment is impact extrusion, in which unmalted barley is gelatinized at pressures of 45 to 75 bar and temperatures of 120 to 160 °C. In any case, barley of malting quality should be employed [8].

A further area of application for barley both within and outside the realm of the *Reinheitsgebot* is its contribution to color and aroma. On one hand, unmalted barley can be roasted and used as an adjunct, as is often the case with dry stout. On the other hand, roasted barley malt or a liquid coloring agent produced using roasted barley malt conform to the *Reinheitsgebot* can be employed to adjust the color of beer. Liquid coloring agents of this kind can be added at any time in the brewing process, even immediately prior to filling.

2.1.2 OATS (*Avena sativa* L.)

Oats most likely evolved from the wild form of *Avena fatua* [17]. The earliest finds in Central Europe date back to the Bronze Age. The plant belongs to the secondary domesticates, because it was first considered a "weed" that surfaced among the Neolithic founder crops [18]. While oats were one of the most important cereals of the Middle Ages, they now hold sixth place in overall cereal production internationally. Oats are distinguished phenotypically from the other principal cereal grains by their paniculate inflorescences. Oats can be planted in areas where a humid, relatively cool climate allows for early tillage and the autumn frosts are not too early [17].

Up to the Middle Ages, oats were frequently used for malting and beer brewing [19, 20]. Later, oats were considered an inferior brewing grain and were only employed for lower quality beers [21].

The slender oat kernel rapidly absorbs water during steeping and germination [22]. Thus, steeping does not last long and subsequent spraying in the germination box is sufficient to reach the desired moisture content [23]. The germination schedule (temperatures, duration, O_2/CO_2, moisture content of the air and the malt) is similar to that of barley; however, the germinating grains form a less compact couch due to their ample husks.

Oats are known for their high protein, fat and β-glucan content [24]. Malt made from oats is therefore not ideal for beer production. However, special varieties of oats with a low β-glucan content, e.g. Duffy, are perfectly suited as an ingredient for brewing beer [25]. Oats possess a lower level of enzymatic activity than barley. Of note is the characteristic bitter flavor of oats, which does not always blend agreeably with the noble bitterness of hops. The inherent bitterness of oats is one possible reason why malted barley rather than malted oats became the primary brewing grain in the *Reinheitsgebot* along with hops and bottom-fermenting yeast [26].

Wort made from 100 % malted oats is comparable to wort produced from barley malt. Lautering occurs very rapidly due to the high husk content of oats. Compared to barley malt wort, wort produced from malted oats contains more zinc (up to 0.6 mg/l), β-glucans and tryptophan [25, 27]. Such beers exhibit the typical flavor of oats and possess a high reductive capacity.

Beers produced with a high percentage of oats in the grist tend to exhibit a pronounced, sustained turbidity [28]. For this reason, it is hardly possible to produce beer with a fine brilliance using oats. On the other hand, the use of malted oats for stabilizing the turbidity of cloudy, top-fermented beers, even within the confines of the *Reinheitsgebot*, is conceivable [29].

2.1.3 MINOR MILLETS

Sorghum is more related in a botanical sense to maize (corn) than to millet and is known by such names as dari and milo. To distinguish it from small-grained millet, the Food and Agriculture Organization (FAO) has grouped them together under the umbrella term "millets" [30]. A wide variety of millets belong to this category; however, only a few come into question for beer production, and they are discussed below. Table 2 (p. 324) provides information concerning the most common millet in Germany, proso millet.

2.1.3.1 Pearl millet (*Pennisetum glaucum* [L.] R. Br.)

Pearl millet originated in West Africa, where the earliest evidence of this type of millet has been found (1000 BCE on Mauritania). Pearl millet is the most economically significant type of millet as it is cultivated over a total area of 26 million hectares, the majority of which is grown in Sub-Saharan Africa. The optimal annual precipitation for pearl millet amounts to between 200 and 600 mm, making it one of the most drought-resistant types of grain in the world [31].

Pearl millet is employed in the production of some African "opaque beers". Germination is largely complete within 24 h. In various experimental trials, malted pearl millet exhibited β-amylase activity similar to that of sorghum malt [32]. Optimal malting conditions were tested in varietal trials and found to be 25–30 °C and 3–5 days of germination [33]. In Malawi, pearl millet forms the basis of the local beer [34].

2.1.3.2 Foxtail millet (*Setaria italica* [L.] P. Beauv.)

Foxtail millet originated in China, where the oldest evidence of its cultivation can be traced to as far back as 7000 to 8000 years. This type of millet was formerly cultivated in eastern Germany and the Rhineland but ceased altogether in the 20th century. Today the chief producer of foxtail millet is China [26].
Compared to other types of millet, foxtail millet is somewhat of an outlier since it requires a rather prolonged germination period during malting, leading to a general lack of homogeneity among the

germinating kernels. Otherwise, the attributes and performance of foxtail millet are very similar to proso millet (section 2.1.3.6).

2.1.3.3 Fonio (*Digitaria exilis*)

Kernels of fonio are very tiny at a thousand kernel weight of only 0.5 g. It is the oldest native African grain and is assumed to have originated on the dry African savanna where most of this type of millet continues to be cultivated. Fonio is tolerant of extremely variable levels of precipitation and grows in regions with 400 to 3000 mm of annual precipitation. "Opaque beers" produced using unmalted fonio are brewed in Togo and Nigeria.

One study was critical of the high malting losses with fonio, although it has been shown that mixing unmalted fonio with other malted millets produces an appealing beer [35]. Upon completion of comprehensive malting trials with fonio varieties, Nzelibe et al. found the most favorable malting parameters to be four days of germination at temperatures around 30 °C. In order to suppress mold growth, formaldehyde was added to the steeping liquor. The kilning temperature for all types of millet normally does not exceed 50 °C [36].

2.1.3.4 Teff (*Eragrostis tef* [Zucc.] Trotter)

The earliest evidence for teff dates back 5650 years. The name is Amharic in origin, the official language of Ethiopia, and means "to be deprived (of something)", reflecting the fact that the kernels are very small (0.3–0.4 g thousand kernel weight). Teff must have been first domesticated in Ethiopia. The primary areas of cultivation are Ethiopia and Eritrea at higher elevations between 1700 and 2800 m that receive 300 to 2500 mm of annual precipitation. In both countries, unmalted teff and unleavened teff bread are used to produce "opaque beers" [35, 37].

In spite of its small size, teff can be malted without difficulty, assuming that the malting equipment, such as the kiln floors, has been adapted or modified to cope with the small diameter of the teff kernels. Trials have shown optimal malting conditions to be a temperature of 24 °C for steeping and germination, a moisture content of 48 % and a four day germination period [38]. Of the four varieties tested, the resultant malt exhibited remarkably high limits of attenuation (up to 83 %) [39, 40, 41, 42].

2.1.3.5 Finger millet (*Eleusine coracana* [L.] Gaertn.)

The oldest evidence of finger millet has been found in what is now Central Sudan and dates to approximately 3000 BCE. Domestication of the plant also appears to have occurred there. India is now the largest producer of finger millet in the world. ´Opaque beers` are brewed with malted finger millet and unmalted maize (corn) in Kenya. An effort was made in varietal trials to select finger millet with low concentrations of tannic substances and high levels of amylolytic enzymes [43]. Several types of finger millet were able to attain levels of β-amylase activity that rival that of barley malt.

2.1.3.6 Proso millet (*Panicum miliaceum* L.)

This grain is purported to have originated in Northern China [44], and up to the introduction of wheat and barley to the region, it was the most prominent cultivated grain there. Beginning in the Middle Ages, proso millet spread out of Asia to Central and Western Europe. It continued to be cultivated in Europe until the arrival of the potato [45]. What little acreage remained for the cultivation of proso millet in Germany disappeared almost completely prior to World War I. Recently, however, attempts have been made to reintroduce proso millet in Germany. It is widespread in arid and semiarid parts of the world. Concerning its photoperiodism, proso millet is a short-day plant and exhibits a high resistance to drought, preferring a hot climate with generous amounts of sun. Compared to barley and wheat, it is less demanding with regard to the quality of the soil [46].

Comprehensive trials at Weihenstephan, Germany and in Cork, Ireland, have shown that a number of beer styles can be produced using this gluten-free grain [47, 48, 49, 50, 51]. In Germany, top-fermenting yeast must be employed in brewing beers with proso millet, but appealing bottom-fermented beers are also possible. The best conditions for malting have proven to be a five day germination period with a temperature of 22 °C for steeping and germination and a moisture content of 44 % [52, 53]. The small diameter of the kernels (2 to 3 mm) does not mean that proso millet will pass through a standard kiln floor, but the grain bed in the kiln will be approximately two-thirds the height of a similar bed consisting of barley. Through intensive selection of proso millet varieties, the brown wild form emerged as most favorable, in part because it is widely available in Central Europe. Even with a final curing temperature of 80 °C, sufficient amylolytic activity is nevertheless retained in the kernels. Interestingly, starch degradation in proso millet is less dependent upon α-amylase and β-amylase than it is on amyloglucosidase. Thus, the mash program accentuates rests between 40 and 55 °C with a simultaneous pH adjustment to 5.2. If secondary contamination can be avoided, such as yeast propagated in wort containing gluten or dust in the mill consisting of gluten-rich malt flour, it is possible to produce gluten-free beer with proso millet, which is satisfying from a sensory standpoint and is well-received with good levels of acceptance [54].

2.1.4 MAIZE OR CORN (*Zea mays* L.)

Maize, or corn as it is commonly known, is the largest cereal both with regard to the height of the plant and the diameter of the seed. Corn can possess a thousand kernel weight of up to 240 g. The earliest evidence for domesticated corn has been dated to 5000 BCE, though its wild progenitor has yet to be identified. Corn is cultivated in many countries, the most prominent among them being the United States, China and Brazil. Many varieties exist and are grouped into the following general categories: dent corn, flint corn, sweet corn, popcorn, waxy corn, flour corn and pod corn [55].

Corn adjuncts commonly serve as a substitute for malt, though in order to do so, the embryo must be separated from the kernels to a great extent. This ensures that as little fat as possible enters the brewing process. The embryo comprises 5 to 14 % of the total weight of the kernel [56] and contains 85 % of the oil [57]. Embryo removal occurs through a type of milling process. The accumulated corn grits can be added during the mashing process in a manner similar to that for rice adjuncts. Gelatinization takes place during mashing. Flaked corn is pre-gelatinized (similar to rice and barley adjuncts) and is produced by steaming, rolling and drying. Flaked corn continues to be an inexpensive ingredient [8] and can shorten the mashing process. It can be used in proportions comparable to those of rice adjuncts.
In a comparison of all-malt wort with worts produced using corn, sorghum and barley adjuncts, no differences were apparent in the worts if the percentages of the respective adjuncts in the grist were limited to 5 %. If higher percentages were added, the protein fractions in the wort decreased linearly, though corn fared slightly better than the others [58]. In general, the free amino nitrogen (FAN) should not fall below 160 mg/l, otherwise additional nutrients are required for the yeast. Vital and viable yeast should be pitched in the wort, but this is certainly not the most relevant limiting factor. Figure 3 shows that corn adjunct mashes containing 10 % malt can be sufficiently converted at 78 °C. By contrast, if the mash is heated to temperatures just below boiling, then the enzymes will be inactivated too rapidly. Gelatinization of the starch occurs but without complete conversion, causing subsequent starch retrogradation. This results in the starch being less accessible to enzymatic degradation [59]. Also of importance is that substantial additions of corn derivatives greatly impact the character of the beer [60].

Figure 3: Gelatinization behavior of 90 % maize (corn) mashes at various temperature maxima (measured with a rotational viscosimeter manufactured by Newport Scientific/Australia).

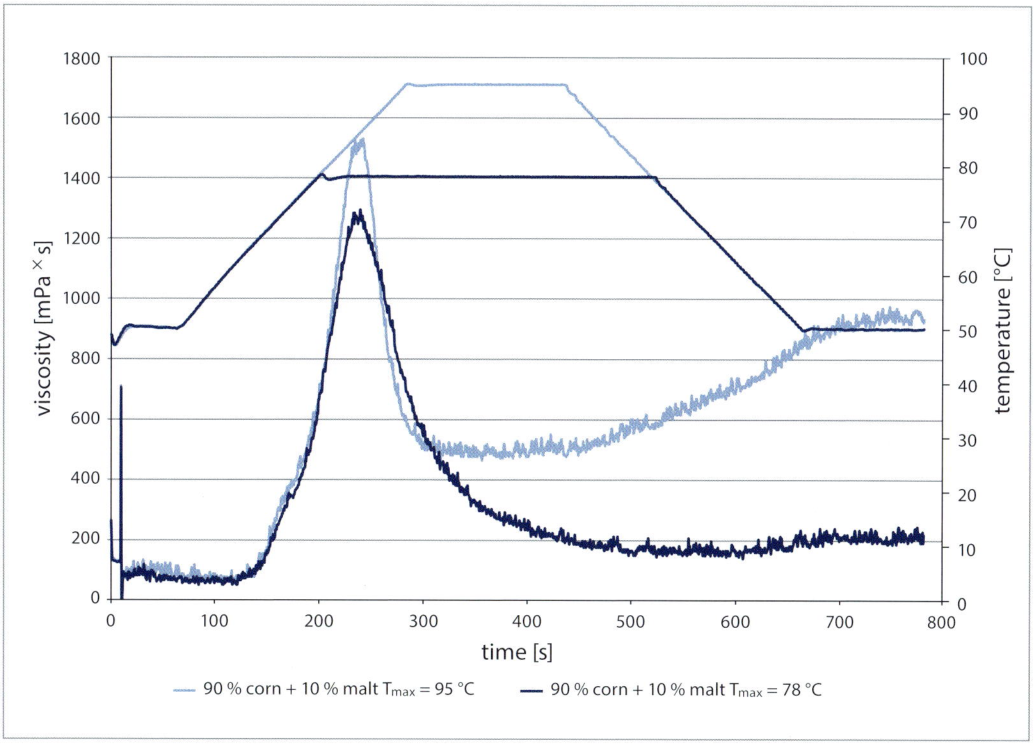

In malting corn, temperatures of approximately 28 °C and two extra days of germination are required, in order to attain a uniform stage of germination among the vast majority of the kernels. As the diastatic power and the extract yield increase so do the malting losses and the risk of mold growth. The standard data for malting in table 2 (p. 324) were obtained from an heirloom variety of blue corn. Except for the concentration of free amino nitrogen, the values are much less favorable compared to barley malt. This is also described in the literature [61].

2.1.5 RICE (*Oryza sativa* L.)

Rice ranks second only to wheat among the world's most important crops. The oldest evidence of rice cultivation was found in southern China and dates to approximately 7000 years ago. The countries where the most rice is cultivated are China, India, most other Asian countries and Brazil. As figure 1 (p. 322) shows, the genus *Oryza* is classified under the family name Poaceae, like barley, wheat and millet. The genus *Oryza* comprises 25 species; of those, only two have been successfully cultivated on a sustained basis: *Oryza glaberrima* Steud. and *Oryza sativa* L. The former is cultivated in Africa, and occurs only in the Niger River delta. The latter is the more successful Asian species, of which there are circa 120,000 known varieties.

Rice can readily adapt to a wide range of environments. There are varieties grown on dry land (mountain rice) that can also thrive in wet fields. However, those that grow in wet fields are not suitable for cultivation on dry land. Aside from the important step of polishing the grains, the fat content can be reduced through repeated washing. This leads to a decline in Y-nonalactone and 1-hexanol, products of fat degradation. Both are aroma compounds, and though it is as yet unclear why, they do not become conspicuous until after fermentation. These compounds have been described as a varietal characteristic of rice and also as indicator substances for foreign seeds [63, 64].

Rice species rich in amylopectin swell intensely, such as the varieties known under the name *Japonica*, which are preferred for beer production because their starch undergoes rapid and extensive gelatinization. Due to their higher gelatinization temperatures, wort production involves the use of "cereal cookers" (fig. 4). However, experimental trials have shown that rice with lower gelatinization temperatures can be added to the mash together with the malt grist. The more gentle application of heat, routine in mashing with barley malt grist, also gelatinizes the rice [64]. In forced aging tests conducted over 15 "warm days" or more, beers brewed with a substantial percentage of rice adjuncts often possess a high colloidal stability.

Malted rice is produced in Nagaland, in Northeastern India, for a beer brewed locally. The rice undergoes three days of wet steeping, followed by three to seven days of germination and is then air-dried [65].

Due to the many varieties of rice in existence, a series of experimental brewing trials was initially limited to a single variety. This variety of rice happens to possess a distinctive attribute. Black rice was chosen, because it has an unusually high concentration of anthocyanidins (50 times more than cold extraction with pilsner malt) and, in an acidic medium, produces a red pigment ordinarily only obtainable with grapes. The malting process with the husked rice is largely unproblematic. Nevertheless, a compromise between germination temperature and growth of fungi associated with grain storage must be made. Therefore, germination occurs over five days at 19 °C and at a moisture content of 43 %. The malt produced according to the standard procedure outlined above is remarkable for its low protein content, though the extract can almost entirely be solubilized, consequently resulting in wort with an abundant amount of fermentable sugar. The beer brewed with this rice has a singular yet attractive aroma profile, which is unlike the aroma one would normally associate with rice. Moreover, the beer possesses a deep, red hue.

Figure 4: A mash schedule for an extremely high proportion of unmalted rice in the grist (up to 90 %)

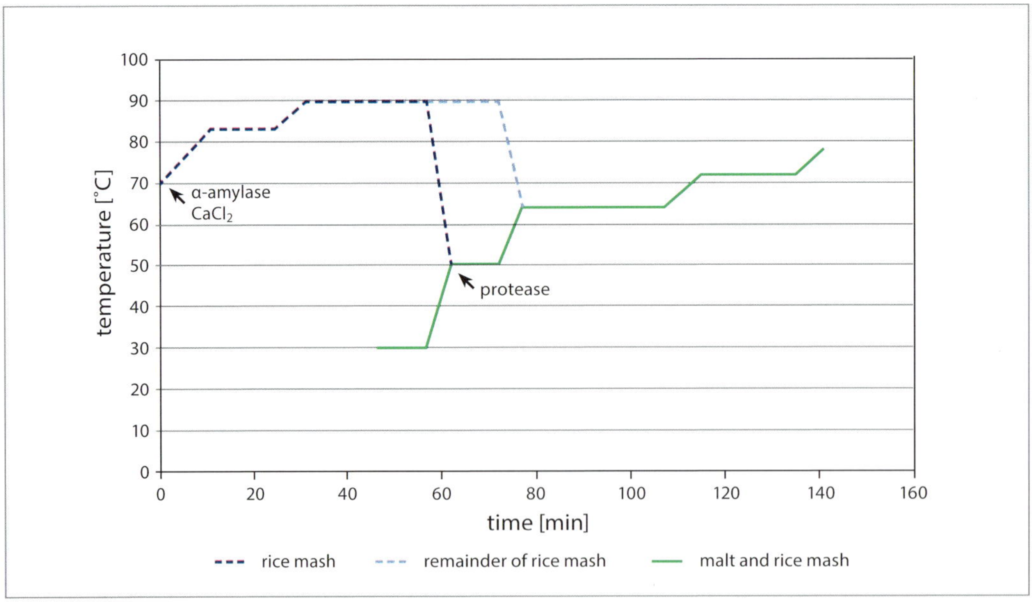

2.1.6 RYE (*Secale cereale* L.)

Rye is a very undemanding cereal and is grown primarily in the northern hemisphere. From both a cultural and historical perspective, it is a very young grain. Before it was surpassed, first by wheat and then by barley in the previous century, rye was the most frequently and continuously cultivated cereal grain in Germany for almost an entire millennium. The oldest evidence of rye, though it was not cultivated there, originates from Stone Age settlements in the Fertile Crescent (approx. 6600 BCE), in present-day northern Syria and Turkey [66].

Rye malt produced using a standard malting process is characterized by an extremely high viscosity due to the substantial amount of pentosans present in the grain. For this reason in particular, oxygen uptake must be avoided throughout the complete brewing process, since the viscosity (and the color!) will otherwise increase rapidly to the point that lautering will be impossible [67]. Hence, it is extremely difficult to achieve absolute brilliance when filtering a rye beer. Immediately noticeable is the dark color of the malt, which is sustained through to the final product. The kinds of beer that can be brewed with rye are therefore limited, especially in Germany where rye beers must be top-fermented. Nonetheless, a pleasant, naturally cloudy, dark, top-fermented beer can be brewed using rye malt.

Roasted rye malt is also available on the market. Because of its darker color (500 – 800 EBC), it has a field of application similar to that of roasted malt produced using spelt and barley.

2.1.7 SORGHUM (*Sorghum bicolor* L.)

Sorghum most likely originated in Ethiopia. Today, it is cultivated over vast expanses of Africa, India, Southeast Asia, Australia and the United States. Though the exact time period is as yet unknown, it was probably first cultivated 3000 to 5000 years ago. It is a drought-resistant plant, and can even flourish in areas too dry for corn [68].

Favorable temperatures for malting sorghum range from 24 to 26 °C [69], but they also encourage mold growth. The addition of a 0.375 % solution of sodium borate and boric acid to the steeping liquor reduces the mold cell count considerably [70]. Extending the length of germination (up to five days) with a sharply increasing moisture content allows the quantity of dissolved pentosans and consequently the viscosity to increase. The five days are nevertheless necessary in order to suffuse the wort with sufficient amounts of soluble nitrogen [71]. The raw protein content and the degree of protein modification in the standard malt are very low, while the values for free amino nitrogen are normal. On the whole, the viscosity is somewhat higher. Characteristic of sorghum malt is the general lack of enzymes, which can also further deteriorate during malt storage [72].

Sorghum is the third most important malt substitute after corn and rice. The composition of sorghum kernels can fluctuate greatly, especially with regard to their phytochemical content. It has been reported that varieties rich in tannins and polyphenols are employed in industrial beer production [73]. Varieties high in amylopectin are preferable for non-alcoholic beer, since the wort exhibits a high proportion of non-fermentable dextrins [74]. These varieties have relatively high gelatinization temperatures, making the use of exogenous enzymes essential. Interestingly, the raw protein content, though comparable to that of barley, only goes into solution to a very limited extent. Similarly, protein degradation scarcely takes place during mashing; very little of the protein is broken down to smaller fragments. Also, the wort color is very pale.

Figure 5: A mash schedule for an extremely high proportion of unmalted sorghum in the grist (up to 100 %) [8]

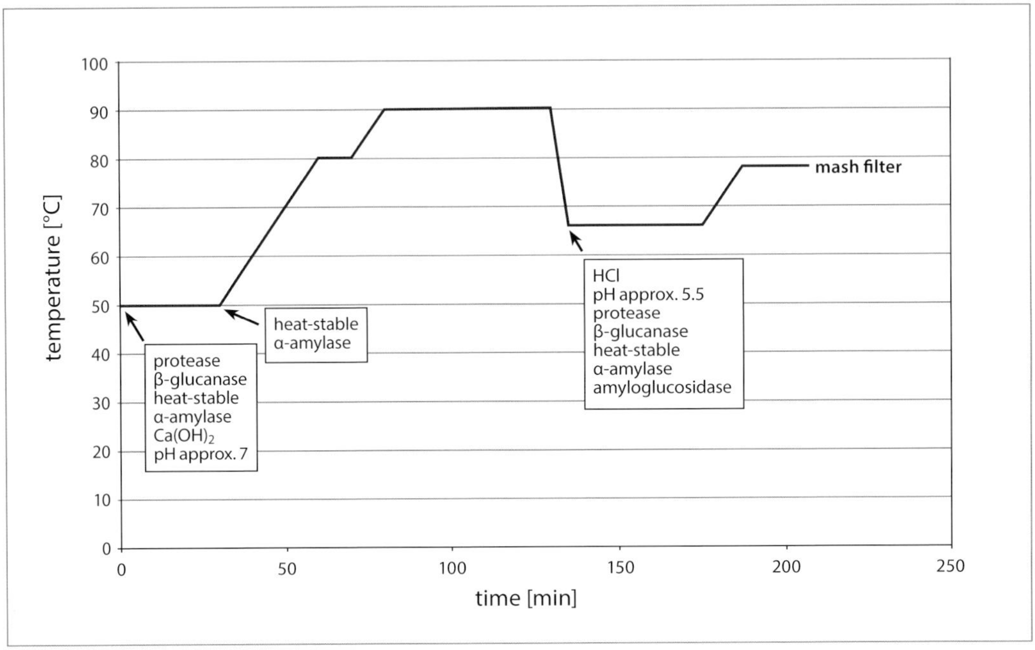

In 1988, the indigenous breweries of Nigeria had to bring their beer production into compliance with a ban on imports of malt, among other goods; therefore, they began brewing solely with unmalted sorghum. Compulsory use of sorghum prompted pioneering research that has yielded some valuable results [8] and a new beer style, for which there is still a ready market even though malt can now be imported into Nigeria again. Figure 5 depicts the mash program that is employed, along with a number of additives and the requisite equipment, i.e. a "cereal cooker" and a mash cooler.

It is recommended that lautering and sparging be carried out in a mash filter due to the absence of husks.

2.1.8 WHEAT SPECIES (*Triticum* L.)

Today, it is a generally held belief that the evolution of wheat species was neither continuous nor directed in any way, but rather it occurred in a similar manner (e.g. spontaneous crosses, mutations) at assorted sites and at different stages. The cultivation of wheat began in the Neolithic. Over the course of roughly 10,000 years, humans have developed the various kinds of wheat now grown as crops from wild wheat species by means of artificial selection. This process is known as domestication. Figure 6 shows the relationships between the artificially selected hybrids.

Figure 6: The ancestors and relatives of wheat (modified from Sears and Feldman [75])

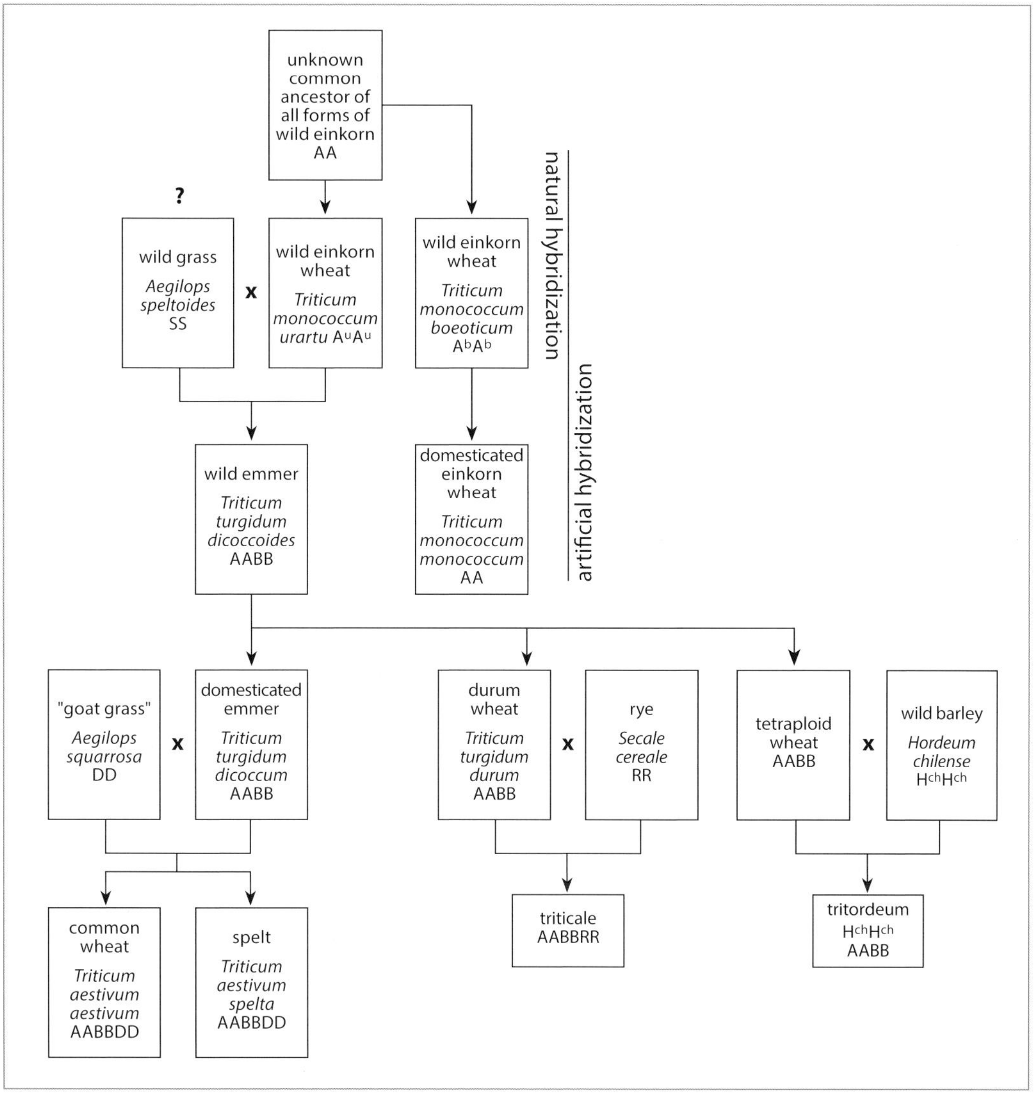

2.1.8.1 Spelt (*Triticum aestivum* ssp. *spelta* Thell. = *Triticum spelta* L.)

Spelt is a type of hexaploid wheat and the closest relative to common wheat also known as bread wheat (fig. 6) [76]. Compared to common wheat, spelt is less demanding during cultivation, less susceptible to lodging and is more resistant to intemperate weather. Spelt can also thrive at higher elevations where wheat can no longer grow. Though spelt originated in Asia, where it was cultivated over 3000 years ago, it was planted over vast areas of Switzerland, Tyrol, Baden-Wurttemberg and Middle Franconia in the Middle Ages. In Germany, it came to be known under the byname "Swabian grain".

The data for spelt malt produced according to the procedure outlined above can be found in table 2 (p. 324), which is comparable in almost every detail to wheat malt. This is also true throughout wort and beer production. Brewing with up to 70 % spelt malt in the grist has proven to be unproblematic. The resultant top-fermented beers have been found to be pleasant and characterful.

The color and aroma of roasted spelt malt is not as intense as more common types of roasted malt or liquid coloring agents produced from roasted spelt malt. Roasted spelt malt possesses a color between 450 and 650 EBC. De-husking the grain imparts a mild taste to the beer. Addition (max. 3 %) should occur shortly before mashing out to avoid any further thermal stress.

2.1.8.2 Einkorn (*Triticum monococcum* L.)

Einkorn kernels are small, and the name of the grain is derived from the fact that there is only one kernel per spikelet on each side of the rachis. This diploid, hulled grain is a relative of common wheat and emmer and also belongs to the genus *Triticum*. The oldest evidence of einkorn has been found in Anatolia and dates to the Pre-Pottery Neolithic [77]. Both einkorn and emmer were grains planted by the Neolithic farmers of Central Europe (Linear Pottery culture). Over the intervening millennia, they have been slowly displaced by barley and free-threshing wheat. This displacement is still occurring today, although einkorn persists, for example, in pockets in southwestern Germany and in neighboring Switzerland [78].

A note concerning the suitability of einkorn for malting: the kernels are planted only with their husks intact, because the embryo is too easily damaged during dehusking. Nevertheless, trials were carried out with dehusked einkorn with a high germinative capacity. The data for einkorn malt produced using the procedure described above were ascertained using a landrace from Germany and appear in table 2 (p. 324). The degree of protein modification and concentration of free amino nitrogen are conspicuously below average despite the fact that sufficient raw protein is present in the einkorn. Notwithstanding its acceptable extract content, the limit of attenuation turns out to be very low. These faults can be corrected with a higher, isothermal germination temperature. The composition of wort produced with einkorn malt is normal. The beer exhibits excellent foam stability and is notable for its satisfying liveliness.

2.1.8.3 Emmer (*Triticum dicoccum* Schübl.)

Finds of emmer date back to the ancient Near East to approximately 8000 BCE. Though it spread into Europe, it began being supplanted by other grains there as early as the Bronze Age; however, emmer held on in southwestern Germany where it was grown as recently as 50 years ago. Recently, there has been renewed demand for it [79]. Emmer is a type of tetraploid, hulled wheat with slender but dense ears, which are characterized by the diverse colors (white, red and black) of the various landraces. Emmer, like einkorn, is also planted with its husks intact. The Schönfeld test for germinative energy has shown that the results can be quite good even with dehusked emmer.

Dehusked emmer malt (table 2, p. 324) is characterized by an extremely high extract yield. The time required for saccharification is also acceptable, though the limit of attenuation is somewhat lacking. The protein content varies within the range normal for wheat, and wort and beer production also progress normally. Sensory evaluation of top-fermented beer brewed with emmer malt reveals a slightly sour, tangy note on the palate, giving a pleasant overall impression.

2.1.8.4 Tetraploid free-threshing (naked) wheat (e.g. durum wheat *Triticum durum* L., rivet wheat *T. turgidum* L. and kamut, either *T. t.* ssp. *polonicum* [L.] Thell. or ssp. *turanicum* [Jakubz.] A. Löve and D. Löve)

These tetraploid wheat species all share a common ancestor in wild emmer, possess 14 chromosome pairs and an alloploid genome, meaning they are hybrids. Of the cultivated varieties, only emmer retains its husks when ripe. The other cultivars of wheat are free-threshing. Individual excavations at archaeological sites in the ancient Near East and Central Asia reveal evidence of wheat in Stone Age strata; however, whether the kernels belong to *T. durum* or *T. aestivum* has yet to be clarified. Modern cultivation of wheat extends from the entire Mediterranean Basin to the Near East, Central Asia down to India and across the USA [80].

In a series of trials, the tetraploid naked wheat cultivar kamut was tested. Kamut originally stems from the Nile River Valley and exhibits conspicuously long kernels, along the length of which modification does not occur uniformly during malting. For this reason, the extract yield is low, and despite its high protein content, the Kolbach index, color and free amino nitrogen are likewise low. The diastatic power is unusually elevated. The top-fermented beers brewed with kamut malt possess a distinct character.

2.1.8.5 Triticale (× *Triticosecale* Wittmack)

Triticale is a new grain cultivar, developed only 100 years ago. It is a hybrid cereal with genetic contributions from female tetraploid and hexaploid wheat as well as male diploid rye. Since the hybrid offspring are sterile, the chromosome sets of the plantlets are artificially doubled in order to make them fertile. To achieve this, they must be treated with colchicine, an alkaloid extracted from autumn crocus. Triticale combines the capacity for productive yield and the grain quality of wheat with hardiness in winter, an undemanding nature and the disease resistance of rye. Only winter triticale is grown in temperate zones, while summer varieties are planted in the tropics and subtropics. The majority of triticale is grown in Poland, France, Australia, Portugal, the USA, Brazil, Germany and the countries formerly belonging to the USSR [81].

The malting properties of triticale, which is primarily grown for feed, have been investigated by various research groups [82, 83]. Creydt et al. [84] have described the varieties and malting technology involved along with particular characteristics, such as the pentosan content, which is manifested in unsuitably high levels of viscosity. The malt displays a close correlation with malted rye. In trials conducted by the authors, favorable results for a wide range of attributes were obtained by modifying the malting parameters to a germination period of five days at a temperature of 15°C and a moisture content of 45 % [85]. The characteristics of color and viscosity did not end up being particularly extreme (table 2, p. 324). The malt was noteworthy for its high diastatic power, while the α-amylase activity was somewhat weak. Similar to rye malt, triticale malt also produces a sustained turbidity in the finished beer [86]. Beers brewed with triticale malt and top-cropping yeast possess a distinctive and prominent aroma.

Unmalted triticale has recently been reported to exhibit a high diastatic power during wort production [87, 88]. If the adjunct mash is not to be boiled, this is particularly advantageous during the mashing process. There are also varieties of triticale with very low gelatinization temperatures. Similar to barley adjuncts, this permits mashing without requiring a separate cereal cooker to boil the adjunct mash.

2.1.8.6 Tritordeum (hexaploid)

Tritordeum is a hybrid created in the 1970s from a species of wild barley *Hordeum chilense* and durum wheat. This grain is very similar to wheat morphologically and agronomically and behaves similarly during bread production [89].

Tritordeum malted according to the process outlined above is remarkable for its extremely high diastatic power as well as its α-amylase activity – the highest of any grain (table 2, p. 324).

2.1.8.7 Wheat (*Triticum aestivum* L.)

Wheat, like spelt, is hexaploid with the difference that it is free-threshing. Wheat is the most important grain in the world, and of the cereals cultivated today, it is the second oldest. Evidence of wheat has been found in the ancient Near East and Central Asia dating to 7800 BCE. In Germany, wheat replaced rye as the chief grain in bread production [90].

Wheat has been malted in Southern Germany, most traditionally in Bavaria, for the production of the wheat beer brewed there, known as weissbier in Bavaria. In contrast to barley, wheat has never been explicitly bred for beer production and thus few standards defining malting and brewing specifications exist (cf. Malt). The information provided in table 2 (p.324) represents average values for commercially available wheat malt.

For many centuries, unmalted wheat has been employed in Belgium in the production of the native wheat beers, *witbier* or *bière blanche*, as well as in the ´spontaneously` fermented beers based on lambic. The grain bills of these beers can comprise up to 50 % unmalted wheat. Since wheat has a low gelatinization temperature, the amylases in the barley can be utilized in the degradation of the wheat endosperm at their customary temperature optima without requiring a cereal cooker to boil the wheat beforehand. Moreover, dextrins and starch are even desirable in the worts of lambic-based beers, since the "wild" flora (e.g. *Dekkera bruxellensis*, *Dekkera lambicus*, *Saccharomyces* sp., *Kloeckera* sp., *Pedioccocus* sp., *Lactobacillus* sp.) found in lambic fermentation can metabolize these and other compounds [92].

Wheat flour from the mill can also be added directly to the mash. Pregelatinized wheat (and barley) can also be puffed (at temperatures over 260 °C) and milled prior to adding it to the grist. Earlier, this only served to intensify the flavor of the beer but today primarily functions as a means for increasing the extract yield [93].

Roasted wheat malt can be employed to increase color and impart aroma to beer. At a color of 800 to 1200 EBC, a mere 5 % can be added to the grain bill to produce dark specialty beers, which must be top-fermented in Germany.

2.2 PSEUDOCEREALS

2.2.1 AMARANTH GRAIN (primarily: *Amaranthus cruentus*, *A. hypochondriacus* and *A. caudatus*)

Amaranth belongs to the family Amaranthaceae and originated in the tropics. These thermophilic plant species date back to at least to 6700 BCE in Mexico. Since then, European cultivars have also been developed. Its cultivation, however, is pursued most earnestly in South America, China, Africa, India and the USA. The most palatable amaranth is *Amaranthus hypochondriacus*, a species descended from wild Mexican ancestors. It is late-maturing and susceptible to frost [94].

Amaranth is notable for its lysine content, which is double or triple the amount found in barley and wheat. During germination, the lysine content also increases. A feature important for the malting process is the extremely low thousand kernel weight of 0.6 g. Floor malting is therefore the preferred method for conducting germination of amaranth. From steeping to kilning, every perforated floor or sieve must accommodate the small diameter of these kernels. In addition, the high specific surface area of the kernels must also be taken into account. Moisture uptake is rapid and volumetrically higher. Given its bulk density, the height of the same amount of amaranth is 80 % that of barley. Steeping and germination temperatures can exceed 30 °C, but the germinating grains have to be washed beforehand to impede mold growth.

Malted amaranth kernels exhibit a reddish hue. They impart a vivid color to fermented foods [96], and the Congress worts are pink in color. The malt (table 2, p. 324) is furnished with ample concentrations of soluble protein and free amino nitrogen. Perhaps the standard methods used for barley malt do not adequately characterize the amylolytic potential of amaranth, but as tested, it has the lowest limit of attenuation of all the malted grains.

In pregelatinized form, the starch from amaranth can be added to a barley malt grist in proportions as high as 20 % without a problem [97].

2.2.2 BUCKWHEAT (*Fagopyrum esculentum* Moench)

Buckwheat belongs to the family Polygonaceae and is botanically unrelated to wheat. Etymologically, the name is derived from the Middle Dutch *boecweite* "beech wheat", due to the similarities buckwheat grains share with the hulled fruit of the beech tree and the pale flour produced from it, which is comparable in color to wheat flour. Zeller counts buckwheat among the underutilized plants of the world and not among those cultivated as crops, despite the fact that 3.2 million tons are produced internationally [98]. Buckwheat most likely comes from Tibet or southern China. From there, it was dispersed along trade routes quite early, reaching Japan around 5000 BCE. In Germany, buckwheat was first mentioned in a document from the Nuremburg Archives dating to 1396, spreading throughout the West well into the 19th century. Buckwheat is planted mainly in China, Russia, the Ukraine and Poland. In Germany, buckwheat is cultivated in isolated pockets for use as a catch crop or a source of green feed.

When malting unshelled buckwheat, one must be aware that the fruit has a tendency to split open while it is steeping, causing mucilaginous deposits of starch to accumulate in the steeping tank. For this reason, abbreviated steeping times are highly recommended, since moisture uptake is rapid. Maximum β-amylase activity is reached at a germination temperature of 20 °C [100]. The distinctive nut-like aroma of the fruit, reminiscent of pistachios, develops during germination and is carried over into the finished beer. The green and kilned malt must be conveyed very carefully, otherwise it may coat the conveyor or crumble. The mash and Congress wort are extremely difficult to lauter or otherwise filter due to their excessive viscosity. This can largely be resolved, however, with an additional separation step upstream or an exogenous enzyme addition. Despite the less than optimal attributes of the malt produced according to the malting process described above, the fact that this pseudocereal is gluten-free and contains some notable phytochemicals makes buckwheat alluring enough to motivate further observation of its behavior during germination. Several research groups have already done so [100, 101, 102, 103, 104, 105]. Through altering the malting parameters it was possible to produce seven times the normal content of rutin, a glycoside of quercetin with antioxidant properties [92], and other tannins.

2.2.3 QUINOA (*Chenopodium quinoa* Willd.)

Quinoa is a pseudocereal and originated in the Andes of South America. It was first cultivated in Peru around 7000 years ago, where it served as the primary source of nutrition for the Inca Empire. It can be grown at elevations as high as 4000 m, is frost-resistant and produces acceptable yields even in areas with little precipitation [106].

For processing after the harvest, it is crucial that the hulls are gently separated from the kernels, in order to remove them and the saponins they contain without damaging the embryos. The saponin content can vary greatly, for instance in bitter quinoa, the content can range from 0.47 to 1.13 g per 100 g while in non-bitter (sweet) cultivars it can be as low as 0.02 to 0.05 per 100 g. More of the saponins are removed by discarding the steeping water. Quinoa very quickly begins germinating in earnest at temperatures as low as 8 °C and is finished after three to four days. The malt produced according to standard malting procedures has already shown promising extract yields and protein modification. The reddish hue of the mash is striking, which is attributable to the pigment β-cyanin (also the case with amaranth) [107], as is the significantly elevated zinc content [108]. Among the pseudocereals, quinoa malt exhibits the highest limit of attenuation [109].

3 SUMMARY

In spite of the fact that barley dominates the brewing process, primarily in its malted form, a number of interesting alternatives are described in this chapter. Due to their singular and distinct quality characteristics, they hold enormous promise for enriching the diversity of beer in the world, even in Germany where the strict regulations of the Bavarian *Reinheitsgebot* apply. On an international scale, alternative grains offer potential advantages to those seeking novelty, not only in malting and brewing but also in the quality they bring to the products created using them.

4 OVERVIEW

What types of grains conform to the *Reinheitsgebot*?
According to the latest information, the following malted grains are permitted in top-fermented beers: barley, einkorn, emmer, kamut, oats, rye, spelt, triticale, tritordeum and wheat.

Without making any operational modifications, what types of grain can be processed in a standard malthouse and brewery for the production of beer?
Cereals of a similar size to barley kernels can be subjected to a standard malting process:

- hulled grains: barley and oats
- naked grains with a portion of malted, hulled grains; this facilitates the formation of a filter bed in the lauter vessel (in Germany, in order for a beer to be named after the type of malt employed in brewing it, at least 50 % of the grist must contain this malt). Naked grains: spelt, emmer, einkorn, kamut, rye, triticale, tritordeum and wheat.
- not permissible according to the *Reinheitsgebot*: sorghum, maize (corn) and rice

What types of grain possess traits that are out of the ordinary?
There are certainly still some pleasant surprises awaiting further research in this area. Increasing the phytochemical content through germination is one current topic of research. Buckwheat has taken the lead in this area; for instance, its rutin content can be increased by a factor of seven through malting.
Beers brewed for those suffering from celiac disease, i.e. gluten-free beverages, can be produced from cereals (minor millets, maize (corn), rice and sorghum) as well as pseudocereals (amaranth, buckwheat and quinoa).

Conspicuous among the malts produced from these alternative starch sources are black rice, amaranth and quinoa, all of which lend beer a reddish tint.

Beer can be brewed from malted oats, rye and triticale, which generate sustained turbidity. The grains all reveal their own flavors and aromas through malting and brewing; however, they also complement, blend well and are amplified by top-fermenting yeast strains, particularly in weissbier.

What kinds of grain are readily available in Germany?
The following malted grains can easily be found in Germany: barley, rye, spelt and wheat. Emmer, einkorn and oats are not as widely available.

5 REFERENCES

[1] Röllig, W.: Das Bier im Alten Mesopotamien. Gesellschaft für die Geschichte und Bibliographie des Brauwesens e.V., Berlin 1970, pp. 10–18.

[2] Demain, A. L., and Solomon, N. A.: Industrial Microbiology. Scientific American 245 (1981), no. 3, pp. 66–75.

[3] Zarnkow, M.: Brautechnologische Erkenntnisse experimenteller Archäologie am Beispiel einer bronzezeitlichen Stadt (Tall Bazi). Lehrstuhl für Technologie der Brauerei I, 3. Rohstoffseminar, Freising/Weihenstephan 2005.

[4] BierStG (i. d. F. vom 29.07.1993) § 9 Abs. 1–8. http://bundesrecht.juris.de/bundesrecht/bierstg/_9.html

[5] Gerstenberg, H.: Vorläufiges Biergesetz. In: Zipfel, W., und Rathke, K.-D.: Lebensmittelrecht. München, Verlag C. H. Beck – Loseblattausgabe, Lfg. 104, Stand: November 1999.

[6] Rehm, S.: Hirsen. Handbuch der Landwirtschaft und Ernährung in Entwicklungsländern. Ulmer, Stuttgart 1971, pp. 281–298.

[7] Hansen: Mitteilung der Deutschen Landwirtschafts-Gesellschaft (1912). pp. 342.

[8] O'Rourke, T.: Adjuncts and their use in the brewing process. Brewer's Guardian 128 (1998), no. 3, pp. 32–36.

[9] Martenka, J.: Reis und Mais in der Brauindustrie. Diplomarbeit, TU München, Lehrstuhl für Technologie der Brauerei I, Freising 2003.

[10] Strasburger, E.: Lehrbuch der Botanik für Hochschulen. Gustav Fischer Verlag, Stuttgart 1998.

[11] Narziß, L.: Abriss der Bierbrauerei (8th ed.). Wiley-VCH, Weinheim 2017. pp. 4.

[12] Körber-Grohne, U.: Nutzpflanzen in Deutschland. Nikol, Hamburg 1995. pp. 46–55.

[13] Narziß, L.: Abriss der Bierbrauerei (8th ed.). Wiley-VCH, Weinheim 2017. p. 13.

[14] Keßler, M., Zarnkow, M., Kreisz, S., Back, W.: Untersuchungen zum Verkleisterungsverhalten unterschiedlicher Zerealien und Pseudozerealien (in Druck).

[15] Meilgaard, M. C.: Wort Composition: With Special Reference to the Use of Adjuncts. Technical Quaterly 13 (1976), no. 2, pp. 78–90.

[16] Müller, K.-P.: Brauversuche unter Einsatz unvermälzter Gerste. Diplomarbeit, TU München, Lehrstuhl für Technologie der Brauerei I, Freising 2005.

[17] Zade, A.: Der Hafer. Eine Monographie auf wissenschaftlicher und praktischer Grundlage. Gustav Fischer, Jena 1918.

[18] Ranhotra, G. S., Gelroth, J. A.: Food uses of oats. In: Welch, R. W.:The Oat Crop. Production and utilization. Chapman & Hall, London 1995.

[19] Both, F.: Gerstensaft und Hirsebier, 5000 Jahre Biergenuß. Isensee, Oldenburg 1998.

[20] Meussdoerffer, F., Zarnkow, M.: Bier – Eine Geschichte von Hopfen und Malz. CH Beck Verlag, München. 2014.

[21] Unger, R.W.: A History of Brewing in Holland 900–1900, Economy,Technology and the State, Brill, Leiden 2001.

[22] Briggs, D. E.: Malting and Brewing. Blackie Academic & Professional, London 1998.

[23] Weichherz, J.: Die Malzextrakte. Springer, Berlin 1928.

[24] Belitz, H.-D., Grosch, W.: Lehrbuch der Lebensmittelchemie (4th ed.). Springer-Lehrbuch, Berlin 1992.

[25] Hanke, S., Zarnkow, M., Kreisz, S., Back, W.: The use of oats in brewing. Brewing Science 58 (2005), no. 3/4, pp. 11–17.

[26] Zarnkow, M.: Das Reinheitsgebot – ein wichtiger Schritt zu unserem heutigen Bier. 49. Technologisches Seminar Weihenstephan. 2016.

[27] Hanke, S., Zarnkow, M., Kreisz, S., Back,W.: Verwendung von Hafer für die Malz- und Bierbereitung. Brauwelt 145 (2005), no. 8/9, pp. 216–219.

[28] Habich, G. E., Schneider, C., Behrend, G.: Die Praxis der Bierbraukunde. Illustriertes Hand- und Hülfsbuch für Brauer. Wilhelm Knapp, Halle a. d. Saale 1883.

[29] Zarnkow, M.: The Use of Response Surface Methodology to Optimise Malting Conditions of Oat (*Avena sativa* L.) as a Raw Material for Alternate Fermented Beverages. International Oat Conference, Minneapolis 2008.

[30] Zeller, F. J.: Nutzung, Genetik und Züchtung kleinkörniger Hirsegräser: 2. Kolbenhirse [*Setaria italica* (L.) P. Beauv.]. Journal of Applied Botany – Angewandte Botanik 47 (2000), no. ½, pp. 50–54.

[31] Zeller, F. J.: Nutzung, Genetik und Züchtung kleinkörniger Hirsegräser: 1. Perlhirse [*Pennisetum glaucum* (L.) R. Br.]. Journal of Applied Botany-Angwwandte Botanik 47 (2000), no. ½, pp. 42–49.

[32] Obilana, A. B.: Millets. In: Belton, P. S., Taylor, J. R. N.: Pseudocereals and Less Common Cereals. Springer, Berlin 2002. pp. 204–206.

[33] Pelembe, L. A. M., Dewar, J., Taylor, J. R. N.: Effect of malting conditions on pearl millet malt quality. Journal of the Institute of Brewing vol. 108 (2002), no. 1, pp. 7–12.

[34] Rao, S. A., Mengesha, M. H., Sibale, P. K., Reddy, C. R.: Collection and evaluation of pearl millet (Pennisetum) Germplasm from Malati. Exonomic Botany 40 (1986), no. 1, pp. 27–37.

[35] Obilana, A. B.: Millets. In: Belton, P. S., Taylor, J. R. N.: Pseudocereals and Less Common Cereals. Springer, Berlin 2002. pp. 210.

[36] Nzelibe, H. C., Obaleye, S., Onyenekwe, P. C.: Malting characteristics of different varieties of fonio millet (*Digitaria exilis*). European Food Research and Technology 211 (2002), no. 2, pp. 126–129.

[37] Zeller, F. J.: Nutzung, Genetik und Züchtung kleinkörniger Hirsegräser: 5. Teff [*Eragrostis tef* (Zucc.) Trotter]. Journal of Applied Botany-Angewandte Botanik 77 (2003), no. 1/2, pp. 47–52.

[38] Obilana, A. B.: Millets. In: Belton, P. S., Taylor, J. R. N.: Pseudocereals and Less Common Cereals. Springer, Berlin 2002. pp. 207–209.

[39] Gebremariam, M. M., Zarnkow, M.; Becker, T.: Effect of Drying Temperature and Time on Alpha-amylase, Beta-amylase, Limit-dextrinase Activities and Dimethyl Sulphide Level of Teff (Eragrostis tef) Malt. Food and Bioprocess Technology: An International Journal (2012). DOI: 10.1007/s11947-012-1025-0.

[40] Gebremariam, M. M., Zarnkow, M.; Becker, T.: Effect of teff (Eragrostis tef) varieties and storage on the malt quality attributes. Journal of the Institute of Brewing (2012). DOI 10.1002/jib.65.

[41] Gebremariam, M. M., Zarnkow, M., Becker, T.: Thermal Stability of Starch Degrading Enzymes of Teff (*Eragrostis tef*) Malt during Isothermal Mashing. Process Biochemistry 48 (2013), no. 12, pp. 1928–1932.

[42] Gebremariam, M.M., Zarnkow, M., Becker, T.: Teff (*Eragrostis tef*) as a raw material for malting, brewing and manufacturing of gluten-free foods and beverages. Journal of Food Science and Technology 51 (2014), pp. 2881–2895.

[43] Zarnkow, M., Almaguer, C., Burberg, F., Back, W., Arendt, E. K., und Kreisz, S.: The Use of Response Surface Methodology to Optimise Malting Conditions of Teff (*Eragrostis tef* L.) as a Raw Material for Gluten-free Foods. In: Brewing Science 61 (2008), no. 5/6, pp. 94–104.

[44] Li, H.: Origin of Cultivated Plants in Southeast Asia. Economic Botany 8 (1970), no. 1, pp. 3–19.

[45] Böckler, W.: Relikte unter den Kulturpflanzen. Zeitschrift für Agrargeschichte und Agrarsoziologie 2 (1954), no. S, pp. 22–40.

[46] Hoffmann-Bahnsen, R., Plessow, J.: Alte Kulturpflanzen neu entdeckt, Rispenhirse (*Panicum miliaceum*) eine ideale Sommerung für den ökologischen Landbau. Mitteilungen der Gesellschaft für Pflanzenwissenschaften 15 (2003). pp. 31–33.

[47] Zarnkow, M. Kreisz, S., Back, W.: Glutenfreie Biere. Brauindustrie 90 (2005), no. 7, pp. 26–28.

[48] Zarnkow, M., Arendt, E. K., Back, W., Burberg, F., Keßler, M., Kreisz, S.: Gluten-free beer from proso millet malt (*Panicum miliaceum*). First International Symposium on Gluten-free Foods and Berverages, Cork 2007.

[49] Zarnkow, M.: Entwicklung von glutenfreiem Bier am Beispiel Rispenhirse. FH Wiesbaden-Geisenheim 2007.

[50] Zarnkow, M., Arendt, E. K., Back, W., Burberg, F., Keßler, M., Kreisz, S.: Gluten-free beer from proso millet malt (*Panicum miliaceum*). 120th Anniversary Convention of the Master Brewers Association of the Americas 2007, p. 24.

[51] Zarnkow, M.: Proso Millet (*Panicum miliaceum* L.) a Sustainable Raw Material for the Malting and Brewing Process. Dissertation, University College Cork, Ireland, 2010.

[52] Zarnkow, M., Keßler, M., Burberg, F., Back,W., Arendt, E. K., Kreisz, S.: The Use of Response Surface Methodology to Optimise Malting Conditions of Proso Millet (*Panicum miliaceum* L.) as a Raw Material for Gluten-free Foods. Journal of the Institute of Brewing 113 (2007), no. 3, pp. 280–292.

[53] Zarnkow, M., Keßler, M., Burberg, F., Back, W., Kreisz, S.: Gluten free beer from malted cereals and pseudocereals. Proceedings of the 30th European Brewery Convention Congress Prague. Fachverlag Hans Carl, Nuremberg 2005, no. 104/1–104/8.

[54] Zarnkow, M., Keßler, M., Burberg, F., Arendt, E. K., Back, W., Kreisz, S.: Optimization of the mashing procedure for malted proso millet. World Grain Summit 2006, 76.

[55] Körber-Grohne, U.: Nutzpflanzen in Deutschland. Nikol, Hamburg 1995. pp. 86–97.

[56] Lásztity, R.:The Chemistry of Cereal Proteins. CRC Press, Boca Raton 1984.

[57] Mounts, T. L., Anderson, R. A.: Corn Oil Production, Processing and Use. In: Barnes, P. J.: Lipids in Cereal Technology. Academic Press, New York 1983. pp. 373–387.

[58] Agu, R. C.: A comparison of maize, sorghum and barley as brewing adjuncts. Journal of the Insitute of Brewing 108 (2002), no. 1, pp. 19–22.

[59] Tegge, G.: Stärke und Stärkedrivate (3rd ed.). B. Behr's Verlag GmbH & Co. KG, Hamburg 2004.

[60] Einsiedler Bierspezialitäten. http://www.beer.ch/ros_biere.html#maisgold. – Aktualisierungsdatum: 22.06.2005.

[61] Eneje, L. O., Ogu, E. O., Aloh, C. U., Odibo, F. J. C., Agu, R. C., Palmer, G. H.: Effect of steeping and germination time on malting performance of Nigerian white and yellow maize varieties. Process Biochemistry 39 (2004), no. 40, pp. 1013–1016.

[62] Ramseyer, U.: Reis. Konsequenzen des Geschmacks. St. Gallen 1988.

[63] Zarnkow, M., Back, W.: Probleme bei der Herstellung von Rohfruchtbieren mit hohen Reisgaben. Brauwelt 144 (2004), no. 13/14, pp. 391–396.

[64] Zarnkow, M., Arendt, E. K., Back, W., Burberg, F., Gastl, M., Herrmann, M., Keßler, M., Kreisz, S.: Influence of Cereal Adjuncts on Beer Flavour and Flavour Stability in Consideration of Rice Adjuncts. Cerevisia – Belgian Journal of Brewing and Biotechnology 32 (2007), no. 2, pp. 110–119.

[65] Teramoto,Y., Yoshida, S., Ueda, S.: An Indigenous Rice Beer of Nagaland, India. Ferment 13, (2000), no. 2, pp. 39–41.

[66] Körber-Grohne, U.: Nutzpflanzen in Deutschland. Nikol, Hamburg 1995. pp. 40–46.

[67] Braun, F.: Mälzungs- und Maischversuche zu Möglichkeiten der technologischen Einflußnahme auf ungünstige Fließeigenschaften bei Roggenmalzwürzen unter besonderer Berücksichtigung des Reinheitsgebotes. Diplomarbeit, TU München, Lehrstuhl für Technologie der Brauerei I, Freising 1998.

[68] Taylor, J. R. N., Belton, P. S.: Sorghum. In: Taylor, J. R. N., Belton, P. S.: Pseudecereals and Less Common Cereals. Springer, Berlin 2002. pp. 25–81.

[69] Palmer, G.: In: Ajerio, K. O., Booer, C. D., Proudlove, M. O.: Aspects of the Malting of Sorghum. Ferment 6 (1993), no. 5, pp. 339–341.

[70] Ajerio, K. O., Booer, C. D., Proudlove, M. O.: Aspects of the Malting of Sorghum. Ferment 6 (1993), no. 5, pp. 339–341.

[71] EtokAkapan, O. U.: Changes in Sorghum Malt During Storage. Journal of the Institute of Brewing 110 (2004), no. 3, pp. 189–192.

[72] Jani, M., Annemüller, G., Schildbach, R.: Untersuchung über den Einfluß der Hirsemalzqualitäten auf die Extraktgewinnung. Brauwelt 139 (1999), no. 23, pp. 1062–1065.

[73] Canales, A. M., Sierra, J. A.: Use of Sorghum. Technical Quarterly 13 (1976), no. 2, pp. 114–116.

[74] Figueroa, J. D. C., Martínez, B. F., Ríos, E.: Effect of Sorghum Endosperm Type on the Quality of Adjuncts for the Brewing Industry. Journal of the American Society of Brewing Chemists 53 (1995), no. 1, pp. 5–9.

[75] Feldman, M., Sears, E. R.: The Wild Gene Resources of Wheat. Scientific American 244 (1981), no. 1, pp. 102–112.

[76] Körber-Grohne, U.: Nutzpflanzen in Deutschland. Nikol, Hamburg 1995. pp. 68–86. [70] Körber-Grohne, U.: Nutzpflanzen in Deutschland. Nikol, Hamburg 1995. p. 324.

[77] Dietrich, O., Heun, M., Notroff, J., Schmidt, K., Zarnkow, M.: The role of cult and feasting in the emergence of Neolithic communities. New evidence from Göbekli Tepe, south-eastern Turkey. Antiquity 86 (2012), pp. 674–695.

[78] Bänninger, A., Jenny, M., Gehring, B., and Bartha, B.: Einkorn/Emmer-Projekt. http://www.sab.ch/v2/merketing/projekte/klettgau.htm. – Aktualisierungsdatum: 21.06.2005.

[79] Körber-Grohne, U.: Nutzpflanzen in Deutschland. Nikol, Hamburg 1995. pp. 326–330.

[80] Körber-Grohne, U.: Nutzpflanzen in Deutschland. Nikol, Hamburg 1995. pp. 28–39.

[81] Schuchert, W.: Triticale: URL http://www2.mpiz-koeln.mpg.de/pr/garten/schau/Triticale/Triticale(d).html. – Aktualisierungsdatum: 24.06.2005.

[82] Malter, R.: Untersuchungen der brautechnischen Einsatzmöglichkeiten von Triticale. Dissertation, TH Köthen, 1992.

[83] Rath, F., Annemüller, G., Creydt, G., Schildbach, R.: Triticale und Triticale-Malze. Monatsschrift für Brauwissenschaft 52 (1999), no. 7/8, pp. 123–130.

[84] Creydt, G., Mietla, B., Rath, F., Annemüller, G., Schildbach, R., Tuszynski,T.: Triticale and triticale malt. Part II: Oriented malting of triticale. Monatszeitschrift für Brauwissenschaft 52 (1999), no. 7/8, pp. 126–130.

[85] Zarknow, M.: Einfluss verschiedener Zerealien und Pseudozerealien auf die Bierbereitung. 5. Rohstoffseminar Weihenstephan 2007.

[86] Zarnkow, M.: Mälzungs- und Braueignung von Triticale. 6. Rohstoffseminar Weihenstephan 2008.

[87] Glatthar, J., Heinisch, J., Senn, T.: A Study on the suitability of unmalted triticale as a brewing adjunct. In: Journal of the American Society of Brewing Chemists 60 (2002), no. 4, pp. 181–187.

[88] Glatthar, J., Heinisch, J., Senn,T.: Unmalted triticale cultivars as brewing adjuncts: effects of enzyme activities and composition on beer wort quality. In: Journal of the Science of Food and Agriculture 85 (2005), no. 4, pp. 647–654.

[89] Martin, A., Alvarez, J. B., Martin, L. M., Barro, F., and Ballesteros, J.: The development of tritordeum: A novel cereal for food processing. Journal of Cereal Science 30 (1999), no. 2, pp. 85–95.

[90] Körber-Grohne, U.: Nutzpflanzen in Deutschland. Nikol, Hamburg 1995. pp. 28–39.

[91] Little, B.T.: Alternative cereals for beer production. Ferment 7 (1994), pp. 163–168.

[92] Zarnkow, M.: Biereinteilung. URL http://www.roempp.com/prod/index1.html. – Februar 2005. Dokumentkennung RD-02-01266.

[93] Britnell, J.: Torrified cereals. Technical Quarterly 10 (1973), no. 4, pp. 176–179.

[94] Aufhammer, W.: Getreide- und andere Körnerfruchtarten. Eugen Ulmer, Stuttgart 1998. p. 162.

[95] Paredeslópez, O., Moraescobedo, R.: Germination of Amaranth Seeds – Effects on Nutrient Composition and Colour. Journal of Food Science 54 (1989), no. 3, pp. 761–762.

[96] Yue, S., Sun, H.: The Development of Food Products of Grain Amaranth. Chinese Cereals and Oils: Proceedings of the International Symposium on New Approaches to Functional Cereals and Oils. Bejing 1997. pp. 188–191.

[97] Fenzl, G., Berghofer, E., Silberhummer, H., Schwarz, H.: Einsatzmöglichkeiten extrudierter, stärkereicher Rohstoffe zur Bierherstellung. Tagungsband Österreichisches Brauereiforum 1997. pp. 1–6.

[98] Zeller, F. J.: Buchweizen (Fagopyrum esculentum Moench): Nutzung, Genetik, Züchtung. In: Bodenkultur 52, no. 3 (2001). pp. 259–276.

[99] Reinhardt, L.: Kulturgeschichte der Nutzpflanzen. Ernst Reinhardt, München 1911.

[100] Wijngaard, H. H., Ulmer, H. M.: The effect of germination temperature on malt quality of Buckwheat. Journal of the American Society of Brewing Chemists 63 (2005), no. 1, pp. 31–36.

[101] Nic Phiarais, B. P.,Wijngaard, H. H., Arendt, E. K.: The impact of kilning on enzymatic activity of buckwheat malt. Journal of the Institute of Brewing 111 (2005), no. 3, pp. 290–298.

[102] Phiarais, B. P. N., Wijngaard, H. H., Arendt, E. K.: Kilning conditions for the optimization of enzyme levels in buckwheat. Journal of the American Society of Brewing Chemists 64 (2006), pp. 187–194.

[103] Wijngaard, H. H., Ulmer, H. M., Arendt, E. K.: The effect of germination time on the final quality of buckwheat. Journal of the American Society of Brewing Chemists 64 (2006), pp. 214–221.

[104] Wijngaard, H. H., Phiarais, B. P. N., Ulmer, H. M., Goode, D. L., Arendt, E. K.: Gluten-free beverages based on buckwheat. Proceedings of the 30th Congress of the European Brewery Convention Prague. Fachverlag Hans Carl, Nuremberg 2005, no. 101/1–101/11.

[105] Wijngaard, H. H., Arendt, E. K.: Otimisation of a mashing program for 100% malted buck- wheat. Journal of the Institute of Brewing 112 (2006), no. 1, pp. 57–65.

[106] Johnson, D. L.: New Grains and Pseudograins. Timber Press: Advances in new crops. Portland, 1990. pp. 122–127.

[107] Mabry, T. J., Taylor, A., Turner, B. L.: The Betacyanins and Their Distribution. Phytochemistry 2 (1963), no. 1, pp. 61–64.

[108] Whali, C.: Quinua hacia su cultivo commercial. Latinreco, Quito 1990.

[109] Zarnkow, M., Geyer,T., Lindemann, B., Burberg, F., Back,W., Arendt, E. K., Kreisz, S.: The Use of Response Surface Methodology to Optimise Malting Conditions of Quinoa (Chenopodium quinoa L.) as a Raw Material for Gluten-free Foods. Brewing Science 60 (2007), no. 9/10, pp. 118–126.

SPECIALTY BEER

1 GERMAN BREWING ON THE WORLD STAGE

This chapter provides information on a selection of beer styles, how they developed and their adoption by brewers around the world. A table (table 2, pp. 354) is provided which includes the following information with reference to these styles: original gravity, alcohol content, apparent attenuation, color, bittering units, percentage of barley malt in the grist, type of fermentation and the most common type of fermentation and typical fermentation temperatures. Aside from the quality attributes of the malt (cf. Malt) and the adjuncts, if applicable, the production of the substrate through the process of mashing greatly influences the style of beer being brewed. Table 1 shows an overview of the possible temperature rests (cf. Mashing).

Table 1: Mash schedule

grist to liquor ratio (1 : 3–5 malt, 1 : 2–3 adjuncts)
Mashing in (contact with brewing liquor, no dust, no clumps)
35–40 °C rest "pre-solubilization of the malt grist" (optional, formerly known as "digestion") degradation of gum substances in plant exudates conversion process: **cytolysis**
45–55 °C rest degradation of proteins and phosphates conversion process: **proteolysis, phosphorolysis**
ca. 62 °C rest optimum temperature for β-amylase (degree of attenuation) ca. 60–65 °C starch gelatinization → without gelatinization, no saccharification à monitoring the gelatinization temperature conversion process: **amylolysis**
ca. 72 °C rest optimum temperature for α-amylase (saccharification rest, iodine test for starch) conversion process: **amylolysis**
ca. 78 °C mash-out retains α-amylase function, lowers viscosity

A brief summary of the most common German bottom-fermented and top-fermented beers is provided first, followed by a short review of various international beer styles.

2 GERMAN BEER STYLES

2.1 BOTTOM-FERMENTED BEERS

The attributes of pale lagers, the most important German beer styles, including pilsner and the beer known as *Helles* in Germany, are listed at the top of table 2. Aside from these more famous pale lagers (the most successful beers worldwide), there are also stronger beers, such as *Export*, pale bock and other specialty beers produced along the same general lines. The darker beer styles include *Märzen*, *Dunkel Export*, *Dunkel Spezial* and *Schwarzbier* as well as dark bock and other strong beers. Aside from bottom-fermented beers, there are also a number of top-fermented beers, such as wheat beer, *Kölsch*, *Altbier* and beers brewed with other cereal grains. A description for each is provided in the following section. According to the *Reinheitsgebot*, bottom-fermented beers may only be brewed from barley malt, while top-fermented beers, by contrast, may be produced from wheat malt as well as from other malted grains.

The high level of diversity found among pale bottom-fermented beers is, in part, due to regional variation. *Helles* is very light in color and has an extremely soft character, which is a result of the high degree of attenuation characteristic of the style. This beer usually possesses less than 20 bittering units. The stronger beers in this general category are more heavily hopped and have a more robust color (e.g. through the addition of specialty malts). Pale German lagers mostly fall into three major categories. In Germany, these are *Hell*, *Export* and *Spezial Festbier* and their respective original gravities are 11–12, 12–13 and 13–14 °P. The brewing liquor is soft, either naturally so or through treatment, which lowers the residual alkalinity to 2–4 °dH. A correction of the mash and wort pH (lowered to between 5.6 and 5.2) is carried out using first wort that has been fermented with lactic acid bacteria known as *Sauergut* (cf. Acidification with Natural Lactic Acid). This produces even softer beers than the use of soft brewing liquor does alone. The mash method is a classic double decoction, though the decoction mash may be held at 96 °C rather than being boiled. Infusion mash methods are also possible with temperatures that correspond to those of the main mash during double decoction (cf. Mashing). Modern wort boiling processes usually last no longer than 60 min. Aroma hops are added in three equal quantities. Bittering or high alpha hops comprise a maximum of one-third of the hops and are apportioned in the first addition approximately 5 min after the beginning of the boil (cf. Hops, Wort Boiling). Conventional fermentation for a pale German lager is carried out at a maximum temperature of 9–10 °C. After primary fermentation, the green beer undergoes a diacetyl rest for 5–7 days until the value for total VDK reaches < 0.12 ppm. A cold lagering period of two to three weeks at 0 to -1 °C ensues. It is important that the yeast be harvested at the appropriate time. Also, during maturation in cylindroconical tanks, yeast should be removed every two days and during cold lagering, once a week. Pilsner is produced in a similar manner, although particular emphasis is placed on creating a beer that is very light in color. Mashing is carried out using a *hoch-kurz* decoction process. *Hoch* is German for "high" and refers to the high mash-in temperature, while *kurz* means "short" and refers to the duration of the process. The hops can be divided over four additions; the last addition occurs in the whirlpool. The bittering units range according to hop variety, etc. from 25 to 45 IBU.

In brewing *Märzen* through to darker beer styles (from dark *Export* to *Doppelbock*), the grist contains an increasingly higher ratio of Munich malt. This imparts a noticeable malt aroma to the beer due to the method by which the malt is produced. The brewing liquor should be somewhat harder (RA of 5 – 7 °dH), and the mash program is generally a triple decoction or an intensive double decoction. An infusion mash with the corresponding temperature rests is also possible; however, in this case, the grain bill should include light and dark caramel malts. Darker beer styles also benefit from a small amount of roasted malt in the grist (0.7 % to 10 %), which imparts more intensity to both the color and flavor. The hop addition is also conducted in three parts with aroma hop varieties. The processes of fermentation, maturation and cold lagering are carried out as described above. The duration of cold lagering is largely dependent upon the strength of the beers and can last from three to 12 weeks.

2.2 TOP-FERMENTED BEERS

In Germany, beers produced using top-fermenting yeast can be divided into Southern German (Bavarian) wheat beer, known as *Weizenbier* or weissbier, and Western German *Kölsch* and *Altbier*, in addition to the beers produced using other cereal grains, such as wheat, rye, triticale, spelt, emmer and others. In order for a beer to be called *Weizenbier* (wheat beer) in Germany, the grain bill must contain at least 50 % wheat malt. These top-fermented beers are very distinctive, largely due to the type of yeast used in their production. The fermentation by-products 4-vinyl guaiacol and 4-vinyl phenol are characteristic for weissbier yeast and lend a clove-like aroma to the beer. Weissbier yeast also produces elevated concentrations of higher alcohols and esters – depending upon the strain and the conditions during fermentation (e.g. temperature, tank geometry, number of batches per tank). Whether the beer is conditioned in the bottle or maturation tank is also of significance. By contrast, the yeast strains used to produce *Kölsch*, *Altbier* and British ale are more neutral in their aroma profiles; however, pronounced ester notes can also arise in those beers.

Kölsch is described as a pale "top-fermented lager", while *Altbier* is darker in color and more highly hopped, especially among the smaller producers. The composition of the malt grist determines the character of the beer to a large extent. The mash program is less intensive, since a higher portion of pilsner malt is employed in the recipe. Also, besides being highly modified, the darker malt still retains a sufficient quantity of enzymes and thus is able to bring about conversion during mashing.

Weissbier (cf. Southern German Wheat Beer), like the pale and dark bottom-fermented beer styles, is brewed to a wide range of original gravities. Depending on the category, gravities can vary from 11.5–11.9 °P, 12.1–12.9 °P, 13–14 °P and can even be brewed to the extract content of bock beer, namely from 16 °P to 20 °P. There are pale and dark wheat beers, which are all unfiltered, as well as *Kristallweizen*, a pale, filtered variety of wheat beer. Certain seasonal *Altbiers* are also brewed to stronger original gravities.

Beers brewed with cereal grains other than barley and wheat (cf. Cereals) are relatively neutral due to their endosperm composition. The malts produced from these grains generally require a higher curing temperature in order to produce more malt aroma. One problem with alternative cereal grains (e.g. rye) is that the level of protein modification far outpaces cytolytic modification during malting. These malts therefore exhibit a high viscosity, which requires an intensive mash program to facilitate solids separation in the brewhouse (lautering). For this reason, a double decoction with a mash-in temperature of 35 °C and rests at 45/50/55 °C or a similarly intensive infusion method with the corresponding temperature rests is warranted. With other grains, such as spelt, the viscosity is lower, and consequently the rests can be shorter. The yeast is an important aspect of the flavor profile, since weissbier yeast will produce a typical weissbier flavor and aroma profile even if the grist does not contain wheat malt. *Kölsch/Altbier* ale yeast strains, on the other hand, bring the character of the substrate to the fore with increased levels of higher alcohols and esters.

3 INTERNATIONAL SPECIALTY BEERS

There are marked differences between British and Continental beers. English, Scottish and Irish brewers have always been able to rely on access to maritime malting barley as a primary raw material for their beer. These barley varieties are not only rich in extract and enzymes, but also possess a low protein content because of the favorable climate and particularly due to the long growing season. Recent examples include the variety Proctor, the winter barley variety Maris Otter as well as their respective descendants. Winter barley varieties have always been valued there because they do not need to be as winter-hardy as they otherwise would for cultivation on the Continent, since the climate is appreciably milder in the

British Isles. These barley varieties are malted quite intensively – after steeping the grain, it spends 8 to 12 days on the germinating floor, is turned only once per day and plowed a mere two to four times per day (English malt plows are thus designed differently). This is followed by a gentle withering period and high temperature curing (90–95 °C). The color of pale malt is similar to that of Vienna malt (5–8 EBC). Due to this very different malting process, mashing can be carried out in a single step – that is, at a constant temperature, e.g. around 65–66 °C. A portion of the brewing liquor is usually added to a vessel with a false bottom (i.e. the "mash tun" in Britain), ahead of time, and the grist is often mashed-in using a pre-masher (equipped with an auger) over the period of one hour. After an additional hour, lautering begins. Because the height of the grain bed is much greater in the mash tun, lautering and sparging can last about six hours or even longer. For this reason, the first wort and sparged wort may be boiled separately in smaller wort kettles or "coppers", as they are known. Sometimes the hops used in the first wort are boiled a second time in the sparged wort.

A historical review: In order to partially inhibit enzymatic activity in the malt, mashing was formerly carried out with falling temperatures: the mash liquor was heated to 78 °C and then added to the mash tun. The grist was wetted in the pre-masher with cold liquor. This resulted in a final mash temperature of 65–66 °C.

Hops were stored in the brewery in a loft, and the first hops in were the first out. Hops from Australia and Tasmania were harvested half a year after the harvest in Britain and shipped to the UK; however, they were often exposed to high temperatures and thus aged prematurely on the journey. The quantities added in the brewhouse were high, and they were also often added on the cold side during primary fermentation or to conditioning casks prior to transport from the brewery.

Unsurprisingly, a "continental climate" prevailed on the Continent, meaning that frosts were more severe in winter and summers were warmer. For spring barley, this often meant a shorter growing season. When weather conditions were less favorable, the barley could even mature unseasonably early. In this type of climate, the resulting malt contained more protein and thus less extract; modification could also be somewhat prolonged despite an intensive malting process: the piece was turned two to three times daily and after four to five days, the grain was allowed to germinate for 24 h without being turned or plowed. After 24 h, the piece was shaken to loosen it and turned; if necessary, the process was repeated. Modification was further enhanced by raising the temperature towards the end of germination to 18–20 °C. The duration of steeping for paler malts was circa three days, and for darker malts, four days. The germination period was eight to ten days, though a one to two-day withering period could supplement the malting process, meaning that the piece was spread upon a well-ventilated floor above the kiln. However, the enzyme content of this green malt was lower than comparable British malt.

Nevertheless, the malt on the Continent was not of a consistent quality. German barley was considered "less refined" and possessed more protein, whereas the Bohemian barley was "finer" and contained less protein. The draft of air rising through Bavarian kilns was not as strong and the green malt underwent a prolonged drying process. The temperature gradient was steep between the heating elements and the kiln floor(s). Very high temperatures, e.g. 100–105 °C, were normally sustained during curing. These somewhat darker malts had therefore lost some of their enzyme activity, which was apparent in the lengthier saccharification rests during mashing and the less favorable degree of attenuation in the finished beer. By contrast, Bohemian barley was more easily modified. The kilns were built so that they channeled a strong draft through the grain bed, allowing the green malt to be dried more rapidly, producing paler malt with a lower degree of thermal stress. The malt was richer in enzymes and led correspondingly to beers that were lighter in color. The brewing water was also quite different – hard in most of Bavaria and soft in Bohemia. The mineral content of the brewing liquor served to further widen the differences between Bohemian and Bavarian beers. Interestingly, both regions originally brewed their beers using a triple decoction mash method. The Bavarian method was based on the physical digestion of the thick mash as well as the enzymatic degradation of the substance groups, primarily of the decoction mash. The Bohemian method, by contrast, strove to inactivate a portion of the enzymes through the "spring" mash method also known as the "decoction mash rest". Each of these mash methods underpins the respective beer style. However, the degrees of attenuation for both of these distinctive beer styles were practically identical. The limit of attenuation for both was between 72–74 %, while the final attenuation of each of the finished beers was about 66 %. The hop additions were different; the Bavarian hops were less refined, and due to the harder water, were used more sparingly to avoid a harsh bitterness. The Bohemian hops were finer and were added in large quantities (for microbiological reasons, among others). Therefore, two very different beer styles developed within a relatively small area of the Continent where the raw materials and equipment originated. In Britain, top fermentation remained widespread, while bottom fermentation was the norm from the 15th century in Bavaria and had spread across most of Central Europe by the middle of the 19th century, though pockets of the former persisted.

A historical review: If craft brewers want to return to the roots of brewing, knowledge of history is worthwhile. With today's raw materials and technology, craft brewers can create beers with much more precision than the brewers of past centuries. An acquaintance with the beer styles described above disseminated outward to other continents with European immigrants or through colonization. The brewers among the immigrants employed their skills to produce German and Anglo-Saxon styles as, for instance, the names of American breweries attest. The beers from these countries were exported for the soldiers and government officials who followed in the wake of colonization. Many of these beers were more heavily hopped and brewed to a stronger gravity in order to increase their stability on the voyage, e.g. India pale ales, imperial stouts and Bavarian *Dunkles*, until breweries could be constructed on-site to quench the thirst of the expatriates at their respective destinations. Malt had to be imported from its nation of origin for beer production at far-flung breweries, because several generations would pass before viable malting barley could be cultivated in those regions. Thus, native raw materials were employed (maize or corn, rice and sugar) to replace a portion of the malt bill. Sugar was not difficult to use and it increased the final attenuation (and therefore the alcohol content). The use of corn was fostered by the starch it contained, which was generally easier to gelatinize than that found in rice. The gelatinization temperatures of rice can vary substantially; for instance, rice grown under hot, relatively dry conditions or cultivated where the growing season is short exhibits a very high gelatinization temperature. Thus, some rice does not gelatinize below the optimal temperatures for amylase enzymes, requiring adjunct mashes to be treated quite differently. These beers gained in popularity, above all in warmer climes. Soon the economic advantages of adjuncts were recognized (simple acquisition, more extract, less expensive than malt), prompting some countries to establish an upper limit for the ingredients utilized as malt replacements.

Approximately 80 years ago, methods were developed, some of them quite complicated, for wort production and above all for fermentation, maturation and lagering that are, in part, still practiced today. The method described here calls for 30 % rice, while 2/3 of the malt bill is two-rowed barley and 1/3 consists of somewhat more protein and enzyme-rich six-rowed barley. About half of the malt bill is added to the "rice mash" in order to facilitate gelatinization and digestion at lower temperatures, around 45 °C. The mash is relatively thin at a grist to liquor ratio of 1 : 4–4.5. This portion of the mash is then heated rapidly to 75–78 °C. The temperature is held for 20 min and then boiled for 20–30 min. Immediately after the adjunct mash program commences, the rest of the grist is mashed in at 45 °C at a grist to liquor ratio of 1 : 3. The two mashes are then combined, bringing the total mash to 68 °C. A rest is held at this temperature until the iodine test is negative or even longer. When this rest is finished, the temperature of the mash is raised to 74 °C to mash out and then pumped into the lauter tun.

Wort boiling originally lasted 90 min, but modern wort boiling systems save time – and therefore the energy involved – in this stage of the process. Aroma hops, primarily from Europe (Hallertau, Saaz) are usually divided into four additions. After the duration of the boiling process is deemed sufficient, the hops and the hot break material are removed in the whirlpool. The temperature of the wort is brought down to pitching temperature, usually in a specially designed wort chiller. The wort is aerated immediately downstream from the chiller. The wort ferments at approximately 15 °C with yeast propagated at the brewery. Once the degree of attenuation reaches the target gravity for primary fermentation, the green beer is moved to the so-called conditioning cellar for two to three days at circa 6 °C, where more yeast sediments out. Subsequently, wort (or sometimes kräusen) is added, and the green beer is pumped into the lager cellar where it remains for 18 to 21 days at this temperature. Tannins are dosed in-line as the beer is transferred to a cold cellar at -1 °C. The tannins are allowed to react with the wort for two days for purposes of stabilization before the beer is filtered and packaged. Several steps in the brewing process were simplified by the introduction of cylindroconical fermentation and lager tanks. Transferring the wort or green beer from tank to tank has largely been eliminated through the ability to purge the cone of sediment. Cylindroconical tanks have also allowed for individual tanks to be cooled independent of others and to be subject to a precisely controlled temperature regimen.

A further example illustrates how problems with malt availability due to a lack of foreign currency resulted in the use of sorghum, both malted and unmalted, for brewing beer. Wort production continued along these lines until a surprisingly good product was developed with top-fermenting yeast. This beer was sufficiently popular that consumers continued to demand it even when malt was reintroduced to the production process.

After planning the beer production process and the means for acquiring the requisite raw materials, brewers aspire to the highest level of profitability in beer production. However, the opposite approach is exemplified by the term "first wort beer". In producing this beer, sparging does not occur, in order to avoid leaching undesirable substances from the grain bed.

The quantity of bitter substances in beer has generally been on the decline worldwide. This has not necessarily been favorable to the overall sensory balance of beer. The reasons behind this decline can, in part, be ascribed to changes in the character of the bitterness in beer, such as the use of higher proportions of bittering (high α) hops. Likewise, a harsh, lingering bitterness is attributable to high molecular weight proteins or dextrins, which produce a positive iodine reaction. The insufficient degradation of these substances is a consequence of poor mashing techniques. Additionally, low quality yeast can leave behind a severe, almost metallic bitterness in beer. The use of tetra extract (tetrahydro iso-α acids) for improving foam stability in "normally hopped" beers leads to an unusual kind of bitterness in beer. The sole use of this extract, either upstream from the filter or from the packaging line prevents, a lightstruck flavor (3-methyl-2-butene-1-thiol) in beer filled in green or clear bottles. The bitterness is less conspicuous when hop extract is exclusively employed to hop the beer.

The increasing uniformity of raw materials, the steady decline in bitterness, the universally available technology used in wort production and advances in brewing techniques, fermentation and maturation, including the production of high gravity beers that are diluted downstream in the process, led to the emergence of ever more homogeneous, characterless, pale lagers. This is true not only in the US but all over the world. This development was reinforced through the introduction of light beers, meaning those brewed to be low in calories. In the 1980s, a development counter to this trend arose with the production of hand-crafted beers, which were based on Anglo-Saxon styles, such as pale ale, brown ale, porter and stout, and on European styles, like *Helles*, pilsner, *Märzen*, weissbier, *Dunkles*, *Spezial* and *Bock*. Craft brewers, as they came to be known, often added hops on the cold side of production as one means of creating more distinctive beers. This practice resulted in highly aromatic beers; however, more hops were usually added throughout the entire process, producing beers with extreme levels of bitterness, in some cases. This occurred in a country where the large brewing conglomerates had reduced the bittering units in their light lagers from circa 14 IBU down to less than 10 IBU! As craft brewers have become increasingly more experimental in their approach, they have found inspiration in Belgian-style ´spontaneously` fermented beers. Similarly, craft brewers utilize a wide array of ingredients to create unique flavors and aromas in their beers. Craft beer has developed in phases and the current most experimental phase has seen the liberal use of non-traditional ingredients (e.g. coffee, fruit, ginger, etc.) often merely to see what flavors and aromas they will produce in beer.

The craft beer wave spilled over into Europe in the early 2000s, where it rapidly found acceptance among an eager drinking public, primarily in Mediterranean countries. In Northern Germany, craft beer was received more readily and openly than in the southern part of the country. However, one must take into account that the number of breweries in the south is much higher than in the north. Many of the breweries in the south are small and have been handcrafting very individual beers for a long time. The fact is that the large breweries (above all, those that export) normally produce 14 or more beer styles, while the smaller ones with their *Märzen*, festbier and other seasonal beers often brew around ten beer styles. Even breweries making weissbier, non-alcoholic beer, low-calorie beer, and lagers from *Helles* and *Pils* to *Bock*, easily produce at least six beer styles. Regardless of this variety, a number of breweries have become specialists in this regard and now excel at producing "sophisticated" craft beers. For example, such beers can sublimely combine caramel notes with a wide range of aromas and flavors derived from hops added both on the hot and cold side of production.

Further reading:

Narziß, L., Back, W., Gastl, M., Zarnkow, M.: Abriss der Bierbrauerei. 8th edition. Wiley-VCH, Weinheim. 2017

Meußdoerffer, F., Zarnkow, M.: Das Bier. Eine Geschichte von Hopfen und Malz. 2nd edition. C.H. Beck, München. 2016

Daniels, R.: Designing Great Beers. Brewers Publications. Division of the Association of Brewers. 1996

Dornbusch, H.: Beer Styles from around the World. Cerevisiae Communications, West Newbury 2016.

Basarova, G., Savel, J., Basar, P., Basarova, P., Lejsek, T.: The Comprehensive Guide to Brewing: From Raw Materials to Packaging. Fachverlag Hans Carl, Nuremberg 2017.

https://www.brewersassociation.org/resources/brewers-association-beer-style-guidelines/

McGreger, C., McGreger, N.: Barrels of Beer (a five-part series of articles). Brauwelt International 34 (2016), no. 3–6 and 35 (2017), no. 1.

4 INTERNATIONAL BEER STYLES

Table 2

	Original gravity	Alcohol	Degree of attenuation (apparent)	Color
	(% w/w)	(% vol.)	(%)	(EBC)
Germany – beer styles, bottom-fermented				
PILSNER	11–12.5	4.8–5.2	78–84	5.5–7.5
DORTMUNDER EXPORT	12.5–14	5–6	78–84	7.7–10
MUNICH HELLES	11.5–13.5	4.8–5.5	78–85	5.5–12
MUNICH DUNKEL	12–14	4.6–5.5	68–74	40–70
VIENNA	12–13	5–5.5	74–78	10–14
OKTOBERFEST/MÄRZEN	13.5–14	5.4–6	75–80	15–20
SCHWARZBIER	11–12.5	4.8–5.2	77–82	60–80
HELLER BOCK/MAIBOCK	16–17	6–7.5	80–84	6.5–10
DARK BOCK	16.1– 18	6–7	70–77	40–70
DOPPELBOCK/EISBOCK	> 18.2 -> 20	7–11.5	70–77	40–80
Germany – beer styles, top-fermented				
DÜSSELDORFER ALT	11.5– 12.5	4.8–5.2	70–80	25–50
KÖLSCH	11.2–12	4.8–5.2	77–82	6–12
WEISSBIER	12–14	5.2–5.8	75–80	8–12
KRISTALLWEIZEN	12–14	5.2–5.8	80–85	6.5–7.5
DARK WEIZENBIER	12.5–14	5.2–5.6	72–77	20–50
DARK WEIZENBOCK	16–17.5	6.2–7.5	77–82	10–24
WEIZENBOCK DUNKEL	16–19	6–8	72–77	30–80
BERLINER WEISSE	7.5–9	3.2–3.7	66–100	5–8

TF = top-fermented; BF = bottom-fermented

Bittering units	Barley malt	Yeast type/ fermentation	Fermentation temperature
(EBC)	(%)		°C
25–45	100	BF	6–12
22–30	100	BF	6–12
15–25	100	BF	6–10
16–25	100	BF	10–12
18–25	100	BF	9–10
22–30	100	BF	8.5–9.5
23–30	100	BF	10–12
22–35	100	BF	7–10
20–30	100	BF	10–12
25–45	100	BF	10–12
30–50	100	TF	15–18
20–30	90-100 (remainder wheat malt)	TF	18–22
10–15	25–50 (remainder wheat malt)	TF	18–22
10–14	25–50 (remainder wheat malt)	TF	18–22
10–15	25–50 (remainder wheat malt)	TF	18–22
18–25	25–50 (remainder wheat malt)	TF	18–22
18–35	25–50 (remainder wheat malt)	TF	18–22
4–8	50–67 (remainder wheat malt)	TF + lactic acid bacteria	16–18.5/17.5–25

	Original gravity	Alcohol	Degree of attenuation (apparent)	Color
	(% w/w)	(% vol.)	(%)	(EBC)
Britain and Ireland				
ORDINARY BITTER	8–9.5	3.2–3.9	60–67	20–31
BEST BITTER	9–10.5	3.6–4.6	61–67	31–37
EXTRA SPECIAL BITTER	10.5–12.4	4.1–5.2	59–64	31–37
PALE ALE (bottled)	10.7–15.9	4–7	65–70	15–37
INDIA PALE ALE	13.5–15.9	5–7	65–70	20–37
MILD ALE	8–9.3	2.8–3.6	55–62	39–91
BROWN ALE	10–11.2	4–5	60–67	39–57
SCOTCH ALE 60′ "Light"	7.5–8.5	3–3.4	56–62	18-34
SCOTCH ALE 70′ "Heavy"	8.5–9.3	3.2– 3.7	56–62	18-34
SCOTCH ALE 80′ "Export"	9.7–10.7	4.1– 4.5	58–64	37–47
SCOTCH ALE 90′ "Wee heavy"	17.0– 21.5	7–9.9	57–70	41–57
PORTER	10–14.7	3.7–5.5	60–70	62–94
SWEET STOUT MILK STOUT LONDON STOUT	9.5–10.5	2.7–3.8	50–58	78–131
OATMEAL STOUT	10–12.2	3.7–5	55–67	91–171
IRISH STOUT DRY STOUT	9.3–10 cask 11.9–14.7 bottle	4.3– 5.4 cask 4.2– 5.5 bottle	60–70	91–184
TROPICAL STOUT	14.7– 19.3	5.75–7.5	65–72	105–158
OLD ALE STRONG ALE	12.4–18.2	4.5–7	55–64	31–78
BARLEY WINE	18.2– 26	7–11.5	55–70	37–78
RUSSIAN STOUT IMPERIAL STOUT	18.2–28	7.2–12.5	60–70	105–211

TF = top-fermented; BF = bottom-fermented

Bittering units	Barley malt	Yeast type/ fermentation	Fermentation temperature
(EBC)	(%)		°C
20–40	70–100	TF	18–24
22–40	80–100	TF	18–24
25–45	85–100	TF	18–24
25–50	85–100	TF	17–19
40–70	85–100	TF	17–19
12–18	75–100	TF	13–16
14–22	85–100	TF	17–19
10–15	90–100	TF	13–16
12–17	90–100	TF	13–16
15–20	90–100	TF	13–16
22–32	90–100	TF	13–16
25–35	92–100	TF	18–24
18–25	90–100	TF	13–21
30–40	90–95	TF	13–21
25–50	94–100	TF	13–21
30–60	90–100	TF	13–16
22–85	85–100	TF	13–16
25–110	85–100	TF	13–16
50–150	90–100	TF	13–16

	Original gravity	Alcohol	Degree of attenuation (apparent)	Color
	(% w/w)	(% vol.)	(%)	(EBC)
Belgium				
ABBEY/TRAPPIST DUBBEL	12.9–15.4	5.5–7	68–70	25.3–65
ABBEY/TRAPPIST SPECIAL	17–19.3	7–9	63–72	14.7–51.8
ABBEY/TRAPPIST TRIPEL	17.5–26	7–10.5	61–70	25.3–65
AMBER ALE	13,2	5,2	16-28	30,6
STRONG PALE ALE	20	8,2	24-42	10,7
FLANDERS BROWN ALE	12–13	4.6–5.6	30-43	38.5–46.5
BRABANT WIT	11– 12.4	4.5–5.2	63– 68	4.1–9.4
STRONG WIT	19,5	8–9	4-8	25.3–38.5
SAISON	12.5–14	5.0–6.0	50–65	15–23
SAISON SPEZIAL	20	7,5	50–65	20–37
LAMBIC GUEUZE blended, aged + young FARO diluted, sweetened KRIEK cherries FRAMBOISE raspberries AARDBEIEN strawberries DRUIVEN muscat grapes PÊCHE peach	12–13	4.4–5.5	70–90	15–31
TRAPPIST (Orval)	11–13.5	5.2–5.7	66	17–23
BIÈRE DU GARDE	14.5–18	5.5–6.5	60–65	15–30
Czech Republic				
PILSNER	12	4.0–4.4	66–70	13
USA				
AMERICAN LAGER	10.5–11.5	4.7–5.0	78–82	5-8
AMERICAN LIGHT BEER	9–9.3	4.0–4.4	85–90	5–7

TF = top-fermented; BF = bottom-fermented

Bittering units	Barley malt	Yeast type/ fermentation	Fermentation temperature
(EBC)	(%)		°C
16–25	90–100	TF	18–29
18–30	90–100	TF	18–29
20–40	90–100	TF	18–29
27	90–100	TF	18–29
52	100	TF	18–29
23–27	100	TF + lactic acid bacteria	changes according to season
14–22	50	TF + lactic acid bacteria	14–21
16–27	50–100	TF	14–21
30–45	90–100	TF	10–12
35–50	90–100	TF	10–12
11–20	60–70 barley malt 30–40 wheat	wild yeast, among other microbes	changes according to season; important in summer
24	85–90 malt + sugar	TF + BF	14–21
24–32	100	TF	14–21
38–40	100	BF	9–10
8–12	65–70	BF	12–14
8–10	65–70	BF	10–12

APPENDIX

1 COMPOUNDS FOUND IN WORT

Note: In the following tables, all standard values provided for these compounds found in wort and beer are based on pale, standard-strength beers unless otherwise indicated.

Table 1: Compounds found in wort

Analysis	Standard value	Evaluation criterion for
limit of attenuation extract (OG)	80–85 % 11–14 % w/w	basic attributes
TBI furfural glutamine	10–50 450–700 µg/ 11–115 mg/	thermal stress
coag. N N (precipitation with MgSO4)	20–40 mg/l 130–180 mg/l	foam
α-amino N zinc	200–250 mg/l 0.15–0.30 mg/l	yeast nutrition, nutrients important for growth
free DMS hexanal hexanol pentanol	20–80 µg/l 10–20 µg/l 1–3 µg/l 2–4 µg/l	indicators for evaporation (during wort boiling)
glucose fructose maltose maltotriose sucrose	7–10 g/l 1–3 g/l 55–65 g/l 3–15 g/l 3–5 g/l	composition of sugars in the wort: yeast nutrition, aroma
linalool	10–60 µg/l	masking effects, hop aroma
vanillic acid protocatechuic acid caffeic acid sinapinic acid ferulic acid catechin	275 ppm 2.4 ppm 2–5 ppm 4 ppm 17–28 ppm 3.6–4.8 ppm	phenolic acids
thiamine (B1) riboflavin (B2) pantothenic acid (B5) pyridoxine (B6) biotin (H) cobalamin (B12) folic acid (B9)	150–190 ppb 150–400 ppb 150–250 ppb 150–800 ppb 5–18 ppb 0–8.6 ppb 0–400 ppb	vitamins
C6–C18:3	3.2–7.1 ppm	free fatty acids in the wort (kettle wort, prior to boiling)
C6–C18:3	1.4–2.5 ppm	free fatty acids in the pitching wort

Table 2: Aroma Compounds in Wort (kettle wort, prior to boiling)

Wort aroma compounds in µg/l		Mean	Maximum	Minimum
3-methylbutanal	(E, W)	274.7	628.8	92.8
2-methylbutanal	(E, W)	100.5	227.5	28.3
methional	(E)	51.6	141.0	20.6
benzaldehyde*	(E)	7.4	35.7	2.9
2-phenylacetaldehyde	(E, W)	122.5	238.7	66.0
Σ Strecker aldehydes		556.8	1072.9	310.0
pentanal	(E)	14.4	25.1	5.3
2-pentanone	(E)	4.3	29.5	0.8
hexanal	(E)	72.5	150.0	21.7
heptanal	(E)	4.4	13.0	0.8
trans-2, cis-6-nonadienal	(E)	1.4	5.0	0.0
2-furfural	(T)	84.2	332.8	21.3
2-acetylfuran	(T)	15.2	25.5	1.1
γ-nonalactone	(E, W)	5.1	31.4	0.8
3-methylbutanol	(E)	135.5	349.6	20.5
2-methylbutanol	(E)	38.1	97.1	7.0
1-pentanol	(E)	31.7	47.1	9.1
1-hexanol	(E)	31.5	107.8	5.0
1-octanol	(E)	2.3	4.9	0.1
1-octen-3-ol	(E)	4.0	7.2	0.7
phenylethanol	(E)	136.3	742.7	37.5

E = indicator for evaporation during wort boiling:
W = indicator for thermal stress in the whirlpool; T = indicator for thermal stress during wort boiling and whirlpool
*benzene derivative (not formed during Strecker degradation reactions)

Table 3: Aroma Compounds in Wort (cast-out wort)

Wort aroma compounds in µg/l		Mean	Maximum	Minimum
3-methylbutanal	(E, W)	34.7	72.3	17.2
2-methylbutanal	(E, W)	11.0	23.8	4.5
methional	(E)	27.7	66.7	12.7
benzaldehyde*	(E)	3.4	7.6	0.7
2-phenylacetaldehyde	(E, W)	91.0	164.8	42.0
Σ Strecker aldehydes		167.5	257.0	90.5
pentanal	(E)	3.1	8.3	0.6
2-pentanone	(E)	4.0	31.3	0.5
hexanal	(E)	3.3	7.1	1.3
heptanal	(E)	1.4	4.7	0.5
trans-2, cis-6-nonadienal	(E)	1.1	2.2	0.5
2-furfural	(T)	197.0	412.0	70.7
2-acetylfuran	(T)	10.4	25.6	1.5
γ-nonalactone	(E, W)	2.9	6.8	0.7
3-methylbutanol	(E)	1.7	6.6	0.2
2-methylbutanol	(E)	0.8	1.3	0.3
1-pentanol	(E)	115.3	186.9	25.9
1-hexanol	(E)	23.6	145.8	1.2
1-octanol	(E)	11.1	50.5	1.3
1-octen-3-ol	(E)	19.7	33.2	1.3
phenylethanol	(E)	3.4	18.1	0.1

E = indicator for evaporation during wort boiling: W = indicator for thermal stress in the whirlpool;
T = indicator for thermal stress during wort boiling and whirlpool
*benzene derivative (not formed during Strecker degradation reactions)

Table 4: Aroma Compounds in Wort (wort, midpoint of chilling)

Wort aroma compounds in µg/l		Mean	Maximum	Minimum
3-methylbutanal	(E, W)	52.0	111.0	18.3
2-methylbutanal	(E, W)	23.0	60.5	7.4
methional	(E)	39.5	103.7	10.9
benzaldehyde*	(E)	2.8	5.8	0.6
2-phenylacetaldehyde	(E, W)	119.5	187.8	71.7
Σ Strecker aldehydes		236.8	388.6	162.2
pentanal	(E)	2.2	8.1	0.7
2-pentanone	(E)	1.8	13.4	0.4
hexanal	(E)	5.5	8.4	2.8
heptanal	(E)	1.7	4.0	0.6
trans-2, cis-6-nonadienal	(E)	0.8	1.3	0.4
2-furfural	(T)	238.5	475.8	125.7
2-acetylfuran	(T)	4.4	8.2	2.5
γ-nonalactone	(E, W)	4.3	8.5	2.2
3-methylbutanol	(E)	32.7	150.2	3.4
2-methylbutanol	(E)	11.1	46.8	1.8
1-pentanol	(E)	6.8	15.5	0.7
1-hexanol	(E)	3.2	10.0	0.5
1-octanol	(E)	1.8	6.7	0.1
1-octen-3-ol	(E)	0.5	1.0	0.1
phenylethanol	(E)	133.0	236.1	70.2

E = indicator for evaporation during wort boiling:
W = indicator for thermal stress in the whirlpool; T = indicator for thermal stress during wort boiling and whirlpool
*benzene derivative (not formed during Strecker degradation reactions)
2 Compounds Found in Beer

2 COMPOUNDS FOUND IN BEER

Table 5

extract alcohol carbon dioxide water	3.5–4.8 % (w/w) 4.4–6.1 % (v/v) 4.2–4.8 % (w/w) 90–93 %
glucose fructose maltose maltotriose sucrose dextrins	0.1–0.5 g/l 0.1–0.2 g/l 2–3 g/l 1–3 g/l 0–0.05 g/l 1.7–3.2 g/l
nitrogenous substances α-amino N	700–800 mg/l 80–150 mg/l
glycerin	1200–1600 mg/l
potassium sodium calcium magnesium sulfate chloride silicon phosphate (organic and inorganic) iron copper manganese zinc aluminum	30–550 mg/l 40–50 mg/l 15–50 mg/l 100 mg/l 30–250 mg/l 100–200 mg/l 20–60 mg/l 370–490 mg/l < 0.1 mg/l < 0.1 mg/l 0.15 mg/l 0.05 mg/l 0.2 mg/l
tannins anthocyanogens tannoids bitter substances	150–200 mg/l 50–70 mg/l 10–40 mg/l 12–50 mg/l
pyruvic acid citric acid malic acid D-lactic acid L-lactic acid oxalic acid	50–70 mg/l 170–220 mg/l 30–110 mg/l 20–60 mg/l 25–50 mg/l 10–20 mg/
thiamine biotin riboflavin pantothenic acid niacin	30 µg/l 10 µg/l 300 µg/l 1500 µg/l 7500 µg/l
SO2	1–10 mg/l
protocatechuic acid gallic acid vanillic acid syringic acid caffeic acid ferulic acid sinapinic acid chlorogenic acid catechin epicatechin	0.2–10 ppm 0–20 ppm 1–12.7 ppm 1–10 ppm 0–10 ppm 1–15 ppm 1–10 ppm 1–10 ppm 1–20 ppm 1–10 ppm
free fatty acids C12–C18:3	0–0.5 ppm

3 AROMA COMPOUNDS AND HIGHER ALCOHOLS FOUND IN BEER

Table 6

Aroma compounds/ higher alcohols in beer		Conc. in weissbier
propanol-1	mg/l	15–30
2-methylpropanol-1	mg/l	20–60
2-methylbutanol-1	mg/l	15–30
3-methylbutanol-1	mg/l	40–100
hexanol-1	µg/l	15–50
heptanol-1	µg/l	10–50
octanol-1	µg/l	10–40
decanol-1	µg/l	5–20
2-phenylethanol	mg/l	15–45
ethyl acetate	mg/l	10–50
acetic acid isobutyl ester	mg/l	0.05–0.8
3-methylbutyl acetate	mg/l	0.5–8
acetic acid hexyl ester	µg/l	3–15
acetic acid heptyl ester	µg/l	2–10
acetic acid octyl ester	µg/l	1.5–5
acetic acid furfuryl ester	µg/l	3–20
acetic acid 2-phenyl ethyl ester	mg/l	0.2–2
butyric acid ethyl ester	mg/l	0.05–0.2
hexanoic ethyl ester	mg/l	0.05–0.3
octanoic ethyl ester	mg/l	0.1–0.4
decanoic ethyl ester	mg/l	0.01–0.05
dodecanoic ethyl ester	µg/l	1–5
isovaleric acid	mg/l	0.2–1
hexanoic acid	mg/l	1–4
octanoic acid	mg/l	2–10
nonanoic acid	mg/l	0.01–0.05
decanoic acid	mg/l	0.1–2
9-decenoic acid	mg/l	0.05–0.3
dodecanoic acid	mg/l	0.01–0.1
γ-nonalactone	µg/l	20–50
4-vinyl guaiacol	mg/l	0.5–4

Average in weissbier	Conc. in pale, standard-strenght lager	Average in pale, standard-strenght lager
21	5–20	10
40	10–20	8.8
18	10–20	15
61	30–50	41
27	10–30	24
32	10–30	12
15	20–40	20
7.5	5–20	3.8
31	10–30	20
29	5–20	9.7
0.25	0.05–0.1	0.05
3.6	0.5–2	0.7
6.5	3–15	3.3
9.4	2–6	2.2
2.6	1–5	1.5
8.4	10–30	10
1	0.1–0.3	0.25
0.18	0.05–0.15	0.05
0.14	0.05–0.3	0.06
0.21	0.1–0.5	0.15
0.01	0.01–0.05	0.01
2.4	2–10	2.1
0.35	0.2–1	0.30
1.7	0.4–2	1.5
3.9	2–10	4
0.01	0.01–0.07	0.01
0.65	0.1–2	0.46
0.2	0.05–0.2	0.09
0.04	0.01–0.07	0.03
36	20–40	13
2	0.1–1	0.3

4 INDICATOR COMPOUNDS FOR AGING [µg/l]

Table 7

		Average value for fresh beer	Average value for artificially aged beer (forced aging)
2-acetylfuran	(A)	4.9	7.5
2-furfural	(T, A)	11.4	95.4
2-furfuryl ethyl ether	(A)	0.5	3.6
hexanal		0.6	2.4
heptanal		0.3	1.1
2-phenylacetaldehyde	(O, A)	4.9	8.6
2-propionylfuran	(A)	2.1	2.8
2-methylbutanal		3.1	5.6
3-methylbutanal	(O, A)	6.2	11.7
5-methylfurfural	(A)	5.4	6.3
benzaldehyde	(O, A)	0.6	0.9
diethyl succinate	(A)	1.1	1.6
γ-nonalactone	(T, A)	27.5	40.2
nicotinic acid ethyl ester		6.9	13.3
ethyl phenylacetate	(A)	0.7	0.8

A = aging indicator
T = thermal stress indicator
O = oxidation indicator

5 CONVERSION OF UNITS

Length

Name of unit	Symbol	Definition	Relation to SI units
metre (SI base unit) (meter)	m	≡ Distance light travels in 1/299792458 of a second in vacuum. ≈ 1/10000000 of the distance from equator to pole.	≡ 1 m
inch (International)	in	≡ 2.54 cm ≡ 1/36 yd ≡ 1/12 ft	≡ 0.0254 m

Mass

Name of unit	Symbol	Definition	Relation to SI units
kilogram (SI base unit) (kilogramme)	kg	≡ mass of the prototype near Paris ≈ mass of 1 litre of water	≡ 1 kg
grain	gr	≡ 1/7000 lb av	≡ 64.79891 mg
bushel (U.S.) of barley	bu	≡ 48 lb	= 21.7724 kg
bushel (U.S.) of malt	bu	≡ 34 lb	= 14.4221 kg

Volume

Name of unit	Symbol	Definition	Relation to SI units
cubic metre (SI unit)	m^3	≡ 1 m × 1 m × 1 m	≡ 1 m^3
litre (liter)	L or l	≡ 1 dm^3	≡ 0.001 m^3
hectoliters		≡ 100 l = 100 dm^3	≡ 0.1 m^3
barrel (imperial)	bl (imp)	≡ 36 gal (imp)	= 0.16365924 m^3
barrel (US fluid)	fl bl (US)	≡ 31 ½ gal (US)	= 0.119240471196 m^3
pint (imperial)	pt (imp)	≡ 1/8 gal (imp)	= 568.26125 × 10^{-6} m^3
pint (US fluid)	pt (US fl)	≡ 1/8 gal (US)	= 473.176473 × 10^{-6} m^3
gallon (imperial)	gal (imp)	≡ 4.54609 L	≡ 4.54609 × 10^{-3} m^3
gallon (US fluid; Wine)	gal (US)	≡ 231 cu in	≡ 3.785411784 × 10^{-3} m^3
bushel (imperial)	bu (imp)	≡ 8 gal (imp)	= 0.03636872 m^3
bushel (US dry level)	bu (US lvl)	≡ 2150.42 cu in	= 0.03523907016688 m^3

source: https://en.wikipedia.org/wiki/Conversion_of_units#Length

Temperature

Name of unit	Symbol	Definition	Relation to SI units
kelvin (SI base unit)	K	≡ 1/273.16 of the thermodynamic temperature of the triple point of water.	≡ 1 K
degree Celsius	°C	[°C] ≡ [K] − 273.15	[K] ≡ [°C] + 273.15
degree Fahrenheit	°F	[°F] ≡ [°C] × 9/5 + 32	[K] ≡ ([°F] + 459.67) × 5/9
degree Réaumur	°Ré		[K] = [°Ré] × 5/4 + 273.15

Speed or velocity

Name of unit	Symbol	Definition	Relation to SI units
metre per second (SI unit)	m/s	≡ 1 m/s	= 1 m/s
foot per minute	fpm	≡ 1 ft/min	= 5.08 × 10^{-3} m/s
inch per second	ips	≡ 1 in/s	= 2.54 × 10^{-2} m/s
mile per hour	mph	≡ 1 mi/h	= 0.44704 m/s

Flow (volume)

Name of unit	Symbol	Definition	Relation to SI units
cubic metre per second (SI unit)	m^3/s	≡ 1 m^3/s	= 1 m^3/s
cubic inch per second	in^3/s	≡ 1 in^3/s	= 1.6387064 × 10^{-5} m^3/s
gallon (US fluid) per minute	GPM	≡ 1 gal/min	= 6.30901964 × 10^{-5} m^3/s

Energy

Name of unit	Symbol	Definition	Relation to SI units
joule (SI unit)	J	The work done when a force of one newton moves the point of its application a distance of one metre in the direction of the force.	"= 1 J = 1 m·N = 1 kg·m^2/s^2 = 1 C·V = 1 W·s
British thermal unit (ISO)	BTUISO	≡ 1.0545 × 10^3 J	= 1.0545 × 10^3 J
calorie (International Table)	callT	≡ 4.1868 J	= 4.1868 J

source: https://en.wikipedia.org/wiki/Conversion_of_units#Length

Density

Name of unit	Symbol	Definition	Relation to SI units
kilogram per cubic metre (SI unit)	kg/m^3	$\equiv kg/m^3$	$= 1\ kg/m^3$
gram per millilitre	g/mL	≡ g/mL	$= 1000\ kg/m^3$
pound (avoirdupois) per gallon (US fluid)	lb/gal	≡ lb/gal	$\approx 119.8264273\ kg/m^3$

Pressure or mechanical stress

Name of unit	Symbol	Definition	Relation to SI units
pascal (SI unit)	Pa	$\equiv N/m^2 = kg/(m \cdot s^2)$	= 1 Pa
bar	bar		$\equiv 10^5$ Pa
atmosphere (standard)	atm		≡ 101325 Pa
atmosphere (technical)	at	$\equiv 1\ kgf/cm^2$	$= 9.80665 \times 10^4$ Pa
pound per square inch	psi	$\equiv 1\ lbf/in^2$	$\approx 6.894757 \times 10^3$ Pa

Brewery

Colour of wort and beer	ASBC (SRM)	Lovibond (Malt)	EBC
	SRM = 0,508 × EBC	°L = (SRM + 0,76)/1,3546	EBC = 1,97 × SRM
Bitterness	ASBC IBU	IBU (ASBC) ≈ BU (EBC)	EBC BU
Filterability F_{spez}			$hl\ m^{-2}\ h^{-1}$

source: https://en.wikipedia.org/wiki/Conversion_of_units#Length

6 LIST OF ABBREVIATIONS

3-MBT	3-methyl-2-butene-1-thiol
a.d.	air-dried
ABTS	2,2′-azino-bis (3-ethylbenzothiazoline-6-sulfonic acid)
acetyl-S CoA	coenzyme A
AN	aniline number
ARB	antiradical behavior
ASBC	American Society of Brewing Chemists (bitter units, colour units, turbidity units)
ATP	adenosine triphosphate
BCE	before the Current (Christian) Era
BF	bottom fermented
BM	bypass membrane filter system
BU	betamyl units
CCT	cylindroconical tank
CIP	cleaning in place
CMA	Centrale Marketing-Gesellschaft der deutschen Agrarwirtschaft (marketing society for german farmers and food industry, liquidated in 2009)
COD	chemical oxygen demand
CV	conductometer value
d.m.	dry matter
DEAE	diethylaminoethyl (cellulose)
DGE	German Nutrition Society
DIN	German Institute for Standardization
DLG	German Agricultural Society
DMS	dimethyl sulfide
DMS-P	DMS precursor
DON	deoxynivalenol
DPI	2,6-dichlorophenolindophenol
DPPH	1,1-diphenyl-2-picrylhydrazyl
DU	dextrin units
EBC	European Brewery Convention (bitterness units, colour units, turbidity units)
EE	evaporation efficiency

ESR	electron spin resonance spectroscopy
EU	European Union
FAN	free amino nitrogen
FFC	flow fraction controller (according to ISO 3511)
FI	flow indicator
FR	flow recording (according to ISO 3511)
GC	gas chromatography
GC-FID	gas chromatography flame ionization detector
GC-MS	gas chromatography mass spectrometry system
HIS	hop storage index
HMF	5-hydroxmethylfurfural
HPLC	high performance liquid chromatography
HTW	high temperature wort boiling
IBD	Institute of Brewing & Distilling
ICP	intracellular pH
IKE	isomerized kettle extract
IOB	Institute of Brewing
LMC	concentrated lactose medium
LOX	lipoxygenase
LSKE	light-stable isomerized kettle extract
LTP	lipid transfer protein
MA	mash acidification
MEBAK	Mitteleurpäische Brautechnische Analysenkommission (Central European Brewing Technology Commission)
NBB®	nutrient medium for beer spoilers according to Back
NBB-A	NBB agar
NBB-B	NBB broth
NBB-B-AM	NBB broth
NBB-C	NBB concentrated broth
NDMA	N-nitrosodimethylamine
NIBC	non-isomerized bitter compounds
NIBEM	Nederlands Instituut voor Brouwgerst, Mout en Bier
OG	original gravity

PBN	N-tert-butyl-α-phenylnitrone
PCR	polymerase chain reaction
PIKE	potassium based Isomerized kettle extract
PrA	proteinase A
PU	pasteurization units
PVPP	Polyvinylpolypyrrolidone
RNA	ribonucleic acid
ROS	reactive oxygen species
RV	reference value
SMM	S-methylmethionine
TBI	thiobarbituric acid index
TE	total evaporation
TF	top fermented
TIA	temperature indicator alarm (according to ISO 3511)
TIC	temperature indicator controller
USSR	Union of Soviet Socialist Republics
VIT®	vermicon identification technology (vermicon AG)
w/v %	percentage by volume (g/100 ml)
w/w %	percentage by weight (% mass)
WK	Windisch Kolbach index
XN	xanthohumol
YCC	yeast cell count

7 INDEX

A

B

C

D

E

F

M

N

O

P

T

U

V

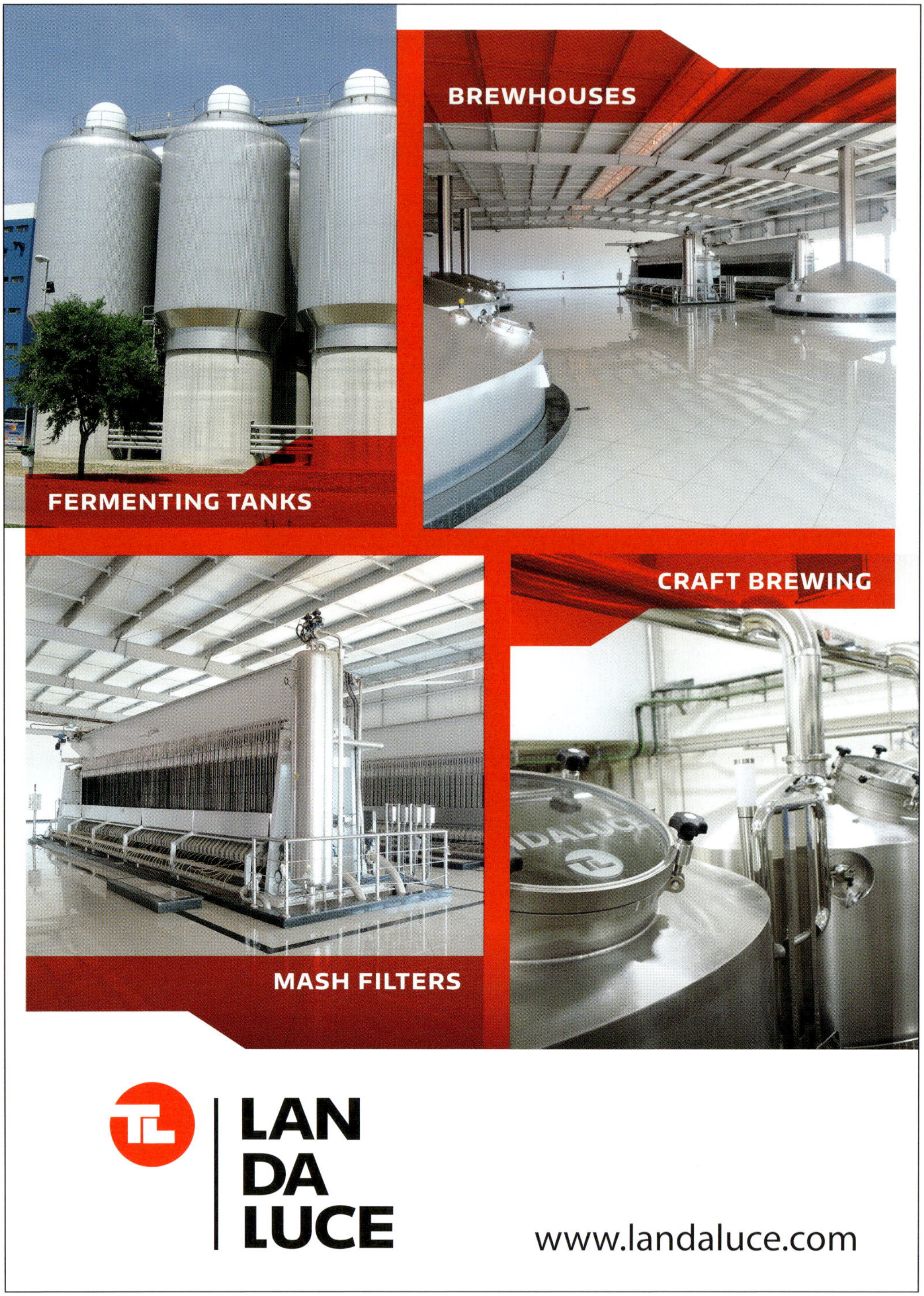
BREWHOUSES
FERMENTING TANKS
CRAFT BREWING
MASH FILTERS
LAN
DA
LUCE
www.landaluce.com